Rules of Inference for '⊃'

Modus Ponens (MP)
One may infer from two previously obtained sentences of the forms

$$p \supset q$$

and

$$p$$

the corresponding sentence

$$q.$$

Modus Tollens (MT)
One may infer from two previously obtained sentences of the forms

$$p \supset q$$

and

$$\sim q$$

the corresponding sentence of the form

$$\sim p.$$

Hypothetical Syllogism (HS)
One may infer from two previously obtained sentences of the forms

$$p \supset q$$

and

$$q \supset r$$

the corresponding sentence of the form

$$p \supset r.$$

Constructive Dilemma (CD)
One may infer from three previously obtained sentences of the forms

$$p \vee q$$
$$p \supset r$$

and

$$q \supset s$$

the corresponding sentence of the form

$$r \vee s.$$

Rules of Replacement for '⊃' and '≡'

Transposition (Trans)
$p \supset q :: \sim q \supset \sim p$

Implication (Impl)
$p \supset q :: \sim p \vee q$

Exportation (Exp)
$(p \mathbin{\&} q) \supset r :: p \supset (q \supset r)$

Equivalence (Equiv)
$p \equiv q :: (p \supset q) \mathbin{\&} (q \supset p)$
$p \equiv q :: (p \mathbin{\&} q) \vee (\sim p \mathbin{\&} \sim q).$

(Continued inside back cover)

ELEMENTARY
SYMBOLIC
LOGIC

SECOND EDITION

WILLIAM GUSTASON
DOLPH E. ULRICH
Purdue University

WAVELAND

PRESS, INC.

Prospect Heights, Illinois

To our parents

For information about this book, contact:
Waveland Press, Inc.
P.O. Box 400
Prospect Heights, Illinois 60070
(847) 634-0081
www.waveland.com

PREFACE

This textbook is designed for a one-semester or two-quarter introductory course in symbolic logic. No previous training is presupposed, and a deliberate attempt has been made to appeal to students from a wide range of disciplines, who enroll in such a course with varying motives and expectations. For the benefit of students with nontechnical backgrounds, relatively intricate topics — especially those pertaining to quantification — are discussed more extensively than is usual. The emphasis is thus on developing the student's grasp of standard techniques and concepts rather than on achieving the degree of sophistication and rigor that would be appropriate for a more specialized and demanding course.

One of our chief aims has been to provide the instructor with considerable flexibility when choosing an order in which to present standard topics. Accordingly semantic methods (Chapter 2) and natural deduction (Chapter 3) for sentential logic are developed independently, the preliminary discussion of the sentential connectives in Chapter 1 providing a common lead-in to both. Thus the instructor may begin in the customary fashion with truth-tables and then proceed to natural deduction; alternatively, he may begin with deductions and follow up with truth-tables. In either case he may at his option finish off with Chapter 4, where the two approaches are linked by a brief discussion of appropriate metatheoretic questions. This book hence differs from the majority of texts, where one must begin with truth-tables, and from the remainder that require coverage of deductions first.

Moreover, the ten rules of inference and replacement for '&', 'v', and '~' presented in Chapter 3 are by themselves semantically complete, and this makes possible additional variations that may intrigue some instructors. With a little maneuvering it is possible to begin with truth-tables and deductions for just these three connectives, proceed to Chapter 4 for the completeness discussion, and then return to Chapters 2 and 3 to develop (in either order, of course) truth-tables and rules for '⊃' and '≡'. Thus, in addition to the two standard orders

Chapters 1, 2 (last two sections optional), 3, and 4 (optional)
Chapters 1, 3, 2 (last two sections optional), and 4 (optional)

one has on this third approach other options, for example:

Chapter 1 (omitting 1.4 temporarily), Section 2.1, 2.3 (first half), 3.1, Chapter 4, 1.4, balance of Chapters 2 and 3 (in either order).

In approaching quantification theory, similar options are available. After the discussion of symbolization and the associated apparatus in Sections 5.1 and 5.2, one may cover — in either order — semantics (Chapter 6) or natural deduction (7.1 and 7.2, with 7.3 optional), returning eventually to the other. The sections on identity and definite descriptions (5.3 and 7.4) may be covered in their natural order or left till the end if desired; and of course they may be omitted should time run short. Sample orders of presentation include:

$$5.1 - 5.3, 6.1 - 6.2, 7.1 - 7.4$$
$$5.1 - 5.2, 6.1 \ (6.2 \ \text{optional}), 7.1 - 7.3, 5.3, 7.4$$
$$5.1, 5.2, 7.1 - 7.3, 6.1, 5.3, 7.4$$

The bulk of the text, then, is taken up with symbolization, semantics, and natural deduction for sentential and quantificational logic. But further material is available for times when some variety is desired or for background reading — a discussion of functional completeness in Section 2.5 for example, and of semantic completeness and soundness (both weak and strong) in Chapter 4. There is also an appendix introducing mathematical induction, which many will find a useful supplement to Chapter 4, and one on axiomatic systems. In addition, an appendix on normal forms and circuits is provided to augment Chapter 2, and another on truth trees for those who desire a more systematic method than that of Section 2.4. Any of these can be inserted at appropriate points — and all may be omitted — without affecting continuity or the coverage of essentials. Optional sections and certain exercises which might best be regarded as "projects" are marked with stars. We note that the first chapter, concerned mainly with preliminary terminology and notation, offers many temptations for squandering class time. Much of the material there can be mastered by students largely on their own, however, and the authors have found that three or four class meetings are sufficient for covering its essentials.

We have tried to write for the *student*, showing by example how one should think about the problems and methods being treated. Particular emphasis is placed on translation into quantificational notation and strategy in discovering deductions. And since students often want to insert tautologies into deductions we have a rule allowing insertion of those of the form '$p \vee \sim p$'. There are surprising advantages — certain deductions turn out to be more easily constructed than in systems lacking such a rule, and conditional proof turns out to be redundant (though of course it is employed as a practical device). In addition, the strategy of indirect proof emerges in a very natural way. It should also be noted that the student who works through Chapter 4 will be able to *demonstrate* the redundancy of conditional proof by obtaining a deduction theorem for the system developed in Sections 3.1 and 3.2.

In the rules for quantificational logic we have tried to put into sharper focus the role often played by free variables. The letters 'u', 'v', 'u'', 'v'', etc., which cannot appear as bound variables, are made available particularly for use with the rules **EI** and **UG**, which consequently can be stated without contorted clauses concerning freedom and bondage. Some instructors may choose to reserve 'v', 'v'', etc., for use with the former and 'u', 'u'', etc., for the latter, though our rules do not require this, of course. Well-known semantic considerations have led some authors to cast **EI** as a sub-proof rule. Such formulations strike many students as nonintuitive and puzzling, however, so we have preferred what seems a more natural version. Certain arguments require circuitous deductions but their number can be kept to a minimum with judicious symbolization.

While the change of publisher from Holt, Rinehart and Winston to Waveland Press in 1982 afforded us the opportunity to make some alterations in the text, there have been more substantive changes and additions for this second edition. Many of the exercise sets have been considerably revamped and expanded, with some of the new exercises raising additional topics of interest. Sections 1.4 and 2.2 have been extensively rewritten to put into better focus, we hope, the relation between 'if-then' and '⊃'. Many of the worked examples in Chapter Three have been replaced by more illustrative ones, and the discussion of theorems has been enlarged. New material to help students in translating from natural language has been added to Section 5.1. Section 5.2 has been reorganized and Section 5.3 of the first edition has been expanded and made into the new Chapter Six with most of the new material concerned with helping students find interpretations showing invalidity. The old Section 5.4 is now 5.3 and the old Chapter Six is now Chapter Seven containing several new and hopefully improved worked examples. A discussion of computer circuits has been added to Appendix A that should be of interest to more technically oriented students, and a discussion of strong induction has been included in Appendix C that many instructors may wish to make use of in discussing topics considered elsewhere in the book. Finally, at the request of many instructors, selected exercises are provided with solutions in the back of the book. Such exercises are indicated by having their numbers boldfaced.

The second edition has benefitted from many thoughtful suggestions supplied by Professors Marc Cohen of the University of Washington, Graeme Forbes of Tulane University, Donald Henson of Wilkes College, William Lycan of the University of North Carolina and Gerald Vision of Temple University. We adopted those suggestions that could be easily incorporated into our scheme of things, and only wish time had permitted us to work in others that would have required more extensive rewriting and reorganization. Thanks go also to our colleagues William L. Rowe, Rod Bertolet and Dorothy Leland, and of course we thank also our students over the years since the first edition appeared for showing us what most needed revision.

From the first edition, we extend our gratitude once more to our friends and former teachers. Professor Jack Nelson of Temple University must be singled out especially; he used an early draft in his classes and made many valuable and detailed suggestions. Our thanks also extend to Ken Tohinaka, John Tugman, Brian Heald, Marie Allison and Ralph Moon. Finally, we thank Susan Ulrich who helped in countless ways from the time of the book's inception until its publication.

West Lafayette, Indiana   W.G.
                                                            D.U.

CONTENTS

CHAPTER ONE
INTRODUCTION

1.1 Arguments

Like most disciplines, logic treats so wide a range of problems and topics that any attempt to describe its subject matter briefly is bound to be misleading and incomplete. But its most elementary and fundamental concern is with matters relevant to the study of arguments – the primary topic of this book. For our purposes an **argument** may be characterized as a sequence of sentences of which one – the **conclusion** of the argument – is marked off as *following from* the others – which are the **premises** of the argument. Our main job will be to develop some tools and techniques useful in determining whether or not certain sentences follow from certain others, and in explaining why they do or do not.

A number of difficult but interesting and important questions are partly questions about what follows from what; for example:

Do there exist relatively simple physical laws from which all other physical laws follow?

Does there exist a series of relatively simple arithmetic truths from which all other arithmetic truths can be deduced?

Do predictions about what is going to happen tomorrow follow from certain statements about what has happened in the past?

May we infer from certain things we all know, or can readily establish, that there exists a Supreme Being whom we may call 'God'?

Because of the intrinsic complexity of such questions, however – not to mention the highly specialized knowledge of other disciplines that most of them call for – we shall generally be dealing with considerably simpler ones and as a result with arguments that are somewhat less interesting.

1

We need not resign ourselves, though, to a steady diet of banalities like the classic

All men are mortal.
Socrates is a man.
Therefore Socrates is mortal.

We shall find use for such arguments, especially in the early sections of several chapters where it will be helpful to begin with such straightforward examples. But we shall also get a chance to deal with a number of more interesting arguments, some of which touch—as do certain of the following examples—on such questions as those mentioned above.

As a first step, readers are asked to examine the following paragraphs critically, asking themselves which of the arguments presented therein are good and which bad. At the same time, they should give the reasons on which they base their decisions.

(I) If God can create a stone too heavy for Him to lift, there is something He cannot do; and if He cannot create a stone too heavy for Him to lift, there is something He cannot create. If there's something God cannot do He is not omnipotent, and if there's something He cannot create He is not omnipotent. Therefore God is not omnipotent.

(II) If it is morally permissible for me to perform a certain action, then it must be morally permissible for everyone to perform that action. But if the consequences of everyone's performing a certain action are on the whole undesirable, then it is not morally permissible for anyone to perform that action. Therefore, if the consequences of my performing a certain action are, on the whole, undesirable, then it is not morally permissible for me to perform that action.

(III) If we introduce the Arabic notation for numbers, and pile a decimal notation on top of that, we sometimes find ourselves blessed with two different ways of writing names for one and the same number. For example, the number we denote by a vertical stroke (and call 'the number one') is identical with the number whose name is sometimes written '.999 $\cdots$'. For let $n = .999 \cdots$ Then $10 \times n = 10 \times .999 \cdots$, that is, $10 \times n = 9.999 \cdots$ But then $(10 \times n) - n = 9.999 \cdots - n$, so that $9 \times n = 9.999 \cdots - .999 \cdots$ Consequently $9 \times n = 9$, that is, $n = 1$. But $n = .999 \cdots$ and so $1 = .999 \cdots$, as was to be shown.

(IV) If freely falling bodies fall faster in proportion to their weight, then a three-pound body would fall faster than one of two pounds. But suppose the three-pound body consists of a two-pound and a one-pound body fastened together. Then it would fall more *slowly* than a single two-pound body, for the one-pound body would retard the motion of the attached two-pound body. Therefore, freely falling bodies do not fall faster in proportion to their weight.

(V) One Saturday, a prisoner was sentenced to be hanged. "The hanging will take place at 8:00 AM," the judge told him, "on Monday, Tuesday, Wednesday, Thursday, or Friday of next week. But you will not have any idea which day it will be until you are so informed at dawn on that day."

Alone later in his cell, the prisoner began to think about the judge's remarks. Obviously they could not wait until the *last* day to hang him, without violating the judge's decree. "If the first four mornings went past and I were still alive," he thought to himself, "I'd know for sure by that fourth afternoon when the hanging would be — there would only be one day left when it *could* be! But the judge said that I *would not* be able to tell in advance like that. So the *last* day is absolutely ruled out as the day for the hanging; it can only take place on Monday, Tuesday, Wednesday or Thursday." So Thursday, he knew, was the last day they could pick for the hanging. But of course they could not wait until the *last* day .... "Thursday's not a real possibility either, then," thought the prisoner. "Were I still alive at dusk on Wednesday, I would know that Thursday had to be the day. Friday has already been ruled out."

"So Wednesday seems to be the latest day the judge could pick. But for the same reasons, Wednesday can be eliminated as a real possibility... and then Tuesday, and finally Monday. They cannot possibly hang me in accordance with the judge's decree!"

Does the reader need to be told that the prisoner was quite surprised when the man in the black hood came to call on Wednesday, just at dawn?

(VI) It is sometimes supposed that (declarative) *sentences* are the things that may properly be said to be 'true' or 'false' — when one says, for example, that it is true that snow is white, it is the sentence "Snow is white" that is being said to be true.

Suppose Downs, in Detroit one bright morning, looks quickly out his window on which a lawn sprinkler is playing, erroneously concludes that it is raining, and utters to his wife the sentence "It is raining," meaning thereby to assert that it is raining that morning in Detroit. And suppose that Smith in Seattle that same morning sees a genuine downpour out his window and utters to his wife the sentence "It is raining," meaning thereby to assert that it is raining that morning in Seattle. Then what Downs says is false, and what Smith says is not.

When someone utters a sentence, and thereby succeeds in asserting something, at least *two* things get done: first, a sentence is uttered, and second, something is said (claimed, asserted) that is either true or false. The considerations advanced in the preceding paragraph suggest that we should not identify what is said, which is either true or false, with the *sentence* uttered in the saying of it.

Now if you reflect on whatever objections you may have raised to these arguments, you will very likely find that they fall into two groups. In the one group are those objections involving a claim that one or more of the premises of a certain argument are false; in the other are those involving the claim that the conclusion of the argument in question simply does not follow from its premises. The distinction between these two sorts of criticism to which an argument may be subjected can be drawn more easily if we look at four comparatively trivial examples, where a more exhaustive discussion is feasible.

(VII) The cube of four is either 16 or 64.
It is not the case that the cube of four is 16.
Therefore the cube of four is 64.

(VIII) Either all the premises of argument (VII) above are true or else some of the premises of argument (VII) are false.
Not all the premises of argument (VII) are true.
So some of the premises of argument (VII) are false.

(IX) Either litmus turns blue in a basic solution or else litmus turns blue in an acid solution.
Litmus turns blue in a basic solution.
Therefore litmus turns blue in an acid solution.

(X) Either the moon is larger than the earth or else the moon is larger than the sun.
The moon is larger than the earth.
Therefore the moon is larger than the sun.

Which of these arguments are good arguments, and which are bad ones? If asked to label them, most would probably write

(VII) Good

(VIII) Bad

(IX) Bad

(X) Bad

though some might put 'Good' beside (VIII) instead.

(VII) is pretty clearly a good argument; its premises are both true, and its conclusion certainly follows from them. (X) is just as clearly bad. Its second premise is false; its first premise is also false; and even if both premises were true, the conclusion still would not have to be, for it does not follow from them.

And now we can notice that (IX) suffers from one of the defects of (X) and (VIII) from the other. What's wrong with (IX) is, to put it simply, that its conclusion just does not follow from its premises. (We may not yet be able to *explain* exactly what we mean when we say that a particular sentence does not follow from certain others, but in such an easy case as this one we can certainly *recognize* what's going on.) Incidentally, this is just what's right about (VIII), and what probably explains why some labeled (VIII) 'Good' rather than 'Bad'. Argument (VIII) is bad in one sense, because its second premise is not true. But if the premises of (VIII) *were* both true, the truth of its conclusion would then be guaranteed—the conclusion of (VIII) certainly does follow from its premises.

All this suggests a way for us to improve a little on our vague 'Good'-'Bad' terminology. Why not check each of our arguments from *two* standpoints? First ask: *are all its premises true?* Then ask: *regardless* of whether or not the premises of this argument are all true, *does its conclusion follow* from them?

(VII) Good All premises are true
Conclusion follows from them

(VIII) Bad Not all premises true
Conclusion does follow from them, though

(IX) Bad All premises are true
Conclusion, however, does not follow from them

(X) Bad Not all premises true
Conclusion does not follow from them—even if they *were* all true, the conclusion wouldn't have to be

It is customary to use some special terms in discussing these matters. An argument is said to be **valid** if its conclusion follows from its premises, and **invalid** if its conclusion does not follow. Alternatively, we shall sometimes say that the premises **imply** the conclusion when the latter follows, and that they do *not* imply it when it does not follow. So every valid argument with true premises must have a true conclusion, while an invalid argument can have true premises but a false conclusion. When an argument is valid and in addition all its premises are in fact true, it is called

sound. Arguments which are not sound—arguments, that is, which either are invalid or have at least one false premise—are **unsound.**

Returning to the old examples with our new terminology, we have:

(VII)	Good	All premises are true Conclusion follows from them	Sound
(VIII)	Bad	Not all premises true Conclusion does·follow from them, though	Unsound (though valid)
(IX)	Bad	All premises are true Conclusion, however, does not follow from them	Unsound (because invalid)
(X)	Bad	Not all premises true Conclusion does not follow from them—even if they *were* all true, the conclusion wouldn't have to be	Unsound (on *both* counts)

The terms 'good' and 'bad', which were never very useful ones, may now be discarded. If someone gives an argument whose conclusion we think is suspect, there are really only two legitimate ways we can attack it. Either we must challenge one or more of the premises, or else we must claim that the conclusion does not follow from them. If we can do neither of these things—if we admit that all the premises of the argument are true, and also admit that the conclusion follows from them—then of course we must admit that the conclusion is true.

Of course, the premises of different arguments concern topics in a wide variety of disciplines, and as a result the job of checking them for truth or falsity falls most often within the province of a specialist in some area other than logic—a philosopher, a chemist, a mathematician, a literary scholar, or the like. It is clearly not the business of logic to dip litmus paper into lye, for example, nor to measure the circumference of the moon. Our concern in the remainder of the text will not be with soundness and unsoundness, then, but with validity and invalidity—features of an argument which, as we shall see shortly, are independent of the specific subject matter of the sentences comprising it.

DEFINITION   An argument is **valid** (its premises **imply** the conclusion) if and only if its conclusion follows from its premises; that is, if and only if it is impossible for all its premises to be true but its conclusion false.

DEFINITION   An argument is **invalid** (its premises do not imply the conclusion) if and only if its conclusion does not follow from its premises; that is, if and only if it is at least possible for all its premises to be true but its conclusion false.

The usefulness of these definitions is limited, of course, by the vagueness of such terms as 'follow from', 'not follow from', 'possible', and 'impossible'. It will be the task of the rest of this book to offer some analyses of these terms as they might be applied to a wide range of arguments found in everyday and technical discourse.

For now, we only note that the terms 'possible' and 'impossible' must be sharply distinguished from the terms 'probable' and '(highly) improbable' — a distinction paralleling that between **deductive** methods for evaluating arguments and **inductive** methods. Only deductive methods will be developed in this book. We mention inductive methods merely to contrast the two. The following arguments, for example, are most naturally assessed from an inductive standpoint:

(XI) Jones badly needed money to pay off a gambling debt.
Jones was having an affair with Smith's wife, who would collect on Smith's insurance policy in case of his death.
Smith was shot with a .38 caliber pistol belonging to Jones.
Two reliable witnesses saw Jones leaving Smith's house on the night of Smith's murder about fifteen minutes after the coroner's estimated time of death.
Jones's fingerprints were found on the murder weapon.
Smith's wife testifies that she conspired with Jones to murder her husband.
Therefore Jones murdered Smith.

(XII) Jones badly needed money to pay off a gambling debt.
Jones was having an affair with Smith's wife, who would collect on Smith's insurance policy in case of his death.
Smith was shot with a .38 caliber pistol belonging to Jones.
Two reliable witnesses saw Jones leaving Smith's house on the night of Smith's murder about fifteen minutes after the coroner's estimated time of death.
No fingerprints were found on the murder weapon.
Smith's wife admits an affair with Jones, but denies any murder plot.
Therefore Jones murdered Smith.

In neither case does the conclusion follow from the premises. It is at least *possible*, albeit not *probable* nor perhaps even very likely, for them all to be true without Jones having murdered Smith. From the point of view of deductive logic, then, each is invalid and there is little more to be said.

But the methods of inductive logic employ different criteria in investigating the relation between the premises and conclusions of such arguments. The assessment is made according to the relative degree of *evidence* the premises provide for the conclusion—the greater the support or evidence, the stronger would the argument be said to be. Depending on the extent to which the premises confirm or disconfirm the conclusion, we would say that the conclusion is *probable* or *improbable* with respect to them. When an argument is assessed as being *strong*, there is a high probability that its conclusion is true, and when assessed as *weak* its probability is low (relative to the evidence supplied by the premises in both cases).

Deductive evaluations are all-or-nothing affairs. When an argument is valid, it is *impossible* for its premises to be true while its conclusion is false. In contrast, inductive evaluations are matters of degree, ranging from very low to very high. When an argument is strong it is *highly improbable* that its premises are true but its conclusion false. The degree would presumably be high in the case of argument (XI) above (perhaps to the extent that a jury would consider the conclusion to follow *beyond reasonable doubt*), fairly low in the case of argument (XII) (the evidence is slim, and mostly circumstantial).

Inductive inference is very common in science, law, and everyday life; for that reason, it is no less important a topic for study than deductive inference. We omit further discussion of inductive inference hereafter mainly because a thorough and systematic study of it is best undertaken only after a familiarity with deductive methods has been acquired.

EXERCISES

I.   Construct a brief, clear-cut example of your own of:
1.   an invalid argument whose premises are all true and whose conclusion is true as well;
2.   an invalid argument whose premises are all true but whose conclusion is false;
3.   an invalid argument whose premises are all false and whose conclusion is true;
4.   an invalid argument whose premises and conclusion are all false;
5.   a valid argument whose premises are all false but whose conclusion is true;

6. a valid argument whose premises are all false and whose conclusion is false as. well;
7. a valid argument whose premises are all true.

II. Explain why any correct answer to (7) above consists of an argument whose conclusion is true.

*III. Construct arguments of your own concerned with:
1. the existence or nonexistence of numbers with some property or properties you regard as interesting;
2. the existence or nonexistence of a machine capable of thinking;
3. the existence or nonexistence of human beings other than yourself;
4. the speed of light;
5. the effect of a species' environment on its development over the centuries;
6. the effect of a human being's environment on his or her cultural development.

Which of your arguments are valid and which invalid? Which are sound and which unsound? Which might be more fruitfully studied from the point of view of *inductive* logic?

1.2 'And', 'Or', and 'Not'

Arguments are our prime concern, but because they are themselves sequences of *sentences* we must begin by looking at the latter. We intend to restrict our attention to those contexts in which each sentence that occurs is either true or false. In so restricting ourselves we do not of course mean to suggest that every sentence is either true or false — imperatives and interrogatives, for example, are clearly neither — nor to deny that a sentence may be used in different contexts to say different things. But the tools and techniques to be developed in this book are only intended to be applicable in those contexts in which each sentence encountered is used unambiguously to say something either true or false.

Strictly speaking, considerations similar to those advanced in example (VI) of the preceding section suggest that it may be improper to apply the terms 'true' and 'false' to *sentences*. Suppose, however, that we introduce a technical term and speak instead of the **truth-value** of a sentence in a particular context (provided that the sentence is used unambiguously in that context and, in all its occurrences, is used to make just one claim). A sentence will be said to **have the truth-value T** in a particular context if and only if it is used unambiguously there and expresses something *true,* and will be said to **have the truth-value F** in that context if and only if it is used unambiguously to express something *false.* We can then restore our original way of speaking: for the sake of *convenience* only, we can agree to call a sentence with the truth-value **T** in a particular context 'true' (in that context), and one with the truth-value **F** 'false'. Nothing in our manner of speaking has changed, but everything is different; when *precision* is more important than convenience, such talk can always be replaced by talk about the truth or falsity of that which, in that context, the sentence expresses.[1]

It will be convenient also to borrow a pair of terms from the grammarians, though our definitions will differ slightly from theirs, and distinguish **compound sentences** from **simple sentences**. The former contain *other* sentences — more precisely, other independent clauses — as proper parts; the latter do not. Thus 'Socrates is wise and Plato is wise', 'Either Socrates is wise or Plato is wise', and 'It is not the case that Socrates is wise' are all compound sentences since each contains at least one other

---

[1]Depending upon the reader's philosophical predilection, these things that sentences may be used to express which are true or false in the primary sense may be thought of as statements, propositions, or whatever. Readers without such predilections may acquire some by reading P.F. Strawson, *Introduction to Logical Theory,* London: Methuen, 1952, Chapters 1 and 6; or W.V. Quine, *Philosophy of Logic,* 2nd ed., Cambridge: Harvard University Press, 1986, Chapter 1. See also R.L. Cartwright, "Propositions," in R. Butler (Ed.), *Analytical Philosophy,* First Series, New York: Barnes and Noble, 1966.

sentence — 'Socrates is wise' or 'Plato is wise' — as a proper part. Since the latter contain no other sentences, however, they themselves are simple.

Simple sentences can be regarded, then, as the building blocks from which compound sentences are constructed, and such expressions as 'and', 'either ... or ---', and 'it is not the case that ...' may be thought of as the mortar used to hold them together. These three expressions will be of particular interest to us because our initial concern will be with compound sentences that contain them and whose truth-values are determined solely by the truth-values of their component sentences. Such sentences suffice for the construction of a number of easily analyzed arguments that will serve as our starting points in the next two chapters.

When one sentence can be obtained from another by prefixing 'it is not the case that ...' to it, the former is called the **negation** of the latter. Thus

(1) It is not the case that God is omnipotent

is the negation of

(2) God is omnipotent;

and

(3) It is not the case that it is not the case that God is omnipotent

is, in turn, the negation of (1).

If two sentences are so related that they must have opposite truth-values, they are said to be **contradictories**, and each is a **contradictory** of the other. (1) and (2) are contradictories, then, as are (1) and (3). Indeed, given our restriction to sentences that are either true or false, an important general point emerges: a sentence and its negation will always be contradictories. Where 'p' represents any sentence, its negation — the corresponding sentence of the form 'it is not the case that p' — will be *false* if 'p' is true and *true* if 'p' is false.

In many cases, of course, English allows us to get the effect of prefixing 'it is not the case that ...' to a sentence without going to quite so much trouble. Sometimes we can simply prefix 'not' to the sentence ('*Not* all men are mortal') or insert it into the verb ('God is *not* omnipotent'). Certain verbs require 'does not' instead ('Litmus *does not* turn blue in an acid solution') and still other variations are possible (such as 'Jim is *un*married', 'Argument (VII) is *in*valid', 'Hercules is *im*mortal', and 'Smedley is *no* ladies' man').[2]

Suppose now that we begin with (2) and

(4) God is benevolent

and combine them into a single compound sentence by inserting the word 'and' between them:

(5) God is omnipotent and God is benevolent.

Clearly *both* (2) and (4) must be true for (5) to be true — if God is not omnipotent, for example, or if He is not benevolent, (5) is false. Compound sentences like (5), which would be true if both components were true but false otherwise, are called **conjunctions** and their components are called their **conjuncts**.

Again, English syntax provides alternative constructions. In neither

(6) God is omnipotent and benevolent

nor

(7) God and UNIVAC are omnipotent,

for example, does 'and' serve to connect two sentences. But it is clear that (6) will be true if and only if (2) and (4) both are, and that (7) will be true if and only if (2) and 'UNIVAC is omnipotent' are, so we will count them as conjunctions as well.

Of course we might also combine (2) and (4) into a single sentence by inserting the word 'either' before the first and 'or' between them:

(8) Either God is omnipotent or God is benevolent.

---

[2] It is natural to extend our terminology slightly and call these sentences 'negations' as well. But one should not presume that such options are always open to us. If we try inserting 'not' into the verb of 'Some dogs are spaniels', for instance, we produce 'Some dogs are not spaniels', and this clearly does not have the same force as 'It is not the case that some dogs are spaniels'. Indeed, since 'Some dogs are spaniels' and 'Some dogs are not spaniels' are *both* true, they are not contradictories and so neither could be the negation of the other.

(As with negations and conjunctions, alternate forms of speech are common, such as 'God is omnipotent or benevolent'.) When we try to state the conditions under which (8) would be true, however, a complication arises. It is clear that it will be true if just one of its components is true and false if neither of them is true, but what about the case where *both* components are true? What if God is not only omnipotent but benevolent as well?

In most contexts a person who uttered (8) might justifiably say that in doing so he or she was only claiming that *at least* one of its components is true because he or she was using 'either... or ---' in the sense of 'either... or --- or both'. In such a case we would surely count (8) as true. That is, of course, a very common way of using 'either... or ---'; a sentence in which it is used in this way will be called a **nonexclusive disjunction**, and its components its **disjuncts**. In almost any context one can imagine, the sentence "All who are over 65 or who require major surgery will receive hospital benefits" would involve a clear case of nonexclusive disjunction, for the intent would surely be that a person who was *both* over 65 *and* required major surgery would receive benefits just as would anyone meeting only one of those conditions. And one who says "The senator's remarks reveal that he is either stupid or ignorant of the issues" does not mean to rule out the possibility that those remarks result from both stupidity and ignorance.

On the other hand one can imagine a different context in which (8) was uttered in order to sternly quash the supposition that God was both omnipotent and benevolent. An age-old argument, for instance, contends that the two qualities are incompatible in a Supreme Being. Were He benevolent He would want a perfect universe and were He all-powerful He could have one; since the universe is not perfect—so the argument goes—He clearly cannot be *both* omnipotent and benevolent. In the context of such a debate (8) would perhaps have to be understood as synonymous with a sentence of the form 'either ... or --- *but not both*'. Sentences of this sort will be termed **exclusive disjunctions**, and they are clearly false when both their *disjuncts* are true. Examples of exclusive disjunctions seem harder to come by than examples of nonexclusive ones, but think of a weary husband who has been on a day-long shopping trip with his wife. In response to her remark that they still must buy groceries and visit her relatives he might say in exasperation "*Either* we buy groceries *or* we visit your relatives." He obviously does not intend to do both, and if both are done he will have to add the utterance of a false sentence to his woes.

To determine in context the truth-value of a sentence containing 'either ... or ---', then, it is sometimes necessary first to determine whether it is being used in the sense of 'either ... or --- (*or both*)' or in the sense of 'either ... or --- (*but not both*)'. In the first case the sentence would be a nonexclusive disjunction and so would be true if at least one disjunct were true and false only if both disjuncts were false. In the latter case, however, it would be an exclusive disjunction and so would be true if and only if its disjuncts differed in truth-value and false if their truth-values coincided. Both kinds of disjunction are false when both disjuncts are false and true when just one disjunct is true, but nonexclusive disjunctions are true — and exclusive ones false — when both disjuncts are true.

Of course it often makes no difference. In such examples as

Either $2 + 2 = 4$ or $2 + 2 = 5$
Either Jones is now in New York or he is now in Paris
Either Simmons is a bachelor or he is married

the component disjuncts are incompatible — they could not both be true. It would be a mistake, though, to conclude that these examples must be construed as exclusive disjunctions. The fact that each could be followed by the phrase 'but not both' without any change in truth-value need not at all be the result of using 'either ... or ---' in the exclusive sense, but rather the result of disjoining two sentences whose incompatibility precludes any question of joint truth — a situation entirely unlike the remark of the weary husband. Indeed, these sentences may be interpreted in either sense — their disjuncts by themselves exclude one another, and it makes no difference to the truth-values of these disjunctions whether we read the 'either ... or ---' as reinforcing this exclusion or not.[3]

We've looked at three kinds of compound sentences: negations, formed typically by prefixing 'it is not the case that ...' to a sentence; conjunctions, formed often by inserting 'and' between two sentences; and disjunctions, formed by wrapping 'either ... or ---' around a pair of sentences. It is customary to call such compounding expressions **sentential connectives**, and important to notice that the truth-value of a compound sentence built up with the aid of the particular connectives we've discussed can often be determined once we know the truth-values of its components. Consider, for example,

(9) Either it is not the case that all human beings are mortal or it is not the case that both all human beings are men and all men are mortal.

---

[3]Cf. W.V. Quine, *Methods of Logic* (ed. 4), Cambridge: Harvard University Press, 1982.

Since 'All human beings are mortal' is true, its negation—the first disjunct of (9)—is false. Since 'All human beings are men' is false and 'All men are mortal' true, their conjunction is false. So the negation of that conjunction—which is the second disjunct of (9)—is true. So (9) has exactly one true disjunct and it is true whether construed exclusively or non-exclusively.[4]

EXERCISES

I. If we replace '*p*' with a true sentence, '*q*' with a false one, and '*r*' with a sentence whose truth-value is unknown to us, which of the following would be true? Which false? Which cannot be determined?
1. Either *p* or both *q* and *r*
2. *p* and either not-*q* or *r*
3. Either not-*p* and *q*, or *p* and not-*r*
4. Either not-*p* or *q,* and either not-*q* or both *r* and *p*
5. *r* and not-*r*
6. Either *q* and *r* or not both *p* and *r*

II. What if we replace '*p*' and '*r*' with false sentences, and '*q*' with a true one?

---

[4]From now on, we shall use the simple term 'disjunction' to mean 'nonexclusive disjunction'. When we want to talk about exclusive disjunctions, we shall use the term 'exclusive disjunction'.

1.3 '&', 'v' and '∼'

We now introduce some abbreviatory devices. To represent *simple* sentences we shall use capital letters of the English alphabet, calling them **sentence letters** to remind ourselves of their function. Using '*O*' for 'God is omnipotent' and '*B*' for 'God is benevolent', for example, the conjunction

(5) God is omnipotent and God is benevolent

can be shortened to '*O* and *B*'. It will be helpful also to introduce some symbols to serve as *sentential connectives*—symbols that perform the same jobs as 'and', 'or', and 'not' by combining with sentence letters to produce compound sentences. In place of the word 'and' in (5) it is natural, for example, to use the **ampersand,** '&', abbreviating (5) most fully as

(5′) *O* & *B*.

Since (5′) abbreviates a conjunction, we shall extend our terminology slightly and regard (5′) itself as a conjunction, and '*O*' and '*B*' as its conjuncts. If '*K*' represents a third sentence (say, 'God is omniscient'), '(*O* & *B*) & *K*' would also be a conjunction, with '*O* & *B*' and '*K*' as conjuncts. In such a case it is sometimes convenient to treat '*O*' and '*B*' as conjuncts of '(*O* & *B*) & *K*' as well.

For 'either ... or ---' in the *nonexclusive* sense—that is, for 'either ... or --- (*or both*)'—we will use the **wedge** symbol, 'v'.[5] Construed nonexclusively, then,

(8) God is omnipotent or God is benevolent

can be written

(8′) *O* v *B*.

Like (8), which it abbreviates, (8′) is a nonexclusive disjunction—or for short just a *disjunction*—and the expressions flanking the 'v' its disjuncts.

---

[5]Unlike English, Latin had two distinct words for 'or': *vel*, used for nonexclusive disjunctions, and *aut*, used for exclusive ones. Our symbol 'v' is derived from the first letter of the former.

As in the case of conjunctions, we will regard 'O' and 'B' individually as disjuncts of '$(O \lor B) \lor K$', along with '$O \lor B$' and 'K'. As will be seen shortly, no special symbol for·'either ... or ---' in the exclusive sense will be needed.

The sign '$\sim$', called the **curl** or the **tilde**, will abbreviate the phrase 'it is not the case that'. Thus

(1) It is not the case that God is omnipotent

can be abbreviated

(1') $\sim O$,

and "It is not the case that it is not the case that God is omnipotent" can be written '$\sim \sim O$'.

Parentheses must be included in our notation so that we can distinguish, for example, between

(10) $\sim (O \, \& \, B)$

and

(10') $\sim O \, \& \, B$.

(10) goes into English most idiomatically as 'God is not both omnipotent and benevolent', while (10') could be translated 'God is not omnipotent and He is benevolent'. Clearly, denying that God is *both* omnipotent and benevolent is not at all the same as affirming that he lacks the first of these attributes but possesses the second. Were God neither omnipotent nor benevolent, then (10), being the negation of a false conjunction, would be true, but (10'), a conjunction with a false right conjunct, would be false.

In addition to sentence letters, sentential connectives, and parentheses, it will be convenient to use small letters from the middle of the alphabet — 'p', 'q', 'r', 's' (adding, if the need arises, numerical subscripts as in 'p_1' and 'q_3') — as **sentential variables**. Unlike sentence letters, sentential variables will not be used to represent particular sentences but rather may be thought of as standing in a given content for *any* arbitrarily selected sentence. In the expression '$p \lor q$', for example, we do not have an abbreviation for a particular disjunction but something more like a schematic picture of disjunctions generally. In effect, '$p \lor q$' presents us with an overall *pattern* or *form* shared by many particular sentences; for this reason expressions like '*p* v *q*' containing sentential variables are called **sentence forms**.[6] Sentence forms are not themselves sentences, and they do not have truth-values; rather they are employed to represent indiscriminately all sentences of the form depicted.

The sentences '*O* v *B*', '*O* v ~*K*', and '(*O* & *B*) v *K*' are all derivable from '*p* v *q*' by making appropriate substitutions for '*p*' and '*q*'. Similarly, the third of these sentences — but not the first two — is also derivable by substitution from '(*p* & *q*) v *r*', and the second — but neither the first nor the third — is derivable from '*p* v ~*q*'. It is helpful to remember that a sentence form like '*p* v *q*' stands to such sentences as '*O* v *B*' and '(*O* & *B*) v *K*' as '$x + y$' stands to '$3 + 1$' and '$(3 - 1) + 2$' in arithmetic. It is precisely because an expression like '$x + y$' signifies no particular number that it is useful in discussing general properties shared by all numbers; and it is because sentence forms stand for no particular sentences that they will be useful in discussing features common to *all* conjunctions or to *all* disjunctions, and so forth.

Reflection on our earlier discussion of the phrases 'and', 'either ... or --- (or both)', and 'it is not the case that' enables us to define in their contexts of use our three sentential connectives '&', 'v', and '~' by specifying the conditions under which sentences containing them will be true or false.

Conjunction   A sentence of the form '*p* & *q*' is true if and only if both conjuncts are true, and false otherwise.

---

[6]More precisely, if less lucidly, we define the notion of a **sentence form** recursively as follows:

  i. Each sentential variable ('*p*', '*q*', '*r*', '*s*', 'p_1', 'p_2', etc.) is a sentence form;
  ii. the result of prefixing '~' to any sentence form is a sentence form;
  iii. the result of inserting '&' or 'v' between any two sentence forms and enclosing the whole in parentheses is a sentence form;
       nothing is a sentence form unless it can be shown to be so on the basis of the above clauses.

Thus '*p*' and '*q*' are sentence forms (clause i) so '~*p*' and '~*q*' are also (clause ii) as are '(*p* & *q*)' and '(*p* v ~*q*)' (clause iii). Then '~~*p*' and '~(*p* v ~*q*)' are also sentence forms (clause ii again), and so on.

It is customary to follow the practice, employed throughout this text, of omitting *outer* parentheses when no ambiguity would result, writing '*p* v *q*' for '(*p* v *q*)' for example, and to replace parentheses with brackets and braces when readability is thereby promoted (for example, '~{[~(*p* & *q*) v *r*] & *s*}' to replace '~((~(*p* & *q*) v *r*) & *s*)'). Of course the parentheses in '~(*p* & *q*)' *cannot* be omitted.

We can now give a technically satisfying definition of a sentence (in our notation): a **sentence** is an expression obtainable from a sentence form by replacing the sentential variables in the latter with sentence letters.

Disjunction   A sentence of the form '*p* v *q*' is true if and only if at least one of its disjuncts is true, and false otherwise.

Negation   A sentence of the form '~*p*' is true if and only if its component is false, and false if its component is true.

We have already noted that

(10) ~(*O* & *B*)

must be distinguished from '~*O* & *B*'. We can notice now, however, that (10) and

(11) ~*O* v ~*B*

may be used interchangeably, and more generally a sentence of the form '~(*p* & *q*)' will always convey precisely the same information as could be conveyed by the corresponding sentence of the form '~*p* v ~*q*'. A conjunction, remember, is true *only* when both components are true and so (10), the negation of a conjunction, will be *false* only when '*O*' and '*B*' are both true. Now (11), a disjunction, is false only when both disjuncts, '~*O*' and '~*B*', are false; hence (11) is false only when their contradictories, '*O*' and '*B*', are *true*. Thus (10) and (11) have the same truth-value regardless of the truth-values of their components: they will both be true so long as '*O*' and '*B*' are not both true, and they will both be false if '*O*' and '*B*' are both true.

Let us now consider:

(12) ~*O* & ~*B*

and

(13) ~(*O* v *B*).

Again we have two sentences that have the same truth-value under all conditions, but here both will be true only when '*O*' and '*B*' are both *false*. Since (12) is a conjunction, it will be true only when '~*O*' and '~*B*' are both true, hence only when '*O*' and '*B*' are both false. And since '*O* v *B*' is false only when both disjuncts are, (13) — the negation of '*O* v *B*' — will be true only when 'O' and 'B' are both false. More generally, any two sentences of the forms '$\sim p$ & $\sim q$' ('not-p and not-q', in idiomatic English) and '$\sim(p \lor q)$' ('it is not the case that either p or q') will be synonymous, and it should be apparent that English contains yet another and more frequently used expression equivalent to both of these: 'neither p nor q'. Using any of the three locutions, one will have said something true if and only if both 'p' and 'q' are false, so sentences of this sort may be symbolized either in the manner of (12) or of (13).

This should give some idea of how conjunction, disjunction, and negation intermesh,[7] but there are other features of our sentential connectives that warrant brief attention. First, both '&' and '$\lor$' are *commutative*—that is, the order of the component sentences in conjunctions and disjunctions does not affect the truth-value of the whole. Compare the operation of addition in arithmetic: just as '3 + 2' and '2 + 3' signify the same number, so must 'A & B' and 'B & A' have the same truth-value, as must '$A \lor B$' *and* '$B \lor A$'. Moreover, conjunction and disjunction are *associative*—that is, the placing of parentheses in compounds with three or more components does not affect the truth-value of such compounds. Thus '$(A$ & $B)$ & C' will be true under the same conditions as 'A & $(B$ & $C)$', since each will be true only when all three conjuncts are true. The same considerations hold for '$A \lor (B \lor C)$' and '$(A \lor B) \lor C$', since each of these is false only if all three disjuncts are false. Again there is a parallel with arithmetic, for '$(3+2)+6$' signifies the same number as '$3+(2+6)$'.

On the other hand, just as no number is uniquely signified by '$6-5+1$', so there is a similar ambiguity with 'A & $B \lor C$'. Here parentheses are essential, for without them we would have no idea whether this is supposed to be a disjunction whose first disjunct is a conjunction, or a conjunction whose second conjunct is a disjunction. That the difference is important can be seen by letting 'A' and 'B' represent false sentences and 'C' a true one. If the above expression is construed as the disjunction '$(A$ & $B) \lor C$', it will be true because its second disjunct is true. Construed as the conjunction 'A & $(B \lor C)$', however, it will be false owing to the falsity of its first conjunct. These two sentences, therefore, do not have the same truth-value under all conditions, and parentheses are indispensable for distinguishing them. We shall call the connective which determines the basic form of a sentence its **main connective**. Thus, '$(A$ & $B) \lor C$' has '$\lor$' as its main connective, while the main connective in 'A & $(B \lor C)$' is '&'. But '$\sim(A$ & $B)$' is a negation, so its main connective would be '$\sim$'.

Ambiguities of these kinds occur less often in natural languages

---

[7]Traditionally, these equivalences are known as **De Morgan's Laws**, and they will be used extensively in Chapter 3.

such as English. From the relative positions of 'either', 'or', and 'and', for example, we can tell that an English sentence of the form 'either p and q or r' is a disjunction with the corresponding sentence 'p and q' as first disjunct. But one of the form 'either p or q and r' would require the same kind of clarification as does 'A v B & C', for one cannot tell from its structure whether it is a disjunction with a conjunction as a component disjunct or a conjunction with a disjunction as a component conjunct. In conversation, such ambiguities are usually alleviated by emphasis or inflection on the part of the speaker. In printed contexts, commas serve the same purpose; we would recognize 'Either Al will go, or else Bill will go and Charlie will go' as a disjunction, for example.

It is worth noting that the latter sentence could be expressed in English by saying instead 'Either Al or Bill will go, and either Al or Charlie will go'. The more general fact is often put by saying that 'v'*distributes* over '&', that is, that a sentence of the form 'p v $(q$ & $r)$' always has the same truth-value as the corresponding sentence of the form '$(p$ v $q)$ & $(p$ v $r)$'.

To see that this is so, assume first that 'p v $(q$ & $r)$' is true. If the first disjunct, 'p', is true, then clearly both 'p v q' and 'p v r' are also true. And if on the other hand 'q & r' is true, then again both 'p v q' and 'p v r' will be, the first because 'q' is and the second because 'r' is.

Conversely, assume that '$(p$ v $q)$ & $(p$ v $r)$' is true. If 'p v q' and 'p v r' are both true because 'p' is, then 'p v $(q$ & $r)$' will be true on the same grounds. The only other possibility is that 'p v q' is true because 'q' is and that 'p v r' is true because 'r' is. But then again 'p v $(q$ & $r)$' will be true, because its right disjunct will be.

Concerning the '~' we confine ourselves here to a brief remark about *double negation*. When the negation of a sentence is itself negated, the resulting sentence is equivalent to the original sentence alone. Since the negation of a sentence is true if and only if that sentence is false, clearly '$\sim\sim O$' as well as 'O' will be true if and only if '$\sim O$' is false. Thus the sign '$\sim$' may be applied to a sentence any number of times and any even number of applications will yield a sentence with the same truth-value as the original.

EXERCISES

I. Translate each of the following English sentences into symbols:
   1. The number four is even and the number five is odd.
   2. The number five is a prime number, and the number four is not.
   3. It is not the case that the numbers four and five are both prime.
   4. The numbers four and five aren't both prime, but the numbers five and seven are.
   5. Neither four nor six is a prime number.

      6. Five is a prime number, but neither four nor six is.

      7. Either six and four are both prime, or else neither is.

II.   Assign truth-values to '*A*', '*B*', and '*C*' such that '*A* & ~(*B* v *C*)' is true but '(~*A* & ~*B*) v (*C* & *B*)' is false.

III. Which of the following sentences are derivable from the sentence form '~*p* v (*q* & ~(*p* v *r*))'? For those which are, specify the substitutions for '*p*', '*q*', and '*r*'.

      1. ~*B* v (*A* & ~(*B* v *D*))

      2. ~*B* v (*A* & ~(*C* v *D*))

      3. ~*A* v (~(*B* & *C*) & ~(*A* v *D*))

      4. ~~*A* v (*B* & ~ (*A* v *C*))

      5. ~*A* v (*A* & ~(*A* v *A*))

      6. ~~*C* v (*E* & ~(*C* v *F*))

IV. Is the sentence '~~*A* & (~(*B* v *C*) & ~*A*)' derivable from any of the following? Briefly explain in each case.

      1. *p* & *q*

      2. *p* & (~*q* & ~*p*)

      3. ~*p* & (*q* & *p*)

V.   Show that '&' *distributes* over 'v', that is, that a sentence of the form '*p* & (*q* v *r*)' is true under the same conditions as '(*p* & *q*) v (*p* & *r*)'. You can use reasoning similar to that in the text for showing distribution of 'v' over '&'.

VI. Find a sentence in symbols having the same truth-conditions as '*A* & *B*' but which contains no occurrence of the connective '&'.

Of course, many English words besides 'and' can serve to indicate conjunction and consequently be handled with the ampersand. For example, the word 'but' is so used in

(14) The senator attended the rally but did not make a speech,

since (14) would clearly be true if both component sentences were true and false if either or both were false. Using '*A*' for 'The senator attended the rally', then, and '*S*' for 'The senator made a speech', (14) can be symbolized as the conjunction '*A* & ~*S*'. Of course the sentence

(15) The senator attended the rally and did not make a speech

would also be symbolized 'A & $\sim S$', yet (14) and (15) seem to differ subtly in their meanings — the former would be more appropriate than the latter, for instance, if there were an implicit expectation that the senator *would* perhaps make a speech at the rally. Despite such rhetorical considerations, however, (14) and (15) would be *true* or *false* under precisely the same conditions; as a result, arguments containing (14) will be evaluated no differently from ones just like them except for containing (15) in place of (14). For our purposes, then, we may put considerations of nuance and connotation aside; sentences of the form 'p but q' may be rendered as 'p & q'.

Similar remarks apply to a wide assortment of English connectives. Those italicized in the following list, though they may differ in nuance from 'and', all serve to indicate conjunction and may be handled with the ampersand.

Although the weather was bad, the game was not called off.

John, *as well as* Sam, attended the meeting.

Not only will the Mets win the pennant, *but* they will win the Series *as well*.

Johnson was an aggressive president, *whereas* Nixon preferred a low profile.

Despite its controversial status, the bill passed the Senate.

The game was exciting, *albeit* violent.

The notions are simple *yet* the subject is not.

The list is not meant to be exhaustive; indeed, the reader should be able to extend it considerably with a little effort. It should be noted also that conjunction may be indicated not by any particular word or phrase but by punctuation alone: Caesar's "I came, I saw, I conquered" is a three-component conjunction.

And now the task of expressing an exclusive disjunction such as

(16) *Either* we buy groceries *or* we visit your relatives, *but not both*

in terms of the connectives '&', 'v', and '~' is no more than an easy exercise. (16) cannot be put simply as

(16') G v R,

of course, for the latter is true — but the former false — if both disjuncts are true. So to express (16) fully we rule out this possibility by translating 'but' as '&' and incorporating into (16') the expression '... but not both' from (16):

(16″) $(G \lor R)$ & $\sim(G$ & $R)$.

Many words besides 'and', then, may be used to indicate conjunction. But we must not jump to the conclusion that such terms always function in this manner. In "None but the strong survive", for example, we do not have a compound sentence at all, 'but' serving merely as a synonym for 'except'.

To further complicate matters, the word 'and' itself can be used nonconjunctively. 'Mary and Jane are sisters' cannot be construed as a conjunction of 'Mary is a sister' and 'Jane is a sister', for these might be true while the first sentence was false. The force of the first sentence is clearly that Mary and Jane are sisters *of one another*. Here, then, 'and' does not serve to connect two sentences; the role it does play in this example will be studied in Chapter 5.

Even where 'and' appears between two sentences as a sentential connective, we need not have a (mere) conjunction. Genuine conjunctions, recall, are commutative — the order of the two conjuncts within the sentence is irrelevant to the truth-value of the whole. But this is not the case with

(17) Jones bought stock and the market soared

or

(18) The market soared and Jones bought stock,

as those who buy at the peak of a bull market can testify. Here 'and' definitely serves to connect two sentences, but plays a temporal role as well. Normally there would be the distinct suggestion with (17) that the event referred to first occurred before that referred to last. In effect 'and' occurs in such sentences in the sense of 'and then', and that sense is not captured by our '&'.

Using 'J' for 'Jones bought stock' and 'M' for 'The market soared', we may of course capture *part* of the sense of (17) — and the same part of the sense of (18) as well — with

(17') *J* & *M*,

for if the latter were false — if Jones did not buy stock, or if the market did not soar — (17) and (18) would both be false as well. But it is important to remember that (17') does not tell the *whole* story — only *part* of the force of (17) is reflected in our notation.

When we turn to disjunctions, we find that English provides only trivial alternatives to the 'either ... or ---' construction: the 'either' may be omitted, or the 'or' followed by 'else'. It is worth noting, however, that 'unless' seems sometimes to stand to 'v' as 'and then' stands to '&'. If, for example,

(19) Jones did not buy stock unless his left elbow itched

is true, then

(20) Either Jones did not buy stock or else his left elbow itched

will be true as well. For if the latter were false — if Jones did buy stock but his left elbow did not itch — then (19) is clearly false as well. So at least *part*[8] of the sense of (19) can be expressed (using obvious notation) by

(20') ~*J* v *L*.

For many purposes the fact that our symbolic resources are limited is not terribly serious. Thus, we may not be able to translate with full accuracy the premises of the following argument:

Jones bought stock and (then) the market soared.
Jones did not buy stock unless his left elbow itched.
Therefore, Jones' left elbow itched.

We can nonetheless consider a related argument in symbols

---

[8]If 'Jones's left elbow itched' is true, it follows that (20) is true also. It is not clear that (19) follows as well, and this is why it is not clear that we can capture the *whole* sense of (19) with (20). We note, however, that when in exercises later on we use the word 'unless', we do so to avoid terminological monotony and the reader may safely take us to mean no more than 'or' by it.

J & M
$\sim J \vee L$
Therefore L

The latter is clearly valid—if its premises are both true then its conclusion must be also, for it follows from the first premise that Jones did buy stock and from that and the second that his left elbow itched. And since (17) and (19)—the premises of the English argument—can only be true if the premises are, it is clear that the former must be valid as well.

Translating English compound sentences into symbols can sometimes be an arduous task if they have a highly complex structure. But the job can be made easier by proceeding in a systematic, step-by-step fashion. As an example, let us consider the sentence

Either the president will veto the bill and Congress will attempt to override or else he will cut his losses, but neither of them will be happy with the outcome unless the labor unions are mollified.

Our first step is to identify the component simple sentences. In the present case, we have six components:

V - the president will veto the bill
O - Congress will attempt to override
L - the president will cut his losses
P - the president will be happy with the outcome
C - Congress will be happy with the outcome
U - the labor unions are mollified

Next, we replace the English components with sentence letters giving:

Either V and O or L, but neither P nor C unless U

We are now in a better position to recognize the logical structure of the sentence. Our next step is to determine the main connective and then work on the component parts separately. The comma clearly indicates that the disjunction on the left ends at that point, and on the right 'neither' begins a second compound, so the main connective here is 'but'. Replacing it with '&' gives

(either V and O or L) & (neither P nor C unless U)

Now we work on the lefthand component. From the placement of 'and' and 'or' we can see that it is a disjunction whose left disjunct is a conjunction, hence giving '$(V \& O) \vee L$'. On the other side, the claim is that condition U is required for *both* P and C, and hence the main connective within the righthand pair of parentheses is 'unless'. Since that word is translated by 'v' we again have a disjunction: '(neither P nor C) v U'. As we saw earlier, the remaining bit of English can be handled in either of two ways, as '$\sim(P \vee C)$' or as '$\sim P \& \sim C$'. Choosing the former — and using brackets to relieve the monotony of parentheses — we have:

$$[(V \& O) \vee L] \& [\sim(P \vee C) \vee U]$$

EXERCISES

I. Put each of the following sentences into symbols using the suggested notation:

1. The Hope diamond is valuable, but not priceless. (V,P)
2. Either the battery is weak or both the distributor and the starter need repair. (B,D,S)
3. The bill will either be vetoed or else it will pass neither the House nor the Senate. (V,H,S)
4. The next card drawn won't be both a heart and a face card, but it will improve my hand anyway. (H,F,I)
5. For you I'll climb the tallest mountain, swim the widest ocean, cross the burning desert. (M,O,D)
6. Either the tax bill or the voting rights bill will pass, but the President will not support both of them. (T,V,A,R)
7. Withdrawals will not be honored unless presented by the depositor and accompanied by his passbook. (W,P,A)
8. Either the gold and silver issues will decline or the Dow-Jones index will fall sharply, but in no case will the President's or Congress' decision on the bill be affected. (G,S,D,P,C)
9. Although he is willing, the vice-president will not deliver the speech unless it is printed in crayon. (V,D,P)
10. Smith won't be able to deduct the mortgage interest on his new house unless he either takes up residence in it or else sells one of his other houses. (D,R,S)
11. Wilson and Jones are both mathematicians, and at least one of them is a computer programmer as well. (W,J,I,O)
12. While the numbers 9, 21 and 45 are all odd, none of them are primes. (N,T,F,I,W,O)
13. At least two of the numbers 2, 4 and 6 are composite. (T,F,S)

14. At most one of the numbers 2, 5 and 7 is even. (T, F, S)
15. Not only will the Democrats and the Republicans both denounce the president's veto, but neither will support his new defense budget without his cooperation on the foreign aid measure. (D, R, M, P, C)

*II. What can be said about the following argument?
Jones did not bluff on that deal unless he was not dealt a strong hand. Jones was dealt a strong hand on that deal and (then) won the whole pot.
Therefore Jones did not win the whole pot unless he bluffed on that deal.

1.4 Material Conditionals and Biconditionals

We have seen that the English disjunction

(20) Either Jones did not buy stock or else his left elbow itched

is correctly translated as

(20') $\sim J \vee L$.

However, in many contexts (20) might be paraphrased as

(21) If Jones did buy stock then his left elbow itched.

We shall call (21) a **conditional** and, following standard practice, refer to 'Jones did buy stock' as its **antecedent** and to 'Jones's left elbow itched' as its **consequent**. Our contention that (20) and (21) stand or fall together can be borne out by arguing informally that (20) and (21) must have the same truth-value. To this end, suppose first that (20) is *true*. Then either Jones did not buy stock or else his left elbow itched. If Jones *did* buy stock, then (20) leaves us with only one alternative: his left elbow itched. Thus, when (20) is true (21) must be also. Now suppose (20) is *false*. Then its disjuncts, 'Jones did not buy stock' and 'Jones' left elbow itched,' must be false. But (21) then has a true antecedent and a false consequent, and in such a case it is surely false. Should it turn out that Jones bought stock but his left elbow did *not* itch, (21) would scarcely be considered true. When (20) is false, then, so is (21).

Thus the conditions under which (21) would be true or false coincide with those for (20); and this point may be put by saying that (21) would be false if and only if its antecedent were true and its consequent false, and true otherwise. Hence not only (20) but (21) could be translated using (20'). However for several reasons, including the fact that the 'if-then' locution occurs so often in argumentative discourse, it will be much more convenient to introduce a special sign for it in our symbolic notation. We shall use the sign ' $\supset$ ' (called the "horseshoe") for this purpose, and a sentence of the form ' $p \supset q$ ' will be called a **material conditional.**[9] The reason for the qualification "material" is that ' $p \supset q$ ' does not capture the full meaning of all

---

[9] For convenience the term 'material' will be omitted hereafter where the context makes clear that sentences of our symbolic notation are under discussion.

English conditionals, though as we will see in the next chapter enough is captured to serve our purpose of analyzing arguments in which conditionals occur (the term will also be applied to any English conditional whose truth conditions do coincide with those of '$p \supset q$'). So the sentence

(21') $J \supset L$.

will be true or false under the same conditions as (20')—both will be assigned **F** only when 'p' is true and 'q' is false. We thus have:

DEFINITION   A **material conditional** whose antecedent is 'p' and whose consequent is 'q' is a compound sentence built up from 'p' and 'q' (typically, one of the form 'if p then q'), which is false if and only if 'p' is true and 'q' false, and true otherwise.

As with negations and conjunctions, English provides alternate forms of speech. The 'then' may be replaced with a comma ('*If* Jones did buy stock, his left elbow itched') and the order may be reversed ('Jones's left elbow itched *if* he bought stock'). And often in place of a sentence of the form 'if p then q' one uses quite different words: 'provided that p, q', 'in the event that p, q', or 'p only if q' (for example, '*Provided that* the merchandise is returned, the money will be refunded', '*In the event that* it rains, the game will be postponed', and 'Smith works in the Empire State Building *only if* Smith works in New York City'). It is especially important to notice this last example. Obviously

(22) Smith works in the Empire State Building only if Smith works in New York City

is true. It can only be construed, then, as synonymous with

(22') If Smith works in the Empire State Building then Smith works in New York City,

and it would be a mistake to think that it could be paraphrased by

(23) If Smith works in New York City then Smith works in the Empire State Building;

he might work in the Chrysler Building.

Sentences (22′) and (23) are said to be *converses* of one another; that is, (23) is the *converse* of (22′) and vice versa. Clearly they need not have the same truth-value, and hence 'if *p* then *q*' and its converse, 'if *q* then *p*', do *not* convey the same information. On the other hand, 'if *p* then *q*' and 'if not-*q* then not-*p*' do convey the same information; the sentences, 'If everyone participated then everyone benefitted' and 'If not everyone benefitted then not everyone participated', are but two ways of making the same point. A similar relation obtains between

If Jesse lies and Frank cheats, then the James Boys are dishonest
If Jesse lies, then if Frank cheats, then the James boys are dishonest

More generally then, sentences of the forms, 'if *p* and *q*, then *r*' and 'if *p*, then if *q* then *r*' say one and the same thing.

All of these characteristics hold for the ' $\supset$ ' of our symbolic notation as well as for the English 'if-then': '*p* $\supset$ *q*' and its converse '*q* $\supset$ *p*' differ in their truth conditions, for the former is false only when '*p*' is true and '*q*' is false, but the latter is false only when '*q*' is true and '*p*' is false. So unlike '&' and 'v', the connective ' $\supset$ ' is *not* commutative. However, '*p* $\supset$ *q*' and '$\sim$*q* $\supset$ $\sim$*p*' are indeed true under the same conditions: both are false only when '*p*' is true and '*q*' is false. Similarly, we have sameness of truth conditions for '(*p* & *q*) $\supset$ *r*' and for '*p* $\supset$ (*q* $\supset$ *r*)': both are false only when '*p*' and '*q*' are true and '*r*' false.

The methods we will develop for dealing with conditionals are only *fully* applicable to sentences of the form 'if *p* then *q*' which can be understood in such a way that their truth-conditions coincide with those for the corresponding sentences of the form 'not-*p* or *q*'. But since virtually every English conditional implies the corresponding material one, even in those cases where our translation of 'if *p* then *q*' as '*p* $\supset$ *q*' does not capture the full meaning of the English original it nonetheless often captures an important aspect of that meaning.

It would be nicer, of course, if *material* conditionals could simply be identified with conditionals generally, that is, with 'if... then---' sentences (and their idiomatic variants). But there are many sentences containing 'if... then---' whose truth or falsity seems to depend on more than the truth-values of their components. To take but one example (see Section 2.2 for more), the material conditional

This filament is made of wood $\supset$ this filament will conduct electricity

must be counted as true so long as its antecedent, 'This filament is made of wood', is false. So it will be true if the filament in question is made, for example, of steel. It seems plausible to suppose, however, that the English sentence "If this filament is made of wood then this filament will conduct electricity" is false regardless of the actual composition of the filament.

It thus emerges that material conditionals are but one sort of conditional one might study. We single them out for study in this book partly because they yield most readily to elementary methods, and partly because many of their features are shared by most conditionals.[10]

One other sentential connective occurs often enough in arguments to justify introduction of a special symbol: '... if and only if---' ('... just in case---' is a synonym). Of course such sentences as

(24) I visit my friend Smith if and only if I visit New York

and

(25) I visit my friend Smith just in case I visit New York

can be handled with the connectives already introduced—we have only to notice that each may be paraphrased by

(26) If I visit my friend Smith then I visit New York, and If I visit New York then I visit my friend Smith

and so can be rendered as

(26') $(S \supset N) \& (N \supset S)$.

Alternately, we could notice that (24) − (26) may all be paraphrased by

(27) Either I visit my friend Smith and I visit New York, or else I don't visit my friend Smith and I don't visit New York,

---

[10]Material conditionals are also the conditionals commonly employed in certain other disciplines besides logic. In most high-level computer languages that have a connective (often written "IMP") for forming conditionals, for example, the conditionals thus formed are *material* ones.

and so render any of them as

(27′) $(S \mathbin{\&} N) \vee (\sim S \mathbin{\&} \sim N)$.

Instead we will introduce the symbol ' $\equiv$ ' (the **triple bar** or **bars**), using it in effect as an abbreviation for the expression '...if and only if---' as it occurs in (24) so that the latter may be written simply

(24′) $S \equiv N$.

We shall call (24) and (24′) **material biconditionals**, construing them as true when their components have the same truth-value and false when their components differ in truth-value.[11]

In translating conditionals and biconditionals, it is important (as always) to identify the main connective. For example, the sentence

(28) If Smith lives in New York and Jones lives in Tallahassee, then I visit my friend Smith if and only if I do not visit Jones

is clearly a conditional and so goes partially into symbols as

$(N \text{ and } T) \supset (S \text{ if and only if not-}J)$.

Translating the conjunctive antecedent and the biconditional consequent, then, we arrive at:

(28′) $(N \mathbin{\&} T) \supset (S \equiv \sim J)$.

Now of course

(29) Jones lives in Florida if he lives in either Tallahassee or Orlando

is also a conditional—the position of 'either' in

---

[11]To extend the notion of a sentence form, we now add to the definition in footnote 6 a new clause:

   iv.   the result of inserting '$\supset$' or '$\equiv$' between any two sentence forms and enclosing the whole in parentheses is a sentence form.

J if either *T* or *O*

shows that the '*T*' and '*O*' components are the two disjuncts of a disjunction. Recalling that sentences of the form '*q* if *p*' must be translated as '*p* ⊃ *q*', we thus obtain

(29') (*T* v *O*) ⊃ *J*.

EXERCISES

I.   Put each of the following sentences into symbols, using the suggested notation:

1. If the sun's rays hit the polar regions squarely, then the polar regions will turn warm. (*R, W*)
2. If the carburetor is not at fault, then the problem lies with either the fuel filter or the gas line. (*C, F, G*)
3. If Kite misses his tap-in and the spectators groan loudly, Watson's concentration will be disturbed. (*M, G, D*)
4. If time is not finite and not circular, then time has neither a beginning nor an end. (*F, C, B, E*)
5. If his heart stops beating, the patient will live if the doctors apply electric shock. (*S, L, E*)
6. Withdrawals made after the first ten days of an interest-paying month will be honored only if the depositor has given the bank written notice of his intent to withdraw as well as the amount of the withdrawal. (*W, I, A*)
7. Jones will graduate if he pays his tuition and passes the requisite number of courses. (*G, T, C*)
8. Jones will graduate whether or not he pays his tuition. (as above)
9. I'll do the job if, but only if, Nelson won't. (*I, N*)
10. If the job has to be done, then I'll do it for you just in case you don't find anyone else to do it. (*J, I, F*)
11. If the spectators let out a spontaneous shout if Kite misses his tap-in, Watson will win. (*S, M, W*)
12. Both William of Ockham and Willard of Cambridge are wrong if universals exist in reality as well as the mind. (*O, C, R, M*)
13. If you want it, you can have it, but you shouldn't blame me if you get into trouble. (*W, H, B, T*)
14. If neither the pilot nor the co-pilot was ill at the time, then the fault lies with the control panel if and only if the explosion occurred before or during warm-up. (*P, C, F, B, D*)

15. If the chairman attends the meeting but either the CEO or the treasurer does not, the Acme deal will be concluded only if either Baker or Evans takes charge of the situation. (C, O, T, A, B, E)

II.   Show that '$\supset$' is not associative by finding sentences 'p', 'q', and 'r' such that '$p \supset (q \supset r)$' and '$(p \supset q) \supset r$' have different truth-values.

III.   Is '$\equiv$' either commutative or associative?

IV.   Are there sentences 'p' and 'q' such that '$p \equiv \sim q$' and '$\sim(p \equiv q)$' have different truth-values? How about '$p \equiv \sim q$' and 'either p or q, but not both'? Show your work.

1.5 Scope of Our Inquiry

In Sections 1.2 and 1.3 we examined *sentences*, focusing on sentential connectives like 'and' and 'or'. Typically, the truth-values of compounds containing them are determined solely by the truth-values of their components. It was just this characteristic we relied on, in fact, in defining such special connectives as '&' and 'v'.

Compound sentences, whether in English or symbols, whose truth-values depend entirely on the truth-values of the sentences occurring in them—are called **truth-functional**. And it is important to realize that English and other natural languages also contain **non-truth-functional** sentences, compounds whose truth-values are *not* dependent solely upon the truth-values of their components. The connectives used in forming such compound sentences function quite differently, then, from those we have concentrated on in the last two sections; indeed, explaining just how they work is in many cases still a topic for philosophical investigation. 'And then'[12] provides one example of such a connective, and 'because' another. The sentence

The chairman resigned *because* he was in ill health,

for example, is not truth-functional. Even if 'The chairman resigned' and 'The chairman was in ill health' are both true, the truth-value of the compound itself is not determined—it may be true or it may be false, depending on whether ill health was in fact the *reason* for the resignation. This is quite unlike what happens with truth-functional sentences; if the truth-values of 'p' and 'q' are given, the truth-values of the corresponding sentences of the forms '$p \& q$', '$p \vee q$', and so on, are thereby determined.[13]

Verbs of *psychological attitude*—such as 'believes', 'doubts', 'hopes'—afford further examples of non-truth-functional compounds. For the truth-value of the sentence

Chris believes that the world is egg-shaped,

---

[12]The reader interested in acquiring an understanding of how investigations of such connectives proceed might examine G.E.M. Anscombe, "Before and After," *Philosophical Review*, vol. 73 (1964), pp. 3-24, and G.H. von Wright, "And Then," *Comm. Phys. Math. of the Finnish Society of Sciences*, vol. 32, no. 7 (1966).

[13]To count as truth-functional, a compound sentence must be such that its truth-value is determined by *every* combination of truth-values for its components. Thus even though the truth-value of this sentence *is* determined when 'The chairman resigned' is true and 'The chairman was in ill health' is false, the fact that it is not determined when both are true suffices to make it non-truth-functional.

(a compound built up from 'The world is egg-shaped' and 'Chris believes that ...') is in no way determined by the truth-value of the component sentence. That someone does or does not hold a certain belief can hardly be established by merely noting the truth-value of a sentence expressing the belief.

While the study of non-truth-functional compounds would be highly challenging, it presupposes a good deal of training in the more elementary areas of logic and so lies beyond the scope of an introductory text. We mention such sentences here only for contrast with our present concern: truth-functional compounds. Arguments containing the latter form the simplest and most basic subject matter within our ken, and it is with them that we must begin.[14]

It remains only to say a few words about the devices ordinarily used in English for marking off the conclusion of an argument. Often this is done by prefixing the word 'therefore' to the conclusion, as in

If the set of integers is finite, it has a largest member.
The set of integers does not have a largest member.
Therefore the set of integers is not finite.

Many other terms can do the job as well; 'consequently' in the place of 'therefore' would serve the same purpose, as would 'thus' or 'hence' or 'so'. Such terms do not function here as sentential connectives, but serve rather to suggest that a certain relation obtains *between* sentences. Typically, these terms are used to indicate that the writer or speaker is attempting to establish the truth of one claim on the basis of the truth of certain others—more specifically, that the former follows from the latter. We shall call them **conclusion indicators**.

It is also possible to express an argument in a single (compound) sentence, the *premises* being marked off explicitly by a **premise indicator** such as 'for':

The set of integers is not finite, for if it is finite it has a largest member but in fact the set of integers does not have a largest member.

In such cases one or more premises are often left unstated, having been assumed to be sufficiently obvious for readers to supply on their own:

The set of integers is not finite, for it does not have a largest member.

---

[14]Since non-truth-functional compounds are beyond our purview, no special symbols will be introduced for words like 'and then', 'because', and 'believes'. Instead, compounds involving such terms must be represented by single sentence letters, as if they were simple sentences.

Again other terms will do the job, 'for' being replaceable with similar effect here by such words as 'since', 'inasmuch as', and—an old friend, used in a new sense—'because'.[15]

It will be helpful, in what follows, to have one standard way of marking off conclusions. We shall use the standard symbol '∴' for this purpose.

If the set of integers is finite, it has a largest member.
The set of integers does not have a largest member.
∴ The set of integers is not finite.

Custom suggests that '∴' be read 'therefore'.

EXERCISES

Some of the following sentences are truth-functional and some are not. Express in symbols those which are (those which are not must be represented by single sentence letters, e.g., 'A').

1. Even though the roads were icy, Sam arrived on time.
2. He was her man, but he done her wrong.
3. While no one complained, they weren't very happy either.
4. Jim thinks a sea dragon inhabits the Wabash.
5. Swarthmore will not defeat Penn unless God intervenes.
6. Swarthmore will not defeat Penn even if God intervenes.
7. Scott will come as long as he doesn't have to wear a tie.
8. Since no one objected, the bill passed without modifications.
9. Sam and Irma have been man and wife for three years.
10. God knows that $2 + 2 = 4$.
11. The owl and the pussycat went to sea together.
12. If Swarthmore wins, I'll eat my hat.
13. In the event of rain, the game will be postponed.
14. It's necessarily true that $12 - 7 = 5$.
15. 'Twas the night before Christmas, and all through the house, not a creature was stirring, not even a mouse.
16. It is possible but not likely that the Cubs will win.
17. The sentence 'Chicago is in Illinois' is true.
18. Though many have tried none have succeeded, but I feel sure someone will, and soon.

---

[15]Presumably it goes without saying that these terms do not *always* serve to distinguish premises from conclusion, nor to indicate that an argument is being given. In "Tom was so late he missed the train" and "I have not seen him since nine o'clock" no arguments are involved and neither 'since' nor 'so' functions as a premise indicator.

CHAPTER TWO

SENTENTIAL LOGIC
Semantic Methods

2.1 Truth-Tables for '&', 'v', and '∼'

In Chapter 1 we noticed that a sentence of the form 'p & q' is true if and only if both conjuncts are true, a sentence of the form 'p v q' is true if and only if it has at least one true disjunct and one of the form '$\sim p$' is true if and only if the negated sentence is false. Each type of compound was thus explained in terms of the conditions under which it would be true, and these conditions may be set out more explicitly in tabular form. The following **truth-tables** lay out the truth-conditions for conjunctions, disjunctions and negations respectively and therefore may be viewed as formally defining the connectives '&', 'v' and '$\sim$'.

(I)	p	q	p & q
	T	T	T
	T	F	F
	F	T	F
	F	F	F

(II)	p	q	p v q
	T	T	T
	T	F	T
	F	T	T
	F	F	F

(III)	p	$\sim p$
	T	F
	F	T

Table (III) has but two rows, since there are just two truth-values the component of a negation may have: it may be true or it may be false. Conjunctions and disjunctions typically have two components, however, so their truth-tables require four rows: one for each of the possibilities (1) that both components are true, (2) that the first is true but the second false, (3) that the first is false but the second true, and (4) that both are false. As table (I) shows, conjunctions are true only in the first of these cases; but a disjunction is shown by table (II) to be true on each of the first three conditions—it has a **T** in all but the last row of its column where both disjuncts are assigned **F**. Each row of these truth-tables, then, gives: first, an assignment of truth-values to the components of a truth-functional compound; and second, the truth-value of the compound itself *for* that assignment.

Now for any truth-functional compound sentence containing one or more of '&', 'v', and '~', we can construct a truth-table exhibiting its truth-conditions by using the above tables as our guide. Consider, for example, the sentence '~*A* v (*B* & *A*)'; from table (III) we can ascertain for each row the truth-values for the component '~*A*', and table (I) allows us to do the same for the conjunction '*B* & *A*'. Thus having the truth-value assignments in each of the rows for both disjuncts, the assignments for the entire disjunction are obtained using table (II).

A	B	~A	B & A	~A v (B & A)
T	T	F	T	T
T	F	F	F	F
F	T	T	F	T
F	F	T	F	T

A compound with three components will require a truth-table of eight rows to set out the conditions under which it will be true and false, since there are exactly eight possible assignments of truth-values to any triad of sentences. Similarly, a four-component compound requires 16 rows, a five-component compound 32 rows, and so on; the number of rows doubles as the number of components is increased by one.[1]

Let us consider more closely compounds with three components; what applies to them applies to those with four or more. In particular, we will examine two such compounds cited back in Chapter 1: '*A* & (*B* v *C*)' and '(*A* & *B*) v *C*'. There is an easy technique for making sure that the table contains all eight truth-value combinations—that none have been duplicated or omitted. Assign all **T**'s to the first four rows beneath the first simple component ('*A*' in this case) and all **F**'s to the last four. Then alternate two rows of **T**'s and two of **F**'s beneath the second component, and finally alternate **T**'s and **F**'s beneath the last. To determine what truth-values '*A* & (*B* v *C*)' will have for each of these eight cases, we first obtain the assignments for the right-hand conjunct '*B* v *C*', putting a **T** in every row of its column where either or both of '*B*' and '*C*' are assigned **T**, and an

---

[1]In general, then, a sentence with *n* components (*n* a positive integer) requires a truth-table with 2^n rows. Note that whereas '(*A* & *B*) v *C*' has three components and so requires an eight-row table, '(*A* & *B*) v *A*' has but *two* components (though '*A*' has more than one *occurrence*) and requires only a four-row table.

F in the others. We then determine the truth-values for the entire conjunction by placing a **T** in every row of its column where both '*A*' and '*B* v *C*' are assigned **T,** and an **F** in the remaining rows. By the same token, the truth-values for '(*A* & *B*) v *C*' are obtained by first getting the truth-values for the component conjunction '*A* & *B*' and then putting a **T** in just those rows where a **T** has been assigned to at least one of the disjuncts, '*A* & *B*' and '*C*':

A	B	C	B v C	A & (B v C)	A & B	(A & B) v C
T	T	T	T	T	T	T
T	T	F	T	T	T	T
T	F	T	T	T	F	T
T	F	F	F	F	F	F
F	T	T	T	F	F	T
F	T	F	T	F	F	F
F	F	T	T	F	F	T
F	F	F	F	F	F	F

Comparing the columns for '*A* & (*B* v *C*)' and '(*A* & *B*) v *C*', we see that they are assigned different truth-values in rows 5 and 7. Thus, this table shows graphically what was noted in Chapter 1, that these two sentences convey quite different information. When, for example, the sentences represented by '*A*' and '*B*' are false and that by '*C*' true, '(*A* & *B*) v *C*' will be true but '*A* & (*B* v *C*)' false as row seven illustrates.

Let us now construct truth-tables for '*A* v ~*A*' and '~*A* v (*B* v *A*)':

A	~A	A v ~A
T	F	T
F	T	T

A	B	~A	B v A	~A v (B v A)
T	T	F	T	T
T	F	F	T	T
F	T	T	T	T
F	F	T	F	T

The truth-table columns for both sentences contain nothing but **T**'s, and they are thus true under *all* conditions – no circumstance whatever will render them false. Any truth-functional sentence that is true no matter what truth-values we may suppose its simple component sentences to have will be called a **tautology**.

DEFINITION   A truth-functional sentence is a **tautology** if and only if it takes the truth-value **T** for every assignment of truth-values to its component sentence letters.

We shall use the sign ' $\models$ ' (the "double turnstile") in abbreviating the assertion that a given sentence is a tautology.[2] So instead of writing

The sentence 'A v $\sim A$' is a tautology

we shall simply write

$\models A$ v $\sim A$.

Thus while ' $\models A$ v $\sim A$' abbreviates a true assertion, ' $\models A$ & B' abbreviates a false one.

Tautologies obviously form a unique subclass of true sentences, for they are true not because the facts are one way rather than another, but because of their *form*. It is because 'A v $\sim A$' is of the form 'p v $\sim p$'—a disjunction one of whose disjuncts is the negation of the other—that it is tautologous. Since they are true independently of how the world happens to be constituted, tautologies are often said to be *necessarily* true, or in philosopher's jargon, true in "all possible worlds."

It should be obvious that no matter what sentence is substituted for 'p' in the form of 'p v $\sim p$' the result will be a tautology. Hence not only is 'A v $\sim A$' true on purely formal grounds, but so too are 'B v $\sim B$' and (making complex substitutions for 'p') ' $\sim C$ v $\sim \sim C$' and '$(A$ & $B)$ v $\sim (A$ & $B)$'. We shall call any sentence form that yields nothing but tautologies upon the substitution of sentences (simple or compound) for its variables a **tautologous form** (the same sentence must of course replace the same variable throughout). To be a tautology, then, a sentence must be derivable by substitution from at least one tautologous form, but it must not be thought that tautologies are derivable *only* from tautologous forms. For while 'A v $\sim A$' is derivable from the tautologous form 'p v $\sim p$', it is also derivable from the nontautologous 'p v q' (substituting 'A' for 'p' and ' $\sim A$' for 'q'). Whereas tautologous forms yield only tautologies upon sub-

---

[2]The sign '$\models$' should *not* be confused with the sentential connectives that are here our prime objects of study. We use it only to save ourselves verbiage. Readers should compare '$\models$' with the sign '$\vdash$' of Section 3.1.

stitution, nontautologous forms can yield both tautologies and nontautologies.

Since their truth is independent of the facts, tautologies offer us no real information as to how things are (or are not). In uttering "It is now raining," one offers some information (or misinformation) about the environment, but in "Either it is now raining or it is not," no information is given concerning the present state of the weather. All tautologies are "trivial" in this sense owing to their lack of factual content, and the reader may be hard put to imagine an actual context where it would not be pointless to utter one. Trivial though they seem, however, tautologies nonetheless are an important tool in the logician's repertoire. Their necessary truth gives them a theoretical significance which we shall make use of in Section 2.3 and 3.1, as well as in Chapter 4.

One more peculiar sort of sentence reveals itself when we look at the truth-table for the negation of '$A \lor \sim A$':

A	$\sim A$	$A \lor \sim A$	$\sim (A \lor \sim A)$
T	F	T	F
F	T	T	F

As its table shows '$\sim (A \lor \sim A)$' has nothing but **F**'s in the column beneath it—there is no condition under which it could possibly be true. Such sentences will be called **contradictions**. Clearly the negation of any tautology is a contradiction (and vice versa), and just as a tautology is necessarily true so a contradiction is necessarily false—its falsehood being due entirely to its form. Thus sentence forms like '$\sim (p \lor \sim p)$' and '$p \& \sim p$' shall be termed **contradictory forms**. Everything that has been said above concerning tautologies and tautologous forms applies *mutatis mutandis* to contradictions and contradictory forms.

It was noted in the preceding chapter that the sentences '$\sim O \& \sim B$' and '$\sim (O \lor B)$' convey the same information, which may be expressed in English by 'God is neither omnipotent nor benevolent'. A truth-table with four rows will establish this in a more lucid fashion than the argument given in Chapter 1.

O	B	$\sim O$	$\sim B$	$\sim O \& \sim B$	$O \lor B$	$\sim (O \lor B)$
T	T	F	F	F	T	F
T	F	F	T	F	T	F
F	T	T	F	F	T	F
F	F	T	T	T	F	T

The table shows the two sentences to have exactly the same arrangement of **T**'s and **F**'s in their respective columns—thus showing them to be true under just the same conditions.

Sentences true under the same conditions have thus far been characterized as *saying the same thing* or *conveying the same information*. We shall now introduce a special term to replace these rather vague phrases.

DEFINITION   Two truth-functional sentences are **equivalent** if and only if, for each assignment of truth-values to their component sentence letters, both receive the same truth-value.

It follows from this definition that all tautologies are equivalent to one another and to no nontautologies; and similarly that each contradiction is equivalent to all other contradictions but to no noncontradictions.

Moreover, just as it is in virtue of their forms that certain sentences are tautologies, so is it in virtue of their respective forms that certain pairs of sentences are equivalent. Any sentence whatever of the form '~*p* & ~*q*' will be equivalent to the corresponding sentence of the form '~(*p* v *q*)'. For the truth-table in effect shows us that the negation of *any* disjunction will have the same truth-conditions as the corresponding conjunction of the negated disjuncts; the particular sentential content makes no difference, since any sentences whatever could have been used in place of '*O*' and '*B*' in constructing the truth-table. Thus, we shall extend our use of 'equivalent' and say that the sentence forms '~*p* & ~*q*' and '~(*p* v *q*)' are equivalent as well. Plainly, then, when two *sentences* are equivalent, they are derivable from a pair of equivalent sentence forms through the substitution of their component sentences for the variables—the same sentence replacing the same variable throughout.[3] It should now be clear that the use of truth-tables is, from a practical standpoint, unaffected by the type of expression—whether sentential variable or sentence letter—employed in its construction. The following table, for example, differs little in appearance from the preceding one, but it shows the equivalence of the forms '~*p* & ~*q*' and '~(*p* v *q*)':

p	*q*	~*p*	~*q*	~*p* & ~*q*	*p* v *q*	~(*p* v *q*)
T	T	F	F	F	T	F
T	F	F	T	F	T	F
F	T	T	F	F	T	F
F	F	T	T	T	F	T

---

[3]Three important characteristics of *equivalence* among sentences are:
1. Every sentence (sentence form) is equivalent to itself.
2. If one sentence (sentence form) is equivalent to a second, then the second is equivalent to the first.
3. If one sentence (sentence form) is equivalent to a second and the second to a third, then the first is equivalent to the third.

EXERCISES

I. Construct truth-tables to determine which of the following are tautologies, which are contradictions, and which are neither:
   1. $\sim(A \,\&\, \sim A)$
   2. $B \vee (\sim B \,\&\, B)$
   3. $\sim(C \,\&\, D) \vee C$
   4. $(G \,\&\, \sim H) \,\&\, [\sim G \vee (H \,\&\, G)]$
   5. $[(R \,\&\, S) \vee (R \,\&\, \sim S)] \vee [(\sim R \,\&\, S) \vee (\sim R \,\&\, \sim S)]$
   6. $[(\sim A \vee B) \vee C] \vee (A \,\&\, \sim C)$
   7. $[(C \vee D) \,\&\, (C \vee \sim D)] \,\&\, [(\sim C \,\&\, E) \vee (\sim C \,\&\, \sim E)]$
   8. $\sim[(P \vee Q) \,\&\, R] \vee \sim(S \,\&\, Q)$

II. Using a truth-table determine whether the following sentence form is tautologous:

$$(p \,\&\, \sim q) \vee (\sim p \vee q)$$

Are any of the following sentences derivable by substitution from the above form? Are any of them tautologies?
   1. $(A \,\&\, \sim C) \vee (\sim A \vee \sim C)$
   2. $[\sim(B \,\&\, C) \,\&\, \sim A] \vee [\sim \sim(B \,\&\, C) \vee A]$
   3. $[\sim \sim A \,\&\, \sim(C \vee B)] \vee [\sim A \vee (C \vee B)]$

III. Using truth-tables, show that the following pairs of sentence forms are equivalent (the same table may be used for more than one exercise):
   1. $p \,\&\, q$; $q \,\&\, p$
   2. $p \vee q$; $q \vee p$
   3. $p \,\&\, (q \,\&\, r)$; $(p \,\&\, q) \,\&\, r$
   4. $p \vee (q \vee r)$; $(p \vee q) \vee r$
   5. $p \,\&\, (q \vee r)$; $(p \,\&\, q) \vee (p \,\&\, r)$
   6. $p \vee (q \,\&\, r)$; $(p \vee q) \,\&\, (p \vee r)$
   7. p; $\sim\sim p$
   8. $\sim p \vee \sim q$; $\sim(p \,\&\, q)$

IV. 1. "If a conjunction is a tautology, then each of its conjuncts must also be a tautology." True or false? Explain.
   2. "If the *negation* of a conjunction is a tautology and the *negation* of one of its conjuncts is a contradiction, then the other conjunct must be a contradiction." True or false? Explain.

2.2 Truth-Tables for '⊃' and '≡'

In Chapter One English sentences containing 'if... then---' and its idiomatic variants were introduced and *material* conditionals[4] singled out for special study. The latter as we saw are expressed in symbols by the sentential connective '⊃', and are false only when their antecedents are true and their consequents false. We also saw that one reason for counting them false in just such a case is that a material conditional of the form 'if p then q' is similar in meaning to the corresponding disjunction 'either not-p or q', which also of course is false only when 'p' is true and 'q' is false. The following truth-table spells out the truth conditions of sentences in our notation of the form '$p \supset q$' and hence serves to define '⊃' (the column for the equivalent '$\sim p \vee q$' is also displayed):

p	q	$p \supset q$	$\sim p$	$\sim p \vee q$
T	T	T	F	T
T	F	F	F	F
F	T	T	T	T
F	F	T	T	T

So 'if p then q' will be translated '$p \supset q$', but it should not be thought that '⊃' always captures the full meaning of 'if... then---'. Indeed, as we will see in this section, the situation is quite unlike that between, say, the conjunctive use of 'and' and '&'. However, we will also see that for the range of arguments with which we shall be concerned, this disparity in meaning is not one that need deter us. For logic at least, there is enough similarity for the task of determining validity.

Let us first look briefly at some of the problems in translating 'if... then---' as '⊃'. Students are sometimes bothered by the assignment of **T** in the last two rows of the truth-table for '⊃' (that is, where the antecedent is assigned **F**), though we could hardly assign an **F** in them since then '⊃' would mean the same as '&'. Even so, there is reason to feel uneasy, and to see why let us focus on the last row of the truth-table: '$p \supset q$' is assigned **T** when both antecedent and consequent are assigned **F**. In some cases this seems unproblematic, as with

(1) If 3 is a multiple of 2, then 3 is even.

---

[4] Some of the points made in this discussion are borrowed from W.V. Quine. See his *Methods of Logic* (ed.4), Cambridge: Harvard University Press, 1982.

Since (1) is true—multiples of 2 being even without exception—we have no qualms about translating it as, say, '$M \supset E$' and assigning it a **T**. However, the variables in '$p \supset q$' may be replaced by *any* pair of sentences we care to choose no matter how relevant or irrelevant their contents might be to one another. Thus where 'R' represents 'The world is flat' and 'F' represents '$2 + 2 = 5$', the truth-table for '⊃' obliges us to assign '$R \supset F$' a **T**. But what are we to say about the corresponding English sentence?

(2) If the world is flat, then $2 + 2 = 5$

Is (2) true? Is it false? Confronted with such a bizarre claim, most of us would probably refrain from making either judgment, settling instead for asking what the world's shape has to do with what the sum of $2 + 2$ is. And this points up two features of English conditionals uttered in day to day affairs that need not be present in sentences of our symbolic notation containing '⊃'. First, we normally do not use conditionals when the truth-values of the component sentences are a matter of common knowledge (as those of (2), and for that matter (1), are). Second and more importantly, we do not use conditionals unless we at least believe there to be some relevant connection between the two components. For example, when someone says, 'If the governor spoke, the audience was bored', there is uncertainty about who spoke and what the mood of the audience was, but the speaker nonetheless is aware of the governor's lack of oratorical skill (thus providing the relevant link between the components). But the truth-table for '⊃' carries no requirement about relevance; it simply tells us to put down a **T** whenever both components are assigned an **F**. No underlying principle or other knowledge connecting the subject matter of antecedent and consequent is presumed.

English conditionals come in many varieties and may be distinguished according to the type of relevant connection obtaining between antecedent and consequent. Sometimes, as in (1), it is a simple arithmetic principle; sometimes, as in the sentence about the governor, it is some specific knowledge about an individual. But often the connection is best construed in terms of one or another kind of *necessity*. The following conditionals exemplify two varieties of necessary connection:

(3) If all men are mortal and Socrates is a man, then Socrates is mortal

(4) If this match is heated to 150° Celsius in the presence of oxygen, then it will ignite.

(3) might be termed a *logical* conditional, for the connection claimed to hold between antecedent and consequent is one of premise to conclusion in

a valid argument: 'if... then (it follows necessarily that)---'. (5) might be termed a *causal* conditional in that the necessity linking antecedent and consequent is one of cause to effect: 'if... then (necessarily as cause to effect)---'.[5] In each of these cases, the 'if-then' sentence asserts considerably more than does its counterpart in symbols using ' $\supset$ ', in fact so much more that in many of these cases the conditional cannot even be regarded as a *truth-functional* compound (cf. Section 1.5). This is best brought out by comparing sentence (1) with a conditional like, 'If 3 is a multiple of 2, then 3 is less than 2'. Like (1) both components are false, but whereas (1) was adjudged true our inclination here is to render just the opposite assessment. This can never happen with ' $\supset$ ', for its truth-table tells us that once the components' truth-values are determined, the truth-value of the entire compound is *thereby* determined.

But though an English sentence of the form 'if *p* then *q*' often asserts more than its symbolic counterpart '$p \supset q$', the latter nevertheless captures an important element in the meaning of the former (even when non-truth-functional): namely that *any conditional must be adjudged false when its antecedent is true and its consequent false.* This feature holds of each of the English conditionals cited above, and thus when we translate them using ' $\supset$ ' we are abstracting from their full meaning and focussing on this purely truth-functional aspect shared by all. In doing so of course the various types of linkage are suppressed, but this generally turns out to be of little consequence. For our main purpose is to analyze and assess arguments, and if an English argument containing 'if... then---' is in fact valid, its validity is preserved when we translate it into symbols replacing 'if-then' with ' $\supset$ ', and moreover we may then *demonstrate* its validity using the methods to be laid out later in this chapter and in the next. For the range of arguments presently under consideration, that is sufficient reason for translating English conditionals using ' $\supset$ '.

There is another and closely related reason for the translation: we saw in Section 1.4 that important logical properties of the 'if-then' construction are captured by the truth-table column for ' $\supset$ '. For example, we saw that a conditional and its converse do not convey the same information, that is, that 'if *p* then *q*' and 'if *q* then *p*' are not equivalent. It may be true that if the governor spoke the audience was bored, but nonetheless false that if the audience was bored the governor spoke—their boredom being the result of

---

[5]While a study of logical and causal conditionals is a highly interesting philosophical enterprise, it is unfortunately beyond the scope of an introductory text. Readers interested in the analysis of logical conditionals are advised to consult C.I. Lewis and C.H. Langford, *Symbolic Logic,* New York: Dover, 1932, and also A. Anderson and N. Belnap, *Entailment,* Princeton: Princeton University Press, 1975. Those interested in causal conditionals should consult A. Burks, *Cause, Chance and Reason,* Chicago: University of Chicago Press, 1978.

a different though equally ineffective orator. On the other hand, 'if p then q'
is equivalent to 'if not-q then not-p': to say that if the audience wasn't
bored then the governor didn't speak is just another way of saying that if he
spoke boredom was the result. We also saw that sentences of the form 'if p
and q, then r' are equivalent to those of the form 'if p, then if q then r'; for
example, 'If the offense holds steady and the defense improves, the Eagles
will win' says no more nor less than, 'If the offense holds steady, then if the
defense improves, the Eagles will win'.

   What is important here is that these properties (and many others) hold
for '⊃' as well as 'if-then', as the following truth-table shows:

p	q	$p \supset q$	$q \supset p$	$\sim p$	$\sim q$	$\sim q \supset \sim p$
T	T	T	T	F	F	T
T	F	F	T	F	T	F
F	T	T	F	T	F	T
F	F	T	T	T	T	T

The equivalence of '$(p \ \& \ q) \supset r$' and '$p \supset (q \supset r)$' is left as an exercise.

   The *material biconditional* was also introduced in the first chapter,
and the following truth-table formally defines the connective '≡' as well
as demonstrating the equivalence of '$p \equiv q$' and '$(p \supset q) \ \& \ (q \supset p)$':

p	q	$p \equiv q$	$p \supset q$	$q \supset p$	$(p \supset q) \ \& \ (q \supset p)$
T	T	T	T	T	T
T	F	F	F	T	F
F	T	F	T	F	F
F	F	T	T	T	T

Thus, as noticed in Section 1.4, a sentence of the form '$p \equiv q$' is true if and
only if both components have the same truth-value. The problems noted
earlier in translating English conditionals apply *mutatis mutandis* to the
translation of English biconditionals into material ones. No claim is made
that '$p \equiv q$' always captures the full meaning of its English counterpart,
but only that it isolates a distinctive truth-functional aspect so that, in
general, validity is preserved in going from arguments with 'if and only if'
to '≡'. Moreover, the important properties of commutativity and as-
sociativity are shared by 'if and only if' and '≡'; '$p \equiv q$' is equivalent to
'$q \equiv p$', and '$p \equiv (q \equiv r)$' is equivalent to '$(p \equiv q) \equiv r$'.

EXERCISES

I.   Using truth-tables, show the following pairs equivalent:
1. $(p \& q) \supset r$; $p \supset (q \supset r)$
2. $p \equiv q$; $(p \& q) \vee (\sim p \& \sim q)$

II. Is '$p \equiv \sim q$' equivalent to '$\sim (p \equiv q)$'? To the negation of 'either p or q but not both'? Show your work.

III. Determine whether each of the following is a tautology, a contradiction or neither:
1. $(A \& B) \supset A$
2. $[(A \supset A) \supset B] \supset B$
3. $(A \equiv \sim B) \& (A \& B)$
4. $(F \vee E) \supset (E \& F)$
5. $(J \& K) \supset (L \supset \sim K)$

IV. To keep in practice, put each of the following into symbols using the suggested notation:
1. If either Corning or Owens wins the batting crown, both the press and the fans will be happy. (C, O, P, F)
2. Sam will come if Charlie does, but he won't stay long unless Emma is there. (S, C, L, E)
3. If either Baskin or Robbins does not perform up to expectations, then the Reds will win the pennant only if the pitching improves. (B, R, P, I)
4. I will not run if nominated, I will not serve if elected. (R, N, S, E)
5. Cohen will be the next Supreme Court Justice if and only if the President recommends him and the Senate approves, but provided that the Old Guard can muster the votes, the Senate will not approve. (C, P, S, O)

2.3 Testing for Validity: The Truth-Table Method

We are now in a position to test truth-functional arguments for validity through the use of truth-tables. (By a **truth-functional argument** we mean one whose premises and conclusion are either simple sentences or else built up from 'and', 'or', etc.) We begin with an elementary example:

Either the stock market will not improve or else the government will both lower interest rates and increase spending. However, the government either will not lower interest rates or else it will not increase spending. Therefore the stock market will not improve.

We shall first put the argument into symbols, using 'S' for 'The stock market will improve', 'L' for 'The government will lower interest rates', and 'I' for 'The government will increase spending'.

(I) $\sim S \vee (L \& I)$
     $\sim L \vee \sim I$
     $\therefore \sim S$

We know that valid arguments, when their premises are true, always carry us to true conclusions, so that it is impossible for the former all to be true when the latter is false. Moreover, we know that the truth-values of the premises '$\sim S \vee (L \& I)$' and '$\sim L \vee \sim I$', as well as the conclusion '$\sim S$', are determined once we assign truth-values to the letters 'S', 'L', and 'I'. So if we run through *all* possible assignments of truth-values to these letters and there is *no* combination resulting in true premises and false conclusion, the argument must be valid.

S	L	I	$\sim S$	$L \& I$	$\sim S \vee (L \& I)$	$\sim L$	$\sim I$	$\sim L \vee \sim I$
T	T	T	F	T	T	F	F	F
T	T	F	F	F	F	F	T	T
T	F	T	F	F	F	T	F	T
T	F	F	F	F	F	T	T	T
F	T	T	T	T	T	F	F	F
F	T	F	T	F	T	F	T	T
F	F	T	T	F	T	T	F	T
F	F	F	T	F	T	T	T	T

The argument's validity emerges from the fact that no matter what truth-values we may suppose 'S', 'L', and 'I' to have, there is no case in which all premises are rendered true but the conclusion false. The above truth-table of eight rows gives us all the required truth-value combinations for three sentence letters, each row giving us a particular assignment of truth-values. So looking at the columns for the premises and conclusion, the validity of the argument is demonstrated by there being no row of the table wherein both premises are assigned **T** and the conclusion **F**.

Comparing the columns for '$\sim S \vee (L \,\&\, I)$', '$\sim L \vee \sim I$', and '$\sim S$' in the truth-table on the preceding page, we can find no row in which the first two are assigned **T** and the third **F**. Notice that it is only in the last three rows that both premises are in fact assigned **T**, but since the conclusion also has a **T** in these rows, the argument is valid.

Of course an invalid argument can carry us from true premises to a false conclusion, and this means there is at least one truth-value assignment to the component letters making all premises true and the conclusion false. Therefore, in the truth-table for an invalid argument, *at least one* row will have all **T**'s assigned to the premises and an **F** to the conclusion. As an example, let us consider the argument in which premise two of argument (I) becomes the conclusion and vice versa:

(II) $\sim S \vee (L \,\&\, I)$
     $\sim S$
     $\therefore \sim L \vee \sim I$

The invalidity of (II) is shown by the fifth row of the above table. There, both '$\sim S \vee (L \,\&\, I)$' and '$\sim S$' are assigned **T** but '$\sim L \vee \sim I$' is assigned **F**. All that is required is that there be at least one such row, for invalid arguments are those in which a true conclusion does not *invariably* accompany true premises.

We shall apply the term **truth-table valid** (or "valid on truth-table grounds") to any argument whose conclusion is never assigned an **F** in a row where all its premises are assigned **T**'s, and the double turnstile sign '$\models$' of Section 2.1 will be used here to express this notion:

The argument with premises '$\sim S \vee (L \,\&\, I)$' and '$\sim L \vee \sim I$' and conclusion '$\sim S$' is truth-table valid

may be abbreviated by writing

$\sim S \vee (L \,\&\, I), \ \sim L \vee \sim I \models \sim S.$

Thus when no sentence occurs to the left of ' $\models$ ', the entire expression asserts that the sentence to its right is a tautology. But when, as above, one or more sentences are on its left, we have the assertion that the argument with those sentences as premises and the sentence to its right as conclusion is truth-table valid.

The sign ' $\models$ ' is not a sentential connective but merely a shorthand device for expressing the relation between premise(s) and conclusion in a truth-table valid argument. We may now define this notion as follows:

DEFINITION A truth-functional argument is **truth-table valid** if and only if there is no row of its truth-table in which all its premises are assigned the truth-value **T** but its conclusion the truth-value **F**.

We shall sometimes omit 'truth-table' for brevity. An invalid argument thus has at least one row with all premises true and conclusion false. Truth-tables therefore yield a mechanical cut-and-dried test for both validity and invalidity.

Now let us compare argument (I) with the following:

Transylvania will not negotiate unless both the Etruscans pull back their missiles and the Ruritanians stop the flow of weapons to the rebels. However, either the Etruscans will not pull back their missiles or the Ruritanians will not stop the flow of weapons. Therefore Transylvania will not negotiate.

Using fairly obvious abbreviations, this becomes:

(III) $\quad \sim N \vee (E \ \& \ R)$
$\quad \quad \sim E \vee \sim R$
$\quad \quad \therefore \sim N$

The same truth-table may be used to show validity here as was used for (I). The reason, of course, is that though arguments (I) and (III) differ in content — in their component sentences — they nonetheless share the same *form*. It should be obvious that *any* argument of this form is valid, inasmuch as the truth-values of the component sentence letters must correspond to one of the truth-value assignments listed in the rows of the truth-table. The table, then, provides a validity test for any argument of this form, not just a particular case like (I) and (III).

We may display the common form of (I) and (III) as a sequence of sentence forms corresponding to their premises and conclusion, in which each variable is paired with a distinct sentence or sentence letter.

(A)   ~p v (q & r)
      ~q v ~r
      ∴ ~p

This array of forms displays the skeletal structure of arguments (I) and (III) since the variables indicate positions within the structure where sentences or sentence letters may be inserted. We shall call any such sequence of sentence forms the last of which is preceded by '∴' an **argument form**, and (I) and (III) — though not of course (II) — are said to be arguments **of the form** (A) displayed above. More fully, let ϕ be any truth-functional argument and ψ any argument form:

DEFINITION ϕ is **of the form** ψ if and only if ϕ can be obtained from ψ by replacing each sentential variable of ψ with a sentence (simple or compound) — the same sentence for the same variable throughout.

Alternatively, we shall say that arguments (I) and (III) are **derivable by substitution** from (A) through the insertion of sentences or letters for the variables. No matter what sentences or letters replace those of (A), the truth-table shows that the resulting argument must be valid. So the truth-table test is best regarded·as one for the validity of an argument form rather than of particular arguments *having* that form. Let us then re-consider the above truth-table, this time using sentential variables:

p	q	r	~p	q & r	~p v (q & r)	~q	~r	~q v ~r
T	T	T	F	T	T	F	F	F
T	T	F	F	F	F	F	T	T
T	F	T	F	F	F	T	F	T
T	F	F	F	F	F	T	T	T
F	T	T	T	T	T	F	F	F
F	T	F	T	F	T	F	T	T
F	F	T	T	F	T	T	F	T
F	F	F	T	F	T	T	T	T

This table establishes the validity of argument form (A), and hence of (I), (III), and *any* further argument of that form. A truth-table with variables thus exhibits a measure of generality not shared by tables with sentence letters. Whereas our earlier truth-table established one argument's validity by listing all truth-value combinations its component sentence letters could have, the present table establishes the validity of an argument form — hence any argument of that form — by listing all truth-value combinations that *any* set of three sentences may have. For the practical end of determining validity, it is immaterial whether the truth-table uses letters or variables, but the foregoing considerations should make it clear that validity pertains to the form of an argument, not its content. An argument is therefore valid when it is derivable from a valid argument form by substitution of component sentences or letters for the variables of that form.

Now, a given argument is generally derivable from more than one argument form. Argument (I), for example, is derivable not only from (A) but from

(B) $p \vee q$
     $r \vee s$
     $\therefore p$

by substitution of '$\sim S$' for 'p', '$L \& I$' for 'q', '$\sim L$' for 'r', and '$\sim I$' for 's'. But a 16-row truth-table will show that argument form (B) is *invalid*. Thus, though a valid argument must be derivable from a valid argument form, this by no means precludes its being derivable from invalid forms as well. What is important, however, is that *only* valid arguments are derivable from valid forms; no matter what substitutions are made for the variables of a valid form — and no matter whether single letters or compounds are substituted — the resulting argument is valid. So the following argument is valid because like (I) and (III) it is derivable from the valid form (A):

(IV) $\sim(A \& B) \vee [C \& (B \vee D)]$
       $\sim C \vee \sim(B \vee D)$
       $\therefore \sim(A \& B).$

Here the complex substitutions are '$A \& B$' for 'p', 'C' for 'q', and '$B \vee D$' for 'r'. Summarizing:
(a)   To be valid, an argument must be derivable from at least one valid argument form.
(b)   *Only* valid arguments are derivable from valid argument forms.
(c)   But it is not the case that valid arguments are derivable *only* from valid argument forms — invalid forms yield them too.
Conversely, it follows from (b) that invalid arguments are derivable only

from invalid forms. Thus argument (II) is invalid inasmuch as *all* forms from which it results through substitution are invalid.[6]

It will be helpful here to establish via truth-tables the validity of four simple and very fundamental argument forms containing '&', 'v', and '~'. They are associated with the rules of inference commonly known as **Disjunctive Syllogism, Addition, Simplification**, and **Conjunction**:

Disjunctive Syllogism: p v q
    $\sim p$
    $\therefore q$

Addition: p
        $\therefore p$ v q

Simplification: p & q
        $\therefore p$

Conjunction: p
        q
        $\therefore p$ & q

The following elementary truth-tables establish the validity of all four:

p	q	$\sim p$	p v q		p	q	p & q
T	T	F	T		T	T	T
T	F	F	T		T	F	F
F	T	T	T		F	T	F
F	F	T	F		F	F	F

Tautologies and contradictions lead to very special cases of validity. Since an argument is valid just in case no row of its truth-table yields true premises and false conclusion, it follows that any argument with a tautology as conclusion is valid. That is, *a tautology is implied by any sentence whatever.* The argument form

p
$\therefore q$ v $\sim q$

---

[6]The following argument of course is also derivable from (A):

$\sim A$ v $(A$ & $A)$
$\sim A$ v $\sim A$
$\therefore \sim A$.

While such an argument is of little practical interest, it is nonetheless of form (A) through substitution of 'A' for each variable. There is nothing wrong with substituting the same sentence for distinct variables – the practice is no different from deriving '$3 + 3 = 3 + 3$' from '$x + y = y + x$'.

is thus valid no matter how irrelevant the content of the premise may be to the conclusion. For since there can be no assignment of truth-values to make the conclusion false, there can be no case of true premises and false conclusion. By the same token, since a contradiction comes out false upon all assignments of truth-values, it follows that any argument with a contradiction as premise is valid. In other words, *a contradiction implies any sentence whatever.* Hence the following argument form is valid:

p & ~*p*
∴ *q*

But since the premise can never be true, it would of course be absurd for anyone to propound an argument of this form. Even so, complex arguments sometimes have inconsistent premises, though a cursory examination will not reveal it. A set of premises is said to be *inconsistent* just in case there is no assignment of truth-values to their component letters which makes *all* of them true together. Therefore if we were to form the conjunction of these premises, the resulting expression would clearly have a truth-table column containing nothing but F's. So any argument whose premises are inconsistent will be valid in the same curious way as the above example. In the terminology of the first chapter, such arguments are (trivially) valid, but they cannot possibly be *sound.* It is for this reason that they are in practice useless, although it has happened more than once that a careless protagonist has unwittingly put forth such an argument. The following argument form is valid, though the inconsistency of its premises is far from obvious; the construction of its truth-table is left as an exercise.

p v (~*q* & *r*)
q v ~*p*
~(*p* v *r*)
∴ *p* & *q*

EXERCISES

I.   Test each of the following for validity using truth-tables:

1.   ~*A* v (*B* & *A*)
     ~*B*
     ∴ ~*A*

2.   ~*B* v (*D* & ~*R*)
     R v ~*B*
     ∴ ~*B* v *D*

3.   (*E* & ~ *F*) v *E*
     F v ~ *E*
     ∴ ~*F*

4.   ~(*E* & *A*) v *R*
     ~(*R* & *E*)
     ∴ ~*A* v ~*R*

5. Either Senator Hominy will not support the incumbent or else it is not the case both that the incumbent will seek reelection and yet not carry the South. So either the incumbent carries the South and Hominy supports him or else he does not seek re-election and Hominy does not support him. (H, R, S)

II. Assume that $\mathcal{T}$ is a sentence such that

$\mathcal{T}$ v $\sim\mathcal{T}$

$\therefore \mathcal{T}$

is a truth-table valid argument. Must $\mathcal{T}$ be a tautology? Explain.

III. Assume that $\mathcal{P}$ and $\mathcal{C}$ are sentences such that the argument

$\mathcal{P}$

$\therefore \mathcal{C}$

is truth-table valid. Is $\sim\mathcal{P}$ v $\mathcal{C}$ a tautology, a contradiction or neither? Explain.

IV. Where the argument

$\mathcal{P}_1$

$\mathcal{P}_2$

.
.
.

$\mathcal{P}_n$

$\therefore \mathcal{C}$

is truth-table valid, must

$\sim(\mathcal{P}_1 \ \& \ \mathcal{P}_2 \ \& \ \dots \ \& \ \mathcal{P}_n)$ v $\mathcal{C}$

be a tautology? Explain.

Arguments Containing '⊃' and '≡'

There are several fundamental argument forms containing '⊃' whose validity we shall now consider.[7] They are associated with the rules of inference traditionally known as *Modus Ponens, Modus Tollens,* **Hypothetical Syllogism,** and **Constructive Dilemma.**

Modus Ponens: $p \supset q$

p

$\therefore q$

Modus Tollens: $p \supset q$

$\sim q$

$\therefore \sim p$

Hypothetical Syllogism: $p \supset q$

$q \supset r$

$\therefore p \supset r$

Constructive Dilemma: $p \supset q$

$r \supset s$

$p \lor r$

$\therefore q \lor s$

---

[7]Hereafter for brevity we shall describe sentences of the form '$p \supset q$' simply as "conditionals"—leaving it understood that they are *material* conditionals.

The first two are among the simplest and most often encountered forms of argument. The following truth-tables make their validity evident.

p	q	$p \supset q$
T	T	T
T	F	F
F	T	T
F	F	T

p	q	$\sim p$	$\sim q$	$p \supset q$
T	T	F	F	T
T	F	F	T	F
F	T	T	F	T
F	F	T	T	T

The form corresponding to **Modus Ponens** should not be confused with the invalid form

$p \supset q$
q
$\therefore p$

One who uses this form of argument is said to commit the fallacy of *affirming the consequent*. Its invalidity is shown by the third row of the left-hand table; if 'p' is replaced by a false sentence and 'q' by a true one the premises will be true but the conclusion false, as in:

If Abraham Lincoln was the British Prime Minister, he was a politician.
Abraham Lincoln was a politician.
Therefore Abraham Lincoln was the British Prime Minister.

Moreover, the argument form associated with **Modus Tollens** should be distinguished from the invalid form

$p \supset q$
$\sim p$
$\therefore \sim q$

To argue this way is to commit the fallacy of *denying the antecedent*, and its invalidity is also shown by the third row of the above table on the right — and with the same substitutions for 'p and 'q' as before. The validity of the forms associated with **Hypothetical Syllogism** and **Constructive Dilemma** may be established using truth-tables of eight and sixteen rows respectively. Their construction is left as an exercise for skeptical readers.

We saw that '$p \supset q$' is equivalent to '$\sim p \lor q$', and '$p \equiv q$' to

'$(p \,\&\, q) \lor (\sim p \,\&\, \sim q)$'. Thus the connectives '$\supset$' and '$\equiv$' are by no means essential to our symbolic notation, but their use makes symbolization of many arguments far easier and more natural than relying on just the basic connectives '$\&$', '$\lor$', and '$\sim$'. An obvious example would be:

The President's tax bill will become law if and only if it passes both the House and the Senate. But the Senate will not pass it unless the House does. Consequently the President's tax bill will become law only if the House passes it.

Using obvious letters for the three components, we have:

(V) $\;B \equiv (H \,\&\, S)$
$\quad \sim S \lor H$
$\quad \therefore B \supset H$

An eight-row truth-table will show the validity of argument (V). However there is a further advantage in using '$\supset$' and '$\equiv$', superfluous though they be in theory. The connective '$\supset$' in particular helps emphasize the intimate connection between arguments and conditionals, and in so doing, provides us with another and more succinct way of characterizing the concept of (truth-table) validity — one which stresses its relation to the notion of truth under all conditions. Let us consider the conditional whose antecedent is the conjunction of the premises of (V) and whose consequent is its conclusion:

(1) $\;\{[B \equiv (H \,\&\, S)] \,\&\, (\sim S \lor H)\} \supset (B \supset H)$

We shall call (1) the **corresponding conditional** of (V). Every argument has a corresponding conditional, and for an argument of n premises ($n \geq 1$)

$\mathcal{P}_1$
$\mathcal{P}_2$
$\;\;\vdots$
$\mathcal{P}_n$
$\therefore \mathcal{C}$

its corresponding conditional (omitting inessential parentheses for the moment) may be schematized as

$(\mathcal{P}_1 \,\&\, \mathcal{P}_2 \,\&\, \ldots \,\&\, \mathcal{P}_n) \supset \mathcal{C}$

Now a conjunction is true only when all its conjuncts are, and since the conjuncts correspond to premises, the antecedent of the conditional is true just in case all premises of the argument are true. Moreover, a conditional is true in every case except where the antecedent is true and the consequent false, and notice that this corresponds to the case where the argument has all true premises and false conclusion. Therefore, an argument is invalid just in case its corresponding conditional has at least one **F** in its truth-table column. Since a tautology has *no* **F**'s in its truth-table column, it follows that:

A truth-functional argument (argument form) is *valid* if and only if its corresponding conditional is a tautology (tautologous).

More succinctly, $\mathscr{P}_1, \mathscr{P}_2, \ldots, \mathscr{P}_n \models \mathscr{C}$ if and only if $\models (\mathscr{P}_1 \,\&\, \mathscr{P}_2 \,\&\, \ldots \,\&\, \mathscr{P}_n) \supset \mathscr{C}$. Thus, conditional (1) is a tautology, since its truth-table will have no row in which its antecedent (the conjunction of the premises of (V)) is assigned a **T** and yet its consequent (the conclusion) is assigned **F**. The truth-table is displayed below. Note that instead of a separate column for each of the components, truth-values are assigned to them as they occur within the whole, thus simplifying our work:

B	H	S	{[B ≡ (H & S)] & (~S v H)} ⊃ (B ⊃ H)
T	T	T	T T T F T T T
T	T	F	F F F T T T T
T	F	T	F F F F F T F
T	F	F	F F F T T T F
F	T	T	F T F F T T T
F	T	F	T F T T T T T
F	F	T	T F F F F T T
F	F	F	T F T T T T T

Of course for the practical task of establishing validity, a truth-table this elaborate is unnecessary—we only need columns for the two premises and the conclusion. The sole point of the above table is to show how valid arguments and tautologous conditionals go hand in hand.

What conditionals are to validity, biconditionals are to equivalence. We saw that two sentence forms are equivalent just in case they have the same truth-value for each assignment of truth-values to their component letters. Now a biconditional is true just in case both components have the same truth-value. Consequently, if we form the biconditional of two equivalent sentences, its truth-table column must be all **T**'s, since for any given row the two sentences are either both assigned a **T** in that row or else both **F**. The biconditional formed from two nonequivalent sentence forms, however, will have at least one **F** in its column, since there must be a row in which the two forms differ in truth-value assignment. Now plainly we can form a biconditional from any pair of sentence forms, and we shall call it their **corresponding biconditional**. We may now refine our notion of equivalence as follows:

Two sentences (sentence forms) are *equivalent* if and only if their corresponding biconditional is a tautology (tautologous).

The following table establishes again the equivalence of '$\sim(A \& B)$' and '$\sim A \vee \sim B$', this time employing the corresponding biconditional:

A	B	$\sim$	(A & B)	$\equiv$	($\sim A$	v	$\sim B$)
T	T	F	T	T	F	F	F
T	F	T	F	T	F	T	T
F	T	T	F	T	T	T	F
F	F	T	F	T	T	T	T

We have now seen how talk of validity and equivalence may be paraphrased away in favor of talk about tautologies. This is just one illustration of the theoretical importance of the latter concept mentioned back in Section 2.1.

EXERCISES

I. Using truth-tables, show the forms corresponding to **Hypothetical Syllogism** and **Constructive Dilemma** to be valid.

II. Test the following arguments for validity using truth-tables:

1. $A \supset B$
   $\sim B \vee \sim A$
   $\therefore \sim A$

2. $A \equiv \sim B$
   $B \vee \sim A$
   $\therefore \sim A$

3. If the Cubs win, then the other teams have either deteriorated or else sold out to the gamblers. But if they sold out, then the Cubs still will not win. So either the other teams have deteriorated or else the Cubs win. (C, D, G)

4. If Larry confesses, then Moe will turn states' evidence and Curly ends up in the slammer. Either Moe turns states' evidence

or Larry confesses. Moreover, Moe won't turn states' evidence unless Curly ends up in the slammer. So in any case, Curly ends up in the slammer. (*L, M, C*)

5. If either the police or the F.B.I. are not mystified, then Vermicelli was rubbed out by the Mozzarella-Roquefort gang. But if the police are mystified, so is the F.B.I. Hence, provided that the police are still mystified, Vermicelli was not rubbed out by the Mozzarella-Roquefort gang. (*P, F, V*)

6. If God exists, animals will not suffer extreme pain unless God is unable to prevent it. If God is unable to prevent an animal suffering extreme pain, then He is not omnipotent. It is false that animals do not suffer extreme pain. Therefore, either God does not exist or He is not omnipotent. (*G, A, P, O*)

III. Let $\mathscr{P}$, $\mathscr{Q}$, and $\mathscr{R}$ be sentences such that $\mathscr{P}, \mathscr{Q} \models \mathscr{R}$. Must it also be the case that $\mathscr{P} \models \mathscr{Q} \supset \mathscr{R}$? What about $\models \mathscr{P} \supset (\mathscr{Q} \supset \mathscr{R})$? Explain.

IV. Let the sentential connective '$' be defined as follows:

p	q	$p \$ q$
T	T	T
T	F	T
F	T	F
F	F	T

Is the following argument valid?

$p \supset (q \& r)$
$\sim p \$ q$
$\therefore \sim p$

Is the following sentence a tautology?

$p \supset (p \$ q)$

Are either of the following pairs equivalent?

1. $p \$ \sim q$; $\sim q \$ p$
2. $p \$ (\sim q \$ r)$; $(p \$ \sim q) \$ r$

V. Put both of the following into symbols and construct their corresponding biconditional. Then determine whether they are equivalent:

1. Jones will be inducted only if he both passes his physical and is administered the oath of allegiance.
2. Jones will neither be inducted but fail his physical nor will he not be administered the oath and yet be inducted.

*2.4 Another Test: The Method of Assigning Truth-Values

When using a truth-table to evaluate an argument, we in effect need only examine those rows in which the conclusion is assigned an **F**. This suggests an easier and quicker technique than the construction of truth-tables with 16 or more rows. We shall call it **the method of assigning truth-values,** and while it is not always easier than drawing up truth-tables it nonetheless simplifies our work in the majority of cases. The method consists, first, in assigning truth-values to those sentence letters of an argument that suffice to render its conclusion *false.* Then we attempt to make assignments to the remaining letters in such a way as to make all premises true. If such an assignment can be found, the argument is thereby shown invalid; but if for each assignment falsifying the conclusion, *all* attempts to make the premises true end in failure, the argument will then be established as valid.

We need in practice concern ourselves only with those cases where the conclusion is false, inasmuch as validity is established when and only when it has been determined that there is *no* truth-value assignment yielding the true-premise–false-conclusion combination. Thus our new technique of first making the conclusion false, in effect, allows us to consider only those rows of the corresponding truth-table where the conclusion is assigned **F** – ignoring the unproblematic **T** rows.

As might be expected, the truth-value assignment method is used to best advantage in establishing *invalidity*, since only one assignment yielding true premises and false conclusion is required. To show *validity*, one must show that *every* assignment making the conclusion false fails to yield a case of true premises, and this can sometimes be an arduous task. We shall first, however, consider invalid arguments:

(1) $(A \ \& \ B) \supset (\sim C \ \& \ E)$

(2) $D \supset (\sim B \ v \sim C)$

(3) $D \ \& \ A$

$\therefore \sim C$

To falsify the conclusion, we suppose 'C' to be true, and by making assignments to the remaining letters, attempt to render all premises true. Now it is obvious that for (3) to be true, both 'D' and 'A' must be true, so we assign **T** to both. We now take up (2), asking whether it can be true when 'C' and 'D' are true. On these assignments, (2) has a true antecedent, and thus for the whole to be true, its consequent must also be true. Since the consequent is a disjunction with a false disjunct, the other disjunct '$\sim B$' must be true for (2) itself to be true. We therefore assign **F** to 'B'.

Moving now to (1), we shall show invalidity if we can render it true on the basis of the truth-value assignments already made or perhaps, if there are letters as yet unassigned, by making further assignments. In the present case, '*E*' has not yet been assigned a truth-value, but in this instance it is not necessary. For (1) has already been made true by the previous assignments, and thus it does not matter what truth-value '*E*' has. Since '*B*' is false, '*A* & *B*' is false, thus giving (1) a false antecedent. Therefore (1) is true since a conditional is false *only* when its antecedent is true and consequent false. We have thus found an assignment of truth-values showing the argument to be invalid, and we can display our work as follows:

A	B	C	D	E	(A & B) ⊃ (~C & E)	D ⊃ (~B v ~C)	D & A	~C
T	F	T	T	–	T	T	T	F

Here, we have set out the assignments to the component sentence letters that yield true premises and false conclusion (the dash under '*E*' simply means that it is immaterial whether we assign **T** or **F**). It is easy to see the advantages of this method; had we used a truth-table to discover this assignment, it would have required 32 rows.

In thinking our way back through the premises, we implicitly employed certain observations concerning the behavior of the sentential connectives '~', '&', 'v', and '⊃', for example, that for (2) itself to be true, '*B*' must be false given the truth of '*C*'. The following principles are gleaned from a study of the truth-tables for these connectives, and are helpful in making truth-value assignments:

A negation is false if the negated sentence is true and vice versa.
A conjunction with a false conjunct is false.
A disjunction with a true disjunct is true.
A conditional with a false antecedent is true.
A conditional with a true consequent is true.

Thus, even when assignments have been made only to some of the component letters, there is often much to be inferred concerning the truth-values of the premises and conclusion. However, matters are not always as simple as they were for our sample argument. There, once the conclusion was made false, there was only one assignment that could be made to each of '*A*', '*B*', and '*D*' to render the premises true. But in many cases, there are alternative assignments to a given premise that will make it true, and it may turn out that with only one of them can the rest of the premises be rendered true as well.

The truth-value assignment technique sometimes affords us a quick test of validity too. Consider the elementary argument:

A & B
∴ B ∨ ~A

Since the conclusion is a disjunction, it can be falsified only by making both disjuncts false, so we assign **F** to '*B*' and **T** to '*A*'. But now we see immediately that the premise cannot be true on this assignment, since both conjuncts must be true and '*B*' of course is not. We may restate the point as follows: for the conclusion to be false '*B*' must be assigned **F**, but for the premise to be true '*B*' must be assigned **T**. Since, therefore, the attempt to obtain a true premise and false conclusion results in a component letter being assigned a **T** at one point and, *per impossibile*, an **F** at another, the argument must be valid.

This, then, is how validity is determined: where φ is any argument, if each attempt to assign truth-values falsifying the conclusion of φ but making all premises true requires that some sentence letter be assigned a **T** at one point and an **F** at another, then φ is valid. Armed with this knowledge, then, let us tackle a more challenging argument.

(4) A ⊃ B

(5) (A & B) ⊃ C

(6) B ⊃ (D ∨ ~C)

∴ A ⊃ D

There is but one assignment making the conclusion false; '*A*' must be assigned **T** and '*D*' assigned **F**. It is now immediately clear that '*B*' must be assigned **T** to make (4) true. However, this makes the antecedent of (5) true, thus requiring that a **T** be assigned to '*C*'. But since '*B*' has been assigned **T** and '*D*' assigned **F**, it is obvious that '*C*' must be assigned **F** in order to make '~*C*' — and hence the consequent of (6) — true. Since incompatible assignments are required of '*C*', the argument is valid. Our work may be summarized as follows:

For the conclusion to be false, '*A*' must be assigned **T** and '*D*' assigned **F**. For (4) to be true, '*B*' must be assigned **T**. But then, for (5) to be true, '*C*' must be assigned **T** and for (6) to be true, '*C*' must be assigned **F**. Therefore the argument is *valid*.

Now if the conclusion of a valid argument can be made false on more than one truth-value assignment, then of course each of these cases will have to be tested in turn. If there are many such assignments — and in addition many further assignments to be made in the premises — the method of assigning truth-values may turn out to be no less cumbersome than the straight truth-table method. On the other hand, should the argument be invalid, the testing comes to an end as soon as a case of true premises and false conclusion emerges — there is obviously no point in going on to consider any further falsifying assignments. There are a great many arguments — both valid and invalid — that are far more easily handled using the present technique. Consider, for example, the forbidding:

(7) $A \supset (\sim B \& C)$

(8) $D \supset (B \vee \sim C)$

(9) $(\sim E \vee C) \supset D$

(10) $A \supset E$

　　∴ $A \equiv \sim D$

Those who balk at the construction of a 32-row truth-table need instead merely assign truth-values to yield a false conclusion and then test the premises for joint truth. Here, there are two cases where the conclusion will be false: where 'A' and 'D' are both true, and where they are both false.

If 'A' and 'D' are both assigned **T** then the conclusion is false. But for (7) and (8) both to be true, their consequents must be true, and there is no assignment of truth-values to 'B' and to 'C' that will make both '$\sim B \& C$' and '$B \vee \sim C$' true. If 'B' is assigned **T**, then (8) is true but (7) is not. If 'B' is assigned **F**, then for (7) to be true, 'C' must be assigned **F**, and in that case (8) will be false. So on this assignment, not all premises can be made true.

So far, then, we have established neither validity nor invalidity. But if we can do the same as above for the other falsifying assignment, validity will then be established. However, we cannot do this. For when 'A' and 'D' are both assigned **F**, then (7), (8), and (10) are all true owing to a false antecedent. We need only, then, assign **T** to 'E' and **F** to 'C' to make (9) true, and hence make all premises true:

A	B	C	D	E	$A \supset (\sim B \& C)$	$D \supset (B \vee \sim C)$	$(\sim E \vee C) \supset D$	$A \supset E$	$A \equiv \sim D$
F	–	F	F	T	T	T	T	T	F

Again, only one case of true premises and false conclusion is needed to show invalidity, so if such a case had arisen on the first assignment, it would not have been necessary to proceed to the second. But to show validity, *all* possible cases with respect to a false conclusion must be traced through (part of the trick, of course, is determining that you *have* exhausted all possibilities), and hence the truth-value assignment technique is shown to its best advantage in establishing invalidity. The technique is most easily applied to arguments whose conclusions admit of falsification in just one or two cases (for instance, 'A', '$A \supset B$', '$A \vee B$', '$A \supset (B \vee C)$'), but for most conclusions admitting of many such cases — as in '$(A \vee B) \supset (C \& D)$' — one might well be advised to construct the full truth-table unless it is strongly suspected that one of those cases will quickly lead to all true premises and hence invalidity.

EXERCISES

I. Using the method of assigning truth-values, determine whether the following arguments are valid:

1. $E \supset (J \& K)$
    $F \supset (E \vee A)$
    $\sim K$
    $\therefore F \supset J$

2. $A \supset (B \& C)$
    $B \supset (\sim A \& C)$
    $\therefore \sim A$

3. $A \supset E$
    $B \supset F$
    $\sim E \vee \sim F$
    $\therefore \sim A \vee \sim B$

4. $M \vee N$
    $L \vee K$
    $(M \vee L) \supset (N \& K)$
    $\therefore N \& K$

5. $(A \& C) \supset F$
    $(B \& F) \supset [\sim E \supset (D \vee \sim A)]$
    $\therefore (A \& B) \supset [C \supset (D \vee E)]$

6. $A \supset (B \supset C)$
    $D \supset (B \& A)$
    $C \supset D$
    $\therefore C \equiv A$

7. If Irv does not prove the continuum hypothesis, then although Karl will not gloat, Henry will. However, neither Karl nor Henry will gloat. Consequently, if Irv proves the continuum hypothesis only if Karl gloats, then Irv in fact proves the continuum hypothesis.

8. If the President vetoes the trade bill, Congress will either back him up or else succumb to public pressure. Japan will not raise its tariffs unless the protectionists have their way. If the protectionists have their way and Congress succumbs to public pressure, the world markets will be in disarray. The President did in fact veto the trade bill, so if Japan raises its tariffs, then Congress will not back up the President and the world markets will be in disarray.

9. Milo buys eggs in Sicily at 1 cent per egg and sells them to the distributors in Malta at 4¼ cents per egg, and *if* he does both, he clears 3¼ cents per egg. If Milo buys eggs in Sicily at 1 cent per egg and buys them back from the distributors in Malta for 7 cents, then he pays a total of 8 cents per egg. If Milo clears 3¼ cents per egg and sells them to the commissary for 5 cents, then he takes in 8¼ cents per egg. If Milo pays a total of 8 cents per egg and takes in a total of 8¼ cents per egg, then he makes a profit. Consequently, if Milo buys back eggs from the distributors in Malta at 7 cents per egg and sells them to the commissary for 5 cents, then he makes a profit.

II. Adapting the method presented in this chapter, determine whether either of the following are tautologies:
   1. $[A \supset (B \ \& \ C)] \supset \{[(D \ \& \ E) \lor \sim B] \supset (D \supset A)\}$
   2. $[(A \supset B) \supset A] \supset A$

*2.5 Truth-Functions

A *truth-functional* compound sentence, it will be recalled, is one whose truth-value is determined solely by the truth-values of its component sentences. The concept of truth-functionality is a very important one, and we shall here develop it a bit more fully. Let us note a parallel between truth-functional compound sentences (whether in ordinary discourse or in symbols) and various compound expressions of arithmetic.[8] Just as '*A* ∨ *B*' is a compound containing the sentence letters '*A*' and '*B*', so '4 + 3' is a compound containing the numerals '4' and '3'. In the first case, we have a sentence containing further sentences; in the second, a numerical expression containing other numerical expressions. More importantly, just as the *truth-value* of '*A* ∨ *B*' is determined given the truth-values of '*A*' and of '*B*', so the *number* signified by '4 + 3' is determined given the numbers signified by '4' and '3'. This latter circumstance is customarily described by saying that the number signified by the compound is a *function of* the numbers signified by its components; for as we vary the latter numbers, the identity of the former is correspondingly altered. Thus, given any pair of numbers whatever, if we insert names for them in place of '*x*' and '*y*' in the expression '*x* + *y*', the result will be a compound expression signifying a specific number. The expression '*x* + *y*' thus represents a systematic way of correlating the numbers signified by the resultant compounds with the pairs of numbers signified by the inserted expressions. A **function**, for our purposes, is just such a correlation. The numbers signified by a given pair of numerical expressions inserted in '*x* + *y*' are said to be **arguments** of the function (called the *addition* function, of course) and the number signified by the compound thus produced the **value** of that function *for* those numbers as arguments. In the present example, the number 7 — since it is signified by '4 + 3' — is the value of the addition function for the numbers 4 and 3, in that order, as arguments of the function.

The familiar notion of addition, then, may be construed as a function of two arguments, that is, a function which takes a number as its value for an ordered pair of numbers as arguments of the function[9] — such a value being signified by compounds of the form '*x* + *y*'. Similarly, the notion of *squaring* a number is a function of one argument, for when a single numerical expression is inserted in 'x^2', the resulting compound signifies a

---

[8] Hereafter, we shall use the phrase 'compound sentence' to mean 'truth-functional compound sentence' and restrict our attention to the arithmetic of the integers (whole numbers). Throughout the discussion, the reader must take care not to confuse linguistic *expressions* with what they signify — the expression '3' is a *numeral*, not a *number*; but it signifies or denotes the number 3.

[9] The notion of an **ordered pair** of arguments is important here, for there are functions that yield *different* values for distinct orderings of the same arguments. Subtraction is an example: 6 − 2 (that is, 4) is a different number from 2 − 6 (that is, − 4).

unique number: the square of the number signified by the inserted expression. The number squared is thus the argument of the function, its square the value for that argument.

Arithmetic functions are distinguished one from another by their having in at least one case *different* values for the same argument(s). Thus, to show that addition and subtraction are distinct functions, it is sufficient to note that the former has the number 6 as its value for the ordered pair of arguments 4 and 2, taken in that order, whereas the latter takes 2 as its value in that circumstance. But the functions represented by '$(x \div 2) + 3$' and '$(x + 6) \div 2$' will obviously have the same value for any argument; no matter which numerical expression is inserted, the resulting compounds always signify the same number. Thus, the functions are one and the same, and this identity is what is asserted in the equation '$(x \div 2) + 3 = (x + 6) \div 2$'.

Now just as there is a functional correlation between the number signified by a compound arithmetic expression and the numbers signified by its components, so is there a functional correlation between the truth-value of a compound sentence and the truth-values of its components. Indeed, it is in virtue of this fact that such sentences are described as "truth-functional"; the sentential connectives '&', 'v', and '~' of our symbolic notation serve as functional signs in much the same fashion as '+' and '−' do in arithmetic discourse. By inserting sentences into 'p v q' we produce a compound sentence whose truth-value is a function of the truth-values of the inserted sentences. Functions that take truth-values both as their values and as their arguments will be called **truth-functions**. The connectives '&' and 'v' each represents a truth-function of *two* arguments.

Now, arithmetic functions and truth-functions differ significantly in the range of items that serve as their arguments, for while there are infinitely many numbers there are only *two* truth-values. This means that for a truth-function of a given number of arguments, we can always specify exactly the number of possible orderings of those arguments — or in other words, **truth-value combinations** — for which the function takes a value. A truth-function of *two* arguments takes a value for exactly four such combinations, which thus correspond to the four rows of the truth-table:

p	q
T	T
T	F
F	T
F	F

Since a conjunction is true only when both conjuncts are, the form 'p & q' expresses that truth-function of two arguments which has the value **T** for the truth-value combination **T–T** and the value **F** for the remaining three: **T–F**, **F–T**, and **F–F**. Therefore this function (the "*and*-function", let us call it) is completely specified by a truth-table column with a **T** in the first row and **F**'s in all the rest. Similarly, a disjunction is true whenever at least one disjunct is, so '*p* ∨ *q*' expresses that truth-function of two arguments which has the value **T** for each of the first three truth-value combinations and the value **F** for the fourth. This "*or*-function", then, is completely specified by a truth-table column having a **T** in every row but the last.

So, unlike arithmetic functions, a truth-function can be completely specified by a table setting out its value for each truth-value combination. One can always determine the number of truth-value combinations for a truth-function of any number of arguments (and thereby, of course, the number of truth-table rows required):

For a truth-function of n arguments, the number of truth-value combinations is 2^n.

We have already seen that where $n = 2$, there are 4 (or 2^2) combinations of truth-values. Clearly for a truth-function of just one argument, such as that represented by '~', that argument can be only **T** or **F** – since there is but one argument, the notion of a truth-value "combination" is redundant here. Thus, any compound sentence with just one component requires a table with 2^1 rows. For example, '~*p*' expresses that truth-function of one argument (the "*not*-function", to give it a name) which has the value **T** for the argument **F**, and **F** for the argument **T**. This function is thus specified by a truth-table column with an **F** in the first row and a **T** in the second.

For a truth-function of *three* arguments there will be, according to our formula, 2^3 distinct truth-value combinations for which it takes a value. Hence, compounds with three component sentences require, as we have seen, a truth-table of eight rows. Truth-functions of *four* arguments will have 2^4 truth-value combinations, and so compounds with four components require 16-row truth-tables for the specification of their truth-conditions. Functions of five or more arguments clearly require quite sizable – and cumbersome – truth-tables.

To show that two truth-functions are distinct functions, it suffices to exhibit one truth-value combination for which they take different truth-values as values. In the case of the *and*- and *or*-functions, the second and third rows of their truth-tables provide such combinations.

By the same token, truth-functions are identical when, for every truth-value combination, they yield the same truth-value. Since, for example, the columns for '~(*p* ∨ *q*)' and '~*p* & ~*q*' have the same arrangement of **T**'s and **F**'s, sentences derivable from them via the same substitutions are true and false under the same conditions. These forms thus provide different ways of expressing the very same truth-function (the "*neither-nor*–function", as we shall call it). It is that truth-function of two arguments that has the value **T** for the truth-value combination **F**–**F** and the value **F** for the rest. And as we can see from the truth-table in Section 2.1, no special connective need be introduced to express this function—the job may be done either by '~' and '&' or else by '~' and 'v'. *Equivalent* sentence forms, therefore, express one and the same truth-function.

Since there are only two truth-values to serve as arguments for a truth-function, we were able to state a formula by which we could determine for any truth-function the number of truth-value combinations for which it takes a value. We may also determine how many truth-functions there are for a given number of arguments:

The number of truth-functions of *n* arguments is 2^{2^n}.

Consider truth-functions of one argument. Here there are 2^{2^1} or 4 truth-functions, for that is as many distinct arrangements of **T**'s and **F**'s as there are for a two-row truth-table. Thus any sentence form with just one component must express one (and only one) of these four functions. They are specified as follows:

p	$f_1(p)$	$f_2(p)$	$f_3(p)$	$f_4(p)$
T	**T**	**T**	**F**	**F**
F	**T**	**F**	**T**	**F**

We have long since introduced the connective '~' for representing the function listed here as '$f_3(p)$'; $f_3(p)$ is thus the "*not*-function". Note that $f_2(p)$ has the same arrangement of **T**'s and **F**'s as the component 'p' itself; we may thus express $f_2(p)$ simply by 'p'. That is, the truth-value of any sentence is (trivially) a function of itself—a truth-function having the value **T** for the argument **T** and **F** for the argument **F**. The *verum* function $f_1(p)$ has the value **T** for both arguments, and may be expressed by 'p v $\sim p$'. Similarly, the *falsum* function $f_4(p)$ has **F** as its value for both arguments, and is thus expressible by 'p & $\sim p$'.

Now there are 2^{2^2} or 16 truth-functions of two arguments, for there are exactly 16 distinct arrangements of **T**'s and **F**'s for a four-row truth-table, including the limiting cases of all **T**'s and all **F**'s. Any compound sentence of two components must therefore express one of these functions:

p	q	$f_1(p, q)$	$f_2(p, q)$	$f_3(p, q)$	$f_4(p, q)$
T	T	T	T	T	T
T	F	T	T	T	F
F	T	T	T	F	T
F	F	T	F	T	T

p	q	$f_5(p, q)$	$f_6(p, q)$	$f_7(p, q)$	$f_8(p, q)$
T	T	F	T	T	T
T	F	T	T	F	F
F	T	T	F	T	F
F	F	T	F	F	T

p	q	$f_9(p, q)$	$f_{10}(p, q)$	$f_{11}(p, q)$	$f_{12}(p, q)$
T	T	F	F	F	F
T	F	T	T	F	F
F	T	T	F	T	F
F	F	F	T	T	T

p	q	$f_{13}(p, q)$	$f_{14}(p, q)$	$f_{15}(p, q)$	$f_{16}(p, q)$
T	T	F	F	T	F
T	F	F	T	F	F
F	T	T	F	F	F
F	F	F	F	F	F

Many of these are already familiar: $f_2(p, q)$ is of course the *or*-function and expressible by 'p v q'; $f_{15}(p, q)$ is the *and*-function expressible by 'p & q'; the *neither-nor*-function is $f_{12}(p, q)$ and may be expressed by '$\sim p$ & $\sim q$' or equivalently by '$\sim(p$ v $q)$'. Moreover, $f_5(p, q)$ might be termed the "*not-both*-function" since it is expressed by '$\sim(p$ & $q)$' and by '$\sim p$ v $\sim q$'. The "*exclusive-or*-function" — which may be expressed in symbols by '$(p$ & $\sim q)$ v $(\sim p$ & $q)$' — is here displayed as $f_9(p, q)$.

Since the *neither-nor*-, *not-both*-, and *exclusive-or*-functions may all be expressed in terms of '&', 'v', and '$\sim$', the question naturally arises whether these connectives alone are capable of expressing the remaining two-argument truth-functions as well. The answer is: not only can all of them be expressed, but we can dispense with 'v', using only '&' and '$\sim$' ('p v q' is equivalent to '$\sim(\sim p$ & $\sim q)$'). That is, having introduced '$\sim$' for $f_3(p)$ and '&' for $f_{15}(p, q)$, all the remaining 18 functions of one and two arguments may be expressed by sentence forms containing, in addition to the necessary parentheses and sentential variables, one or more occurrences of *just* these connectives. What is more surprising is that '~' and '&' can do the same for *all* truth-functions whatever! Any truth-function of *any* number of arguments may be expressed by a sentence form whose only connectives are '~' and '&'. To fully appreciate the extent of this claim, one should keep in mind that there are 2^{2^3} or 256 truth-functions of three arguments, 2^{2^4} or 65,536 truth-functions of four arguments, and so forth.

To bear out this contention, let us first consider some examples. The function $f_{13}(p, q)$ has the value **T** for just one truth-value combination, that given in row three. In this row, 'p' is assigned an **F** but 'q' a **T**. Since a conjunction is true only when both conjuncts are, '~p & q' takes a **T** in *just* this row and so expresses $f_{13}(p, q)$. More generally, any truth-function specified by a truth-table column having a **T** in just *one* row can·be expressed by a sentence form in which the simple components are conjoined together, but those to which **F** is assigned in that row are negated. So $f_{13}(p, q)$ may be expressed solely in terms of '~' and '&' by '~p & q', and similarly $f_{14}(p, q)$ may be expressed by 'p & ~q'.

But now we must consider functions that have the value **T** for more than one case. The function $f_{10}(p, q)$ has the value **T** for each of the truth-value combinations given in rows 2 and 4. Since it has the value **T** in just these cases, then, it can be expressed by '(p & ~q) v (~p & ~q)'. For this expression is assigned a **T** just in case either 'p & ~q' or '~p & ~q' is assigned a **T**; that is, just in case 'p' is assigned a **T** and 'q' an **F** (row 2) or else 'p' and 'q' are both assigned **F**'s (row 4). By the same technique $f_3(p, q)$ is expressible — dropping unneeded parentheses for the moment — by '(p & q) v (p & ~q) v (~p & ~q)'. In general, a truth-function specified by a truth-table column having **T**'s in two or more rows is expressible by a disjunction each of whose disjuncts is a conjunction of the type described above. If there are k rows with a **T**, there will be exactly k conjunctions disjoined together. The next step is to eliminate the 'v', and this is easily accomplished since we already know that 'p v q' may be equivalently expressed as '~(~p & ~q)'. Applying this knowledge to our examples, $f_{10}(p, q)$ is expressible by '~[~(p & ~q) & ~(~p & ~q)]' and $f_3(p, q)$ by '~[~(p & q) & ~(p & ~q) & ~(~p & ~q)]'. So functions having the value **T** in two or more cases can also be expressed solely in terms of '~' and '&'.

The same pattern of reasoning may be extended to functions of three or more arguments. Two such functions are:

p	q	r	$f_j(p, q, r)$	$f_k(p, q, r)$
T	T	T	F	T
T	T	F	F	F
T	F	T	T	T
T	F	F	F	F
F	T	T	F	F
F	T	F	F	T
F	F	T	F	F
F	F	F	F	F

The function $f_j(p, q, r)$ has the value **T** for just one truth-value combination — that given in row three. It thus has the value **T** in just the row in which 'p & $\sim q$ & r' is assigned a **T**, and is therefore expressible by that form. For $f_k(p, q, r)$ we again form such a conjunction for each of the three rows in which a **T** occurs and then disjoin them, obtaining '$(p$ & q & $r) \vee (p$ & $\sim q$ & $r) \vee (\sim p$ & q & $\sim r)$'. Eliminating '$\vee$' as before, we derive

$$\sim [\sim (p \text{ \& } q \text{ \& } r) \text{ \& } \sim (p \text{ \& } \sim q \text{ \& } r) \text{ \& } \sim (\sim p \text{ \& } q \text{ \& } \sim r)]$$

which contains only the connectives '&' and '$\sim$'. There are simpler expressions for this function, such as '$(p$ & $r) \vee (\sim p$ & q & $\sim r)$', but the important consideration is that the technique sketched here will *always* give the required sentence form if there is at least one **T** in the column.

Plainly the technique may be extended to functions of four or more arguments as well, the only complication being the size of the truth-tables required. For any number n of sentential variables: (1) construct the appropriate 2^n row truth-table; (2) for any arbitrary column containing k number of **T**'s ($1 \leqslant k \leqslant 2^n$), produce a conjunction of the n components for each **T** row, negating each component to which **F** is assigned in that row; (3) if $k > 1$, disjoin them and eventually eliminate '$\vee$'. We thus have a routine by which we can always *find* an expression whose only connectives are '$\sim$' and '&' to represent any randomly constructed column of **T**'s and **F**'s of any truth-table — in effect, to express any truth-function of any number of arguments. The only case left to consider is where only **F**'s occur in the column, as in $f_4(p)$ and $f_{16}(p, q)$. Now any sentence of the form 'p & $\sim p$' must be false no matter what the components' truth-values are, and so $f_4(p)$ may be expressed as 'p & $\sim p$'. To express $f_{16}(p, q)$, we simply write 'p & $\sim p$ & q'. Our routine thus breaks down into three cases:

(a) exactly one **T** in the column: conjoin the components, negating those to which an **F** is assigned in the **T** row;

(b) more than one **T** in the column: produce the conjunctions for each **T** row described in (a), negate each and conjoin them, then negate the whole;

(c) no **T**'s in the column: conjoin all components along with the negation of any one of them.

Thus '$\sim$' and '&' form a **truth-functionally complete** set of connectives — they are capable of expressing all of our truth-functional discourse. But as some readers may already have suspected, '&' can be eliminated in favor of 'v'. Since 'p & q' is equivalent to '$\sim(\sim p$ v $\sim q)$', the connectives '$\sim$' and 'v' form a complete set too. However, '&' and 'v' form an incomplete set, since there are truth-functions which cannot be represented by expressions containing only those connectives. We leave it to the reader to show that they cannot alone express the *neither-nor*-function.

EXERCISES

I. Produce a sentence form containing 'p', 'q', and 'r' whose connectives are '$\sim$' and '&', and which is assigned a **T** *only* when 'q' is assigned a **T** and 'r', an **F**.

II. Let us introduce the connective '$|$' (the *stroke*) for the *not-both*-function, $f_5(p, q)$:

p	q	$p \mid q$
T	T	F
T	F	T
F	T	T
F	F	T

Surprisingly, this connective, *by itself*, can express all truth-functions whatever! Since we have already shown that '$\sim$' and '&' form a complete set, we can establish the completeness of '$|$' if we can produce expressions equivalent to '$\sim p$' and to 'p & q' whose only sentential connective is '$|$'. Find such expressions, using a truth-table to justify your answers.

Hint: for negation, try to find an expression built up from 'p' and '$|$' which has **F**'s when 'p' has **T**'s and vice versa.

III.   We represent the *neither-nor*-function, $f_{12}(p, q)$, by ' $\downarrow$ ' (the *dagger*):

p	q	$p \downarrow q$
T	T	F
T	F	F
F	T	F
F	F	T

This connective can also express all truth-functions. Show the completeness of ' $\downarrow$ ' in the same manner you did for '|' in the preceding exercise. Then express each of the following in terms of ' $\downarrow$ ': $f_2(p, q)$; $f_4(p, q)$; $f_5(p, q)$.

*IV.   There are two robots, Boris and Bela. If you utter a true sentence, Boris always responds "True!", and if you utter a falsehood, he responds "False!" Bela, on the other hand, is a fiend who replies "False!" upon hearing a true sentence, "True!" upon hearing a falsehood. You approach a robot wishing to find out which fork in the road ahead will take you to East Grubbsville, your destination. However, you do not know if the robot in front of you is Boris or Bela. You are allowed to utter just *two* (declarative) sentences to the robot, and must determine from his answers which road will take you to East Grubbsville. What sentences should you utter? Suppose you are allowed to utter just *one* sentence—find one which will do the job!

CHAPTER THREE

SENTENTIAL LOGIC
Deductive Methods

3.1 Rules of Inference and Replacement for '&', 'v', and '~'

It is not hard to see that the argument

Either the number two is composite or else the number two is prime.	$C \vee P$
It is not the case that the number two is composite.	$\sim C$
Therefore the number two is prime.	$\therefore P$

is valid: if its premises are both true, then its conclusion has to be true too. In fact it's obvious that any argument of the form

$p \vee q$
$\sim p$
$\therefore q$

will be valid in that sense, for if a disjunction is true but its first disjunct is not then its second disjunct must be.
    We have an equally easy time with examples like

God is omnipotent and benevolent.	$O \& B$
Therefore God is omnipotent.	$\therefore O$

and

God is omnipotent.	O
Therefore God is omnipotent or omniscient.	$\therefore O \vee K$

These arguments, as well as all others of the forms

p & q
∴ *p*

and

p
∴ *p* v *q*

are clearly valid: whenever a conjunction is true its first conjunct must be, and whenever the first disjunct of a disjunction is true the whole disjunction must be.

Things get a little more complicated with

God is omnipotent and benevolent.	*O & B*
Therefore God is either omnipotent or omniscient.	∴ *O* v *K*

but this argument can still be handled fairly easily if we combine two of the things we've already noticed: it follows from the premise that God is omnipotent, and from that it follows in turn that God is either omnipotent or omniscient.

When, however, we come to such a complicated argument as the following, a bit more work is required:

Either our senses are wholly reliable, or else our sense of sight sometimes deceives us, our sense of touch sometimes misleads us, and our sense of hearing sometimes gives us false impressions.	*R* v [(*S & T*) & *H*]
It is not the case that our senses are wholly reliable.	~*R*
Therefore either our sense of sight sometimes deceives us or else our sense of sight always deceives us.	∴ *S* v *A*

From the two premises of this argument it follows that

Our sense of sight sometimes deceives us, our sense of touch sometimes misleads us, and our sense of hearing sometimes gives us false impressions.	(*S & T*) & *H*

The intermediate argument from those premises to that conclusion is an argument of the form

p v *q*
~*p*
∴ *q*

(substituting '*R*' for '*p*' and '(*S* & *T*) & *H*' for '*q*'), and it is clear, as we've already noticed, that any argument of that form is a valid one.

But of course whenever

Our sense of sight sometimes deceives us, our sense of touch sometimes misleads us, and our sense of hearing sometimes gives us false impressions	(*S* & *T*) & *H*

is true,

Our sense of sight sometimes deceives us and our sense of touch sometimes misleads us	*S* & *T*

will be true, too — any argument of the form

p & *q*
∴ *p*,

we've noticed, is valid.

For the same reason, when

Our sense of sight sometimes deceives us and our sense of touch sometimes misleads us	*S* & *T*

is true,

Our sense of sight sometimes deceives us	*S*

must also be true; and from the latter we can infer

Either our sense of sight sometimes deceives us or else our sense of sight always deceives us,	*S* v *A*

since all arguments of the form

p
∴ *p* v *q*

are valid.

We've shown that the conclusion of our original argument follows from the premises of that argument by *deducing* it from them, taking care at each stage of the game to make *only* inferences that could be justified by reference to certain elementary forms of argument. Since each of the argument forms we relied upon along the way can take us from true sentences only to other true sentences, we could be sure at each point in the deduction that the sentence being inferred at that point, from preceding sentences, had to be true when those preceding sentences were. At the end, then, we could be sure that the conclusion of our original argument had to be true if its premises were, and thus be sure that our original argument was valid.

The whole **deduction** (as we shall call it) can be written up so as to record our train of thought by listing in order the chain of inferences we've put together:

$R \vee [(S \& T) \& H]$
$\sim R$
$(S \& T) \& H$
$S \& T$
S
$S \vee A$

To clarify the whole thing we can number the various lines of our deduction and note, to the right, what is involved at each point. We began with two premises and wanted to deduce a certain conclusion from them; so we can begin by writing

1. $R \vee [(S \& T) \& H]$ Premise

2. $\sim R$ Premise   DEDUCE: $S \vee A$

The first thing we noticed was that '$(S \& T) \& H$' followed from these two premises, so we continue:

3. $(S \& T) \& H$ 1, 2

Since the rule of inference we are relying on here — the rule that permits one to infer from a disjunction and the negation of its first disjunct its second disjunct — has a name, **Disjunctive Syllogism** (usually abbreviated **DS**), we can note that too:

3. $(S \& T) \& H$ 1, 2 DS

We next inferred '*S* & *T*' from (3) on the grounds that any argument whose premise is of the form '*p* & *q*' and whose conclusion is the corresponding sentence '*p*' is valid. The rule that permits one to infer from any conjunction its left conjunct is usually called **Simplification** (abbreviated **Simp**) so we write, as our fourth step:

4. *S* & *T*   3 Simp

We then noticed that '*S*' follows from (4) on the same grounds:

5. *S*   4 Simp

For our final step we recalled that any argument whose conclusion is of the form '*p* v *q*' and whose premise is the corresponding sentence '*p*' is valid, and so deduced '*S* v *A*' from (5). The rule that permits one to infer from any sentence any disjunction whose first disjunct is that sentence is called **Addition** (abbreviated **Add**), so we can finish writing up our deduction by writing:

6. *S* v *A*   5 Add

The completed deduction looks like this:

1. *R* v [(*S* & *T*) & *H*]   Premise

2. ∼*R*   Premise   DEDUCE: *S* v *A*

3. (*S* & *T*) & *H*   1, 2 DS

4. *S* & *T*   3 Simp

5. *S*   4 Simp

6. *S* v *A*   5 Add

To give ourselves some range, we begin with four rules of inference associated with four valid forms of argument, along with a special rule allowing us to introduce sentences of the form '*p* v ∼*p*' into deductions whenever we want:

Rules of Inference	*Justification*

Conjunction (Conj)

In any deduction, one may infer from two previously obtained sentences

$$p$$

and

$$q$$

the corresponding sentence

$$p \ \& \ q.$$

Any argument of the form

$$p$$
$$q$$
$$\therefore p \ \& \ q$$

is valid, for if each of a pair of sentences is true then the conjunction of the two must also be true.

Simplification (Simp)

In any deduction, one may infer from a previously obtained sentence of the form

$$p \ \& \ q$$

the corresponding sentence

$$p.$$

Any argument of the form

$$p \ \& \ q$$
$$\therefore p$$

is valid, for if a conjunction is true then its first conjunct must also be true.

Addition (Add)

In any deduction, one may infer from a previously obtained sentence

$$p$$

any sentence of the form

$$p \lor q.$$

Any argument of the form

$$p$$
$$\therefore p \lor q$$

is valid, for if the first disjunct of a disjunction is true then the whole disjunction must also be true.

Disjunctive Syllogism (DS)

In any deduction, one may infer from two previously obtained sentences of the forms

$$p \lor q$$

and

$$\sim p$$

the corresponding sentence

$$q.$$

Any argument of the form

$$p \lor q$$
$$\sim p$$
$$\therefore q$$

is valid, for if a disjunction is true but its first disjunct is not then its second disjunct must be.

Excluded-Middle Introduction (E-M I)

At any point in a deduction, one may introduce any sentence of the form

$$p \lor \sim p.$$

No matter what sentence is substituted for 'p', the disjunction

$$p \lor \sim p$$

must be true, for if the first disjunct is not true the second is.

EXERCISES

I. Deduce the conclusions of each of the following arguments from their premises:

1. $A \lor B$
    $\sim A \,\&\, \sim C$
    $\therefore \sim A \,\&\, B$

2. $(F \,\&\, G) \lor (G \,\&\, H)$
    $\sim (F \,\&\, G)$
    $\therefore G \lor \sim H$

3. $N \lor (S \lor \sim A)$
    $\sim N \,\&\, \sim P$
    $\sim S \,\&\, \sim Q$
    $\therefore \sim N \,\&\, (\sim S \,\&\, \sim A)$

4. $A \lor B$
    $\sim A \,\&\, C$
    $\therefore (B \lor \sim D) \,\&\, (D \lor \sim D)$

5. $R \,\&\, S$
    $\therefore (R \lor S) \lor (T \lor \sim T)$

6. $R \,\&\, S$
    $\therefore (R \lor \sim S) \,\&\, (T \lor \sim T)$

7. $(\sim E \,\&\, \sim S) \,\&\, \sim V$
    $[E \lor (G \,\&\, B)] \,\&\, (\sim E \lor S)$
    $\therefore (G \lor \sim S) \,\&\, (\sim E \lor S)$

II. Make up an argument such that the deduction of its conclusion from its premises requires that each of our five rules be used at least twice.

* III. Let's use the symbol '$*$' temporarily (for this exercise only) as an abbreviation for the English expression '... and possibly ---'. Then, for example, we can abbreviate the sentence 'Raquel is coming to the party, and possibly Paula is too' by writing '$R * P$'. Which of the following arguments are valid?

$R * P$ R
$\therefore R$ P
                  $\therefore R * P$

Is every argument of the form

$p * q$
$\therefore p$

valid? How about arguments of the form

$$p$$
$$q$$
$$\therefore p * q?$$

*IV.   Make up an argument that is valid but that *cannot* be handled using just our five rules, thus establishing that we need to add a few more rules to our list. (One of the difficulties, of course, is finding a way to show that no combination of our rules will suffice to allow the deduction of your argument's conclusion from its premises.)

Rules of Replacement

Let's examine in detail what happens when we try to treat the clearly valid argument

Fabre's insect studies were either useless, significant, or not unhelpful.	$U \lor (S \lor \sim\sim H)$
Fabre's insect studies were not useless.	$\sim U$
Therefore Fabre's insect studies were either significant or helpful.	$\therefore S \lor H$

From the two premises we can obviously deduce

$$S \lor \sim\sim H$$

by **DS**; and it is clear that

$$S \lor H$$

follows from the former in the sense that '$S \lor \sim\sim H$' cannot be true unless '$S \lor H$' is also. But it turns out that our five rules do not permit us to infer the latter from the former.[1] What we need is a rule that somehow reflects the fact that in contexts such as this one a sentence and its double negation are *interchangeable*.

Notice, by the way, that it will *not* do to add to our list a rule like this:

---

[1] A good exercise (cf. Exercise IV above) would be to try to give some convincing justification for this claim.

At any point in any deduction, one may infer from any previously obtained sentence

p

the corresponding sentence of the form

$\sim\sim p$;

and similarly from

$\sim\sim p$

one may infer

p.

There wouldn't be anything *wrong* with adding such a rule, of course; any argument of the form

p
$\therefore \sim\sim p$

as well as any argument of the form

$\sim\sim p$
$\therefore p$

is valid. It's just that such a rule would *not* help us with the problem confronting us here. *Rules of inference* apply only to entire lines of deductions. **Simp,** for example, can be applied to 'A & B' to obtain 'A', but it cannot be applied *inside* the parentheses of ' $\sim(A$ & $B)$' to get ' $\sim A$'.

To handle the sentence 'S v $\sim\sim H$', then, what we need is a rule that *will* let us go "inside" and replace just the *part* of the form ' $\sim\sim p$' with the corresponding sentence of the form 'p'. So where '...p...' represents any sentence in which 'p' occurs, the rule we need could be formulated this way:

Double Negation (DN)

In any deduction, one may infer from any previously obtained sentence

...p...

containing an occurrence of some sentence, 'p', any corresponding sentence

...$\sim\sim p$...

provided that the latter is *just like* the former *except* for containing an occurrence of the corresponding sentence

$\sim\sim p$ ·

in exactly one place where the former contained an occurrence of 'p'; and similarly, from

$...\sim\sim p...$

one may infer

$...p...$

To avoid such a long statement of this rule, we'll simply write

Double Negation (DN)
$p :: \sim\sim p$

intending the odd mark ':: ' to be read as something like '...can replace or be replaced by, at any point in any line of a deduction...'.

We also need some other rules of a similar sort, for example one that will allow us to replace in any line of any deduction a sentence of the form

$p \& q$

with the corresponding sentence of the form

$q \& p$.

The following longish set of rules (which we shall call **Rules of Replacement**) has proved useful enough in the past to justify its length.

Rules of Replacement

Commutation (Com)
$p \& q :: q \& p$
$p \vee q :: q \vee p$

Association (Assoc)
$p \& (q \& r) :: (p \& q) \& r$
$p \vee (q \vee r) :: (p \vee q) \vee r$

Distribution (Dist)
$p \& (q \vee r) :: (p \& q) \vee (p \& r)$
$p \vee (q \& r) :: (p \vee q) \& (p \vee r)$

Double Negation (DN)
$p :: \sim\sim p$

De Morgan (De M)
$\sim p \& \sim q :: \sim(p \vee q)$
$\sim p \vee \sim q :: \sim(p \& q)$

Intuitive justifications for each of these rules can be supplied by using those given in Section 1.3 for **De Morgan** and the second of the **Distribution** rules as a guide. Going left to right on the first **Distribution** rule, for example, we might think along the following lines. Let 'p', 'q', and 'r' represent any sentences. Assume that 'p' is true and that the disjunction 'either q or r' is also true. Now if 'q' is true, then 'p' and 'q' are both true and so 'either p and q, or p and r' will be true because its *first* disjunct is. On the other hand, if 'q' is *not* true then 'r' must be (since 'either q or r' is); in that case both 'p' and 'r' are true, so again 'either p and q, or p and r' will be true, this time because its *second* disjunct is.[2]

With our full set of ten rules available, we can show a great many arguments valid by deducing their conclusions from their premises that we could not have handled using the five rules of inference alone. Discovering deductions is a skill that comes with practice. One approach that is often useful is to consider the premises of the argument at hand carefully and see what can be inferred immediately from them by our rules. We then consider those sentences in the same way, and continue until we reach our conclusion:

1. $D \& (E \& \sim C)$ Premise
2. $C \lor \sim (A \lor \sim B)$ Premise   DEDUCE: $\sim A \& B$
3. $(D \& E) \& \sim C$ 1 Assoc
4. $\sim C \& (D \& E)$ 3 Com
5. $\sim C$ 4 Simp
6. $\sim (A \lor \sim B)$ 2, 5 DS
7. $\sim A \& \sim \sim B$ 6 De M
8. $\sim A \& B$ 7 DN

---

[2]Going *right* to *left* on this same rule, assume that 'either p and q, or p and r' is true. Then no matter which disjunct is true, 'p' will certainly be true. In addition, if the left disjunct is true 'q' will be true, and if the right disjunct is the true one then 'r' will be true, so one way or the other 'either q or r' will be true as well.

Of course each of our rules may also be justified using truth-tables—see Section 2.1, especially the results of the exercises in Group III there.

In lines (3)-(5) we take (1) apart to get ' $\sim C$ '.[3] The latter can clearly be used with (2) to obtain (6), (6) suggests (7) and (7) gives our conclusion.

Arguments that yield readily to this approach—deducing what we can from the premises, and then looking around for something useful to do next—frequently involve use of **DS** at one or more key points with the early lines setting up the **DS** opportunity and the later ones tidying up afterwards:

1.	$\sim G$	Premise	
2.	$(\sim F \& G) \lor (\sim F \& H)$	Premise	
3.	$F \lor I$	Premise	DEDUCE: $I \& \sim G$
4.	$\sim G \lor \sim \sim F$	1 Add	
5.	$\sim (G \& \sim F)$	4 De M	
6.	$\sim (\sim F \& G)$	5 Com	
7.	$\sim F \& H$	2, 6 DS	
8.	$\sim F$	7 Simp	
9.	I	3, 8 DS	
10.	$I \& \sim G$	9, 1 Conj	

Since (1) tells us that 'G' is not true, we see that the left disjunct ' $\sim F \& G$ ' of (2) cannot be true either. In steps (4)—(6) we obtain its negation, and the ensuing **DS** at line (7) sets up another at line (9).

Since there are infinitely many valid arguments of distinct forms and we're trying to get by on just ten rules, our deductions will sometimes have to be rather lengthy. But it should be noted that we shall often have

---

[3] **Simp** only permits us to infer the *left* conjunct of a conjunction, not the right one. So long as we permit ourselves to use but one rule per line, then, there is no way to move from (3) to ' $\sim C$ ' in a single step. Instead we must first apply **Com** to (3) to get ' $\sim C \& (D \& E)$ ', and **Simp** will then deliver ' $\sim C$ '. Similar problems would come up if we wanted to use **DS** with a pair of sentences such as 'F' and ' $\sim F \lor G$ '; 'F' is not the negation of the left disjunct of ' $\sim F \lor G$ ', so we'd have to use **DN** to get ' $\sim \sim F$ ' before **DS** could be applied. Of course we might justifiably add rules to our system that would permit us to move directly from '$p \& q$' to 'q', and from 'p' and ' $\sim p \lor q$ ' to 'q', but with **Com** and **DN** available they would be more trouble than they're worth.

For now we'll put up with minor inconveniences of the above sort. But later on (see Section 3.4), we shall allow ourselves to combine **Com** or **DN** with any other rule as a single step.

available several different ways of deducing the conclusion of an argument from its premises using our ten rules, and the reader should realize that facility in discovering short deductions can be a great timesaver. For example, if we notice here that premise (2) is of the form '$(p \& q) \vee (p \& r)$' we can apply **Dist** at the start and save several steps:

1.	$\sim G$	Premise	
2.	$(\sim F \& G) \vee (\sim F \& H)$	Premise	
3.	$F \vee I$	Premise	DEDUCE: $I \& \sim G$
4.	$\sim F \& (G \vee H)$	2 Dist	
5.	$\sim F$	4 Simp	
6.	I	3, 5 DS	
7.	$I \& \sim G$	6, 1 Conj	

Deductions that require **Dist** near the end rather than the beginning are very hard to discover using our basic approach of starting with the premises and deducing what we can. An alternate approach is to look closely at the *conclusion* of the argument and think *backwards* from it, looking for an easier sentence to obtain first. For example, confronted with

1.	$J \vee (K \vee L)$	Premise	
2.	$J \vee M$	Premise	DEDUCE: $(J \vee K) \vee (L \& M)$

it is probably best to notice at the start that our conclusion is of the form '$p \vee (q \& r)$' and so could be gotten by **Dist** if we first had the corresponding sentence of the form '$(p \vee q) \& (p \vee r)$'. Our *hope* is that it will be easier to see how to get '$[(J \vee K) \vee L] \& [(J \vee K) \vee M]$' than it was to see how to get the conclusion itself. And in this case, the hope is fulfilled: the left conjunct '$(J \vee K) \vee L$' is easy to get from premise (1) by rearranging it slightly, and the right conjunct '$(J \vee K) \vee M$' can be gotten from premise (2) if we use **Add** on (2) to get '$(J \vee M) \vee K$' and then rearrange that:

3.	$(J \vee K) \vee L$	1 Assoc
4.	$(J \vee M) \vee K$	2 Add
5.	$J \vee (M \vee K)$	4 Assoc
6.	$J \vee (K \vee M)$	5 Com
7.	$(J \vee K) \vee M$	6 Assoc
8.	$[(J \vee K) \vee L] \& [(J \vee K) \vee M]$	3, 7 Conj
9.	$(J \vee K) \vee (L \& M)$	8 Dist

Though its usefulness is not confined to such cases, the "thinking backwards" technique is especially worth considering when the conclusion contains both '&' and 'v' and there is thus a good prospect that **Dist** may be needed to obtain it.

Sometimes, of course, short deductions cannot be found and we have to resign ourselves to finding one fifteen or twenty lines long. But readers will generally find that tricks discovered in the process of working through easier problems can often be modified to help with the harder ones. And the more familiar one becomes with our ten rules and the various ways they can be used in combination, the easier it will be to discover new deductions of reasonable length. The following example, for instance, combines the kind of **Dist** move used on the last couple of problems above with the use of **DS** reminiscent of the deductions with which we began:

1.	$A \vee B$	Premise	
2.	$A \vee \sim B$	Premise	DEDUCE: A
3.	$(A \vee B) \& (A \vee \sim B)$	1, 2 Conj	
4.	$A \vee (B \& \sim B)$	3 Dist	
5.	$(B \& \sim B) \vee A$	4 Com	
6.	$\sim B \vee \sim \sim B$	E-M I	
7.	$\sim (B \& \sim B)$	6 De M	
8.	A	5, 7 DS	

By conjoining (1) and (2) and distributing, we can get the 'B' from (1) and the '$\sim B$' from (2) to come together with an '&' between them as the left disjunct of (5). But of course '$B \& \sim B$' cannot possibly be true, so in steps (6) and (7) we establish its negation. **DS** on (5) and (7) then delivers 'A'.

Having worked through this deduction, prospects are not so bleak as they might have been when we come to the argument

| 1. | $B \vee B$ | Premise | DEDUCE: B |

We want to deduce 'B' from our premise, and there appears to be no obvious way to do it. But the preceding example taught us that it would be almost as good to deduce '$B \vee (B \ \& \ \sim B)$'. Thinking backwards from that, **Dist** comes to mind and we find ourselves needing only '$(B \vee B) \ \& \ (B \vee \sim B)$', which is easy enough to get:

2.	$B \vee \sim B$	E-M I
3.	$(B \vee B) \ \& \ (B \vee \sim B)$	1, 2 Conj
4.	$B \vee (B \ \& \ \sim B)$	3 Dist

.
.
.

From (4), the deduction of 'B' proceeds exactly as in the last few lines of the previous example.

EXERCISES

Symbolize each of the following arguments, using the letters suggested. Then deduce their conclusions from their premises.

1. The number two is not both prime and composite. The number two is prime. Therefore the number two is not composite. (P, C)
2. The number two is not composite. Therefore the number two is not both composite and prime. (C, P)
3. Either the first witness told the truth or the third witness did not. The first witness did not tell the truth, but either the second witness or the third one did. Therefore the second witness told the truth. (F, S, T)
4. Either the first witness or the second witness lied, and either the first witness or the third witness lied. The first witness did not lie. Therefore the second and third witnesses both lied. (F, S, T)
5. Either time had a beginning, or time is not finite. Either time will end, or time is not finite. Therefore either time is not finite, or else time had a beginning and will end. (B, F, E)

6. Either **Conj** is not a rule of replacement but **Simp** is, or else **Conj** is not a rule of replacement but **DN** is. It is not the case that either **Simp** is a rule of replacement or **Add** is. Therefore, neither **Conj** nor **Simp** is a rule of replacement, and **Add** isn't, either. (*C, S, D, A*)

7. I either act to further my own interests, or I act to further the interests of others. Therefore either I act to further my own interests, or else I act to further the interests of others but do not act to further my own. (*M, O*)

8. Either the blood test and the EKG are not both accurate or else the patient has a heart disorder. Either the blood test was accurate or else neither the nurse nor the lab technician acted competently. The nurse acted competently. Therefore either the EKG is not accurate or else the patient has a heart disorder. (*B, E, P, N, L*)

9. You should not open the bidding with "Two no-trump" unless you have 22–24 high-card points and a stopper in every suit. You have 22–24 high-card points but you do *not* have a stopper in every suit. Therefore even though you have 22–24 high-card points, you should not open the bidding with "Two no-trump." (*B,H,S*; recall the discussion of 'unless' in Section 1.3)

10. Either silver stocks are going up or silver bullion prices are not; and either the Dow Jones Transportation Index is going up or the Dow Jones Industrial Index is not. Silver bullion prices are going up and so is the Dow Jones Industrial Index. Therefore silver stocks are going up and so is the Dow Jones Transportation Index. (*S, B, T, I*)

11. The valence of the calcium ion is either +2 or +3, and the valence of the phosphate ion is either −2 or −3. So either the valence of the calcium ion is +3, or the valence of the phosphate ion is −2, or else the valence of the calcium ion is +2 and the valence of the phosphate ion is −3. (*C, D, P, Q*)

12. Either Socrates is mortal, or he's not immortal. Therefore, Socrates is mortal. (*M*)

13. The set of prime numbers is either denumerable or countable. The set of prime numbers is either finite, denumerable or uncountable. Therefore the set of prime numbers is either denumerable or finite. (*D, C, F*)

14. Either Neanderthal man was especially well adapted to living in cold climates, or else he did not live in Europe during the Würm glaciation. Unless the carbon-14 studies of his remains are inaccurate, Neanderthal man did live in Europe during the Würm glaciation. Therefore either Neanderthal man was especially well adapted to living in cold climates, or the carbon-14 studies of his remains are inaccurate. (*A, L, C*)

*15. Either no evil exists in the world, or God is unwilling to prevent its existence, or He is unable to prevent its existence. Either God is both willing and able to prevent the existence of evil, or else He is neither willing nor able to prevent its existence. Therefore either no evil exists in the world or else God is both unwilling and unable to prevent its existence. (E, W, A)

Deductive Validity and Theorems

We have confined our attention in this section to arguments whose validity hinges solely upon the behavior of '&', 'v', and '~'. In connection with such arguments stated in our symbolic language it is useful to define two terms explicitly.

DEFINITION   A **deduction** of the sentence $\mathscr{C}$ from the sentences $\mathscr{P}_1$, ..., $\mathscr{P}_k$ is a finite sequence of sentences beginning with $\mathscr{P}_1$ through $\mathscr{P}_k$ and ending with $\mathscr{C}$ wherein each line after $\mathscr{P}_k$ is either a sentence of the form 'p v $\sim p$' or else follows from earlier sentences in the sequence by one of our rules. (The lines must be numbered consecutively, the first k lines must be labeled as premises, and the line number(s) and rule involved in obtaining each later step must be listed to the right.)

DEFINITION   An argument is **deductively valid** if and only if it is possible to construct a deduction of its conclusion from its premises.

Thus the argument 'The universe is not both non-empty and infinite; it is non-empty; therefore the universe is finite' is deductively valid, as we can show by presenting the following deduction:

1. $\sim(\sim E \,\&\, \sim F)$ Premise
2. $\sim E$ Premise   DEDUCE: F
3. $\sim\sim E \lor \sim\sim F$ 1 De M
4. $E \lor \sim\sim F$ 3 DN
5. $\sim\sim F$ 2, 4 DS
6. F 5 DN

Sometimes it is convenient to employ a sort of shorthand in discussing deductive validity using the sign '$\vdash$' which we call the "turnstile". We write, for example,

$\sim(\sim E \,\&\, \sim F), \sim E \vdash F$

instead of the wordier

The argument with premises '~(~E & ~F)' and '~E' and conclusion 'F' is deductively valid,

that is, instead of writing

'F' is deducible (using our ten rules of inference and replacement) from '~(~E & ~F)' and '~E'.

It should always be kept in mind, of course, that in deducing the conclusion of an argument we are *not* establishing that it is *true* but only that it would have to be true *if* the *premises* were true:

We have already seen that our rules can be used to deduce the conclusions of many arguments from their premises. But sometimes they can be used to do something else. Certain sentences can be deduced, using just our rules, *from no premises at all*. What happens in the case of such a sentence is that we use **E-M I** at the beginning of the deduction, and then go on to deduce the sentence in question from the one **E-M I** starts us off with, for example:

1.	$F \vee \sim F$	E-M I
2.	$(F \vee \sim F) \vee E$	1 Add
3.	$E \vee (F \vee \sim F)$	2 Com
4.	$(E \vee F) \vee \sim F$	3 Assoc
5.	$(F \vee E) \vee \sim F$	4 Com

What has been shown here is very different from what we show when the conclusion of an argument is deduced from its premises. The latter sort of deduction establishes only that a certain sentence must be true *if* certain others are. But here no premises at all are involved; instead, the truth of 'Either the universe is finite, or it is empty, or it is not finite' is established *unconditionally*. The sentence in question has, in effect, been shown to be a *logical truth*, that is, true on logical grounds alone. Again, it is useful to define two terms:

DEFINITION   A **proof** is a finite sequence of sentences wherein each line (including the first) is either of the form '$p \vee \sim p$' or else follows from earlier sentences in the sequence by one of our rules. (The lines must be numbered consecutively, and the line number(s) and rule involved in obtaining each step must be listed to the right.)

DEFINITION   A sentence is a **theorem** if and only if it is possible to construct a proof of which it is the last line.

Here too it is convenient to employ shorthand so we will write

$\vdash (F \lor E) \lor \sim F$

as an abbreviation for

'$(F \lor E) \lor \sim F$' is deducible (using the rules of this section) from no premises at all,

that is, for

'$(F \lor E) \lor \sim F$' is a theorem (of our deductive system).

As theorems become more intricate, we can naturally expect their proofs to get more complicated. To show that '$[(A \& B) \lor (A \& \sim B)] \lor [(\sim A \& B) \lor (\sim A \& \sim B)]$' is a theorem, for example, we need a proof such as this:

1. $B \lor \sim B$	E-M I
2. $A \lor \sim A$	E-M I
3. $(B \lor \sim B) \& (A \lor \sim A)$	1, 2 Conj
4. $[(B \lor \sim B) \& A] \lor [(B \lor \sim B) \& \sim A]$	3 Dist
5. $[A \& (B \lor \sim B)] \lor [(B \lor \sim B) \& \sim A]$	4 Com
6. $[A \& (B \lor \sim B)] \lor [\sim A \& (B \lor \sim B)]$	5 Com
7. $[(A \& B) \lor (A \& \sim B)] \lor [\sim A \& (B \lor \sim B)]$	6 Dist
8. $[(A \& B) \lor (A \& \sim B)] \lor [(\sim A \& B) \lor (\sim A \& \sim B)]$	7 Dist

Theorems, and their proofs, are not only interesting in their own right; they will play an important role in Chapter 4 as well.[4]

---

[4]The reader may wish to compare the sign '$\vdash$' (and the concepts of *deductive validity* and *theorem*) with the sign '$\models$' (and the concepts of *truth-table validity* and *tautology*) introduced in Sections 2.1 and 2.3.

EXERCISES

I. Show that each of the following is a theorem of the deductive system
    developed in this section:
    1. $\sim F \vee F$
    2. $\sim F \vee (\sim E \vee F)$
    3. $\sim (F \& \sim F)$
    4. $(E \& F) \vee (\sim E \vee \sim F)$
    5. $A \vee \sim (A \& B)$
    6. $C \vee [\sim C \& (D \vee \sim D)]$
    7. $[(R \& S) \vee (R \& \sim S)] \vee [\sim (S \vee R) \vee (\sim R \& S)]$
    8. $\sim [(P \vee Q) \& \sim P] \vee Q$

II. Answer the following questions, *explaining* each of your answers:
    1. If $\mathcal{P}$ and $\mathcal{Q}$ are expressions which are theorems of our system,
        must the corresponding expression
        $$\mathcal{P} \& \mathcal{Q}$$
        also be a theorem?
    2. If $\vdash \mathcal{P}$ and
        $$\vdash \sim (\mathcal{P} \& \mathcal{Q})$$
        must $\mathcal{Q}$ be a theorem as well?
    3. If $\mathcal{P}$ and $\mathcal{Q}$ are expressions such that
        $$\vdash \sim \mathcal{P} \vee \mathcal{Q}$$
        must it be the case that
        $$\mathcal{P} \vdash \mathcal{Q}?$$

3.2 Rules of Inference and Replacement for '⊃' and '≡'

Many of the arguments we have looked at thus far seem to have been worded in an especially contrived manner, for example:

Either the blood test and the EKG are not both accurate or else the patient has a heart disorder.	$\sim(B \& E) \vee P$
Either the blood test was accurate or else neither the nurse nor the lab technician acted competently.	$B \vee \sim(N \vee L)$
The nurse acted competently.	N
Therefore either the EKG is not accurate or else the patient has a heart disorder.	$\therefore \sim E \vee P$

It would be far more natural to reword the premises and the conclusion, and present the argument this way:

If the blood test and the EKG are both accurate, then the patient has a heart disorder.	$(B \& E) \supset P$
If either the nurse or the lab technician acted competently, then the blood test was accurate.	$(N \vee L) \supset B$
The nurse acted competently.	N
Therefore if the EKG is accurate then the patient has a heart disorder.	$\therefore E \supset P$

Of course we want to extend our treatment of '&', 'v', and '~' so that we can handle arguments presented in this sort of language, too, and that requires us to find and add to our set of ten rules for '&', 'v', and '~' some rules of inference and replacement for '⊃' and '≡'. We begin by noticing that each of the following arguments is valid in the informal sense—discussed in Chapter 1—that their conclusions must be true if their premises are:

If I think then I exist.
I think.
Therefore I exist.

If all killing is wrong then killing in defense of one's children is wrong.
It is not the case that killing in defense of one's children is wrong.
Therefore it is not the case that all killing is wrong.

If God is omniscient then God always knows what time it is.
If God always knows what time it is then what God knows at any given moment differs from what He knows at any other moment.
Therefore if God is omniscient then what God knows at any given moment differs from what He knows at any other moment.

Either time is infinite or time is circular.
If time is infinite then time had no beginning.
If time is circular then time has no end.
Therefore either time had no beginning or time has no end.

These arguments are, respectively, of the forms:

$p \supset q$	$p \supset q$	$p \supset q$	$p \lor q$
p	$\sim q$	$q \supset r$	$p \supset r$
$\therefore q$	$\therefore \sim p$	$\therefore p \supset r$	$q \supset s$
			$\therefore r \lor s$

Clearly, no argument of any of these forms could have true premises but a false conclusion. Accordingly, we add to the five rules of inference that we already have (**Conjunction, Simplification, Addition, Disjunctive Syllogism,** and **Excluded-Middle Introduction**) the following four:

Rules of Inference   ***Justification*** [5]

***Modus Ponens* (MP)**

In any deduction, one may infer from two previously obtained sentences of the forms

$$p \supset q$$

and

$$p$$

the corresponding sentence

$$q.$$

Any argument of the form

$$p \supset q$$
$$p$$
$$\therefore q$$

is valid, for if a conditional is true and its antecedent is also, then its consequent must be.

***Modus Tollens* (MT)**

In any deduction, one may infer from two previously obtained sentences of the forms

$$p \supset q$$

Any argument of the form

$$p \supset q$$
$$\sim q$$
$$\therefore \sim p$$

---

[5] These justifications may be spelled out in more detail by referring explicitly to the definition of a material conditional given in Section 1.4 (see Exercise IV below). They are elaborated in a different way, by truth-tables, in Section 2.3.

and

$$\sim q$$

the corresponding sentence of the form

$$\sim p.$$

is valid, for if a conditional is true but its consequent is not then its antecedent cannot be.

Hypothetical Syllogism (HS)

In any deduction, one may infer from two previously obtained sentences of the forms

$$p \supset q$$

and

$$q \supset r$$

the corresponding sentence of the form

$$p \supset r.$$

Any argument of the form

$$p \supset q$$
$$q \supset r$$
$$\therefore p \supset r$$

is valid — if the first two conditionals are true then the third one must also be true.

Constructive Dilemma (CD)

In any deduction, one may infer from three previously obtained sentences of the forms

$$p \lor q$$
$$p \supset r$$

and

$$q \supset s$$

the corresponding sentence of the form

$$r \lor s.$$

Any argument of the form

$$p \lor q$$
$$p \supset r$$
$$q \supset s$$
$$\therefore r \lor s$$

is valid, for if one or the other of two true conditionals has a true antecedent then one or the other of them has a true consequent.

By way of contrast, it should be clear that *neither* of the following forms of argument is valid and hence they should never be employed in deductions:

$$
\begin{array}{ll}
p \supset q & p \supset q \\
q & \sim p \\
\therefore p & \therefore \sim q
\end{array}
$$

The first is sometimes confused with **Modus Ponens** and the second with **Modus Tollens**. If we replace '*p*' with the false sentence 'More than twelve planets orbit the sun' and '*q*' with the true sentence 'More than eight planets orbit the sun', however, both resulting arguments have true premises but false conclusions:

If more than twelve planets orbit the sun, then more than eight planets orbit the sun.	If more than twelve planets orbit the sun, then more than eight planets orbit the sun.
More than eight planets orbit the sun.	Not more than twelve planets orbit the sun.
Therefore more than twelve planets orbit the sun.	Therefore not more than eight planets orbit the sun.

We will need some additional rules of replacement, as well, and the considerations of Section 1.4 suggest the following:

Transposition (Trans)

$p \supset q :: {\sim}q \supset {\sim}p$

Exportation (Exp)

$(p \,\&\, q) \supset r :: p \supset (q \supset r)$

Implication (Impl)

$p \supset q :: {\sim}p \lor q$

Equivalence (Equiv)

$p \equiv q :: (p \supset q) \,\&\, (q \supset p)$

$p \equiv q :: (p \,\&\, q) \lor ({\sim}p \,\&\, {\sim}q).$[6]

More thoroughgoing justifications for these rules are provided by the methods of Section 2.2.

With these additional rules we can now handle such arguments as:

1. $C \supset ({\sim}P \supset A)$	Premise	If *abstract* entities are those that cannot be perceived with the senses, then if the number 89 cannot be perceived with the senses then the number 89 is an abstract entity.
2. $N \supset (A \supset E)$	Premise	If the number 89 exists, then if it is an abstract entity then of course at least one abstract entity exists.

---

[6]The 'if ... then ---' locution is undoubtedly used, like most locutions, in a variety of different senses in ordinary English discourse. With a single set of rules like ours, of course, we get at only one of the ways in which it can be used. Presumably *other* sets of rules could be constructed that would capture *other* senses of the locution in question. The reader interested in attempts to construct other such sets of rules might begin with the references cited in Section 2.2, footnote 5.

3. $\sim N \supset \sim B$ Premise   If the number 89 does not exist, then there are no prime numbers between 84 and 96.

4. B Premise   There is a prime number between 84 and 96.

∴ $C \supset (P \vee E)$ Therefore if abstract entities are those that cannot be perceived with the senses then either the number 89 can be perceived with the senses or else at least one abstract entity exists.

We first use **MT** to play (3) and (4) off against one another

5. $\sim\sim B$ 4 DN

6. $\sim\sim N$ 3, 5 MT

and notice that **DN** will permit us to use **MP**:

7. N 6 DN

8. $A \supset E$ 7, 2 MP

If the consequent of (1) were the same as the antecedent of (8) we could use **HS**; of course this is not the case, but if we use **Exportation** on (1) first, we'll be able to use **HS** anyway:

9. $(C \& \sim P) \supset A$ 1 Exp

10. $(C \& \sim P) \supset E$ 9, 8 HS

We could now use **Implication** on (10) to get '$\sim(C \& \sim P) \vee E$'; **De Morgan** would then lead us to '$(\sim C \vee \sim\sim P) \vee E$', **DN** and **Assoc** would give us '$\sim C \vee (P \vee E)$' and **Implication** would finish the job. But we can save a couple of steps by using **Exportation** on (10) first:

11. $C \supset (\sim P \supset E)$ 10 Exp

12. $C \supset (\sim\sim P \vee E)$ 11 Impl

13. $C \supset (P \vee E)$ 12 DN

It is generally good strategy to use **MP, MT, HS,** and **CD** as often as possible, often going out of one's way via **DN, Implication, Exportation, Transposition,** and the like to develop opportunities to apply such rules. Presented with

1. $A \supset (B \& C)$	Premise	
2. $D \supset E$	Premise	
3. $A \lor D$	Premise	
4. $B \supset F$	Premise	
5. $(\sim E \supset F) \supset (\sim G \supset \sim H)$	Premise	DEDUCE: $H \supset G$

it is probably best to notice first that (3) is the disjunction of the antecedents of (1) and (2):

6. $(B \& C) \lor E$	1, 2, 3 CD

The conclusion we're after, '$H \supset G$', is concealed in the consequent of (5), and one might notice this next:

7. $(\sim E \supset F) \supset (H \supset G)$	5 Trans

So we need '$\sim E \supset F$'. Since no obvious way of obtaining it suggests itself, we look around for something else to do. If we commute (6) we'll be able to use **Dist,** of course, and experience in the preceding section has surely taught us to give this route a chance:

8. $E \lor (B \& C)$	6 Com
9. $(E \lor B) \& (E \lor C)$	8 Dist

Now which half of (9) is most likely to get '$\sim E \supset F$' for us? Let's see what we can do with

10. $E \lor B$	9 Simp

first—if it leads nowhere we can always turn to '$E \lor C$' later. (10) is a disjunction but there's no way to use it in a **Constructive Dilemma**; so instead of leaving it as it is let's use **Implication** to turn it into a conditional:

11. ~~E ∨ B   10 DN

12. ~E ⊃ B   11 Impl

That was the right road to take! (12) has the same consequent as (4) has antecedent, so **HS** and **MP** finish our deduction:

13. ~E ⊃ F   12, 4 HS

14. H ⊃ G   13, 7 MP

You will often be tempted to use the rule **Implication** to transform conditionals into complex disjunctions, hoping the familiar rules of Section 3.1 can be employed. And it must be admitted that there are a few problems where that strategy is useful (see Exercises 3 and 4 in Group III below, for example). But on most arguments this is an inadvisable approach that leads to long, complicated deductions where much simpler ones can be found with a bit of ingenuity. Consider again the argument about diagnosing heart disorder given at the start of this Section:

1. (B & E) ⊃ P   Premise
2. (N ∨ L) ⊃ B   Premise
3. N   Premise   DEDUCE: E ⊃ P

Since (3) is not the antecedent of (2), we cannot do **MP** immediately. But if we apply **Add** to (3) first we'll get '*B*' quickly:

4. N ∨ L   3 Add
5. B   2, 4 MP

(5) is just "half" the antecedent of (1), and one might consider using **Implication** to turn (1) into ' ~ (B & E) ∨ P' so that an eventual **DS** can be found. But a much shorter deduction is available if we simply use **Exportation** on (1) instead:

6. B ⊃ (E ⊃ P)   1 Exp
7. E ⊃ P   6, 5 MP

There are even arguments where it is best to use the rules **Implication** and **Equivalence** to transform appropriate sentences involving '&', 'v' and '~' into conditionals and biconditionals so that we may make maximum use of the new rules for '⊃' and '≡'. Exercise 9 in Group II below is one example (it is the argument about Neanderthal man from Section 3.1); here's another:

1. ~E v [(W & A) v (~W & ~A)]   Premise

2. E   Premise   DEDUCE:

3. E ⊃ [(W & A) v (~W & ~A)]   1 Impl   (~W v A) & (~A v W)

4. E ⊃ (W ≡ A)   3 Equiv

5. W ≡ A   2, 4 MP

6. (W ⊃ A) & (A ⊃ W)   5 Equiv

7. (~W v A) & (A ⊃ W)   6 Impl

8. (~W v A) & (~A v W)   7 Impl

It will later be convenient to add to our system one last rule, the rule of **Conditional Proof**, which will correspond to a very natural way of arguing and will have the advantage of shortening many deductions considerably. But it will turn out (see Chapter 4) that we *need* not add that rule; we shall be able to prove nothing with it that we could not have established using only the rules already presented in this chapter.

EXERCISES

I.   Finish annotating the following two deductions:

    1.   1. (A & B) ⊃ C   Premise

          2. A   Premise

          3. C ⊃ (D & E)   Premise   DEDUCE:

          4. A ⊃ (B ⊃ C)   ~B v (D & E)

          5. B ⊃ C

          6. B ⊃ (D & E)

          7. ~B v (D & E)

    2.   1. ~(J & K)   Premise

          2. ~(H v ~I) ⊃ (J & K)   Premise

          3. H ⊃ ~F   Premise

          4. ~I ⊃ ~G   Premise

          5. ~F ⊃ G   Premise   DEDUCE:

          6. ~~(H v ~I)   F ≡ ~G

          7. H v ~I

          8. ~F v ~G

9. $F \supset \sim G$
10. $\sim G \supset \sim\sim F$
11. $\sim G \supset F$
12. $(F \supset \sim G) \& (\sim G \supset F)$
13. $F \equiv \sim G$

II. Deduce the conclusions of each of the following arguments from their premises:

1. $(D \supset C) \& (A \supset B)$
   $C \& A$
   $\therefore A \& B$
2. $A \supset R$
   $\sim(R \& S)$
   $\therefore S \supset \sim A$
3. $(F \vee G) \supset H$
   $I \supset J$
   $F \& \sim J$
   $\therefore H \& \sim I$
4. $(O \& P) \supset Q$
   $R \supset (Q \supset S)$
   $O \& R$
   $\therefore P \supset S$
5. $S \supset (M \supset W)$
   $P \supset (M \& S)$
   $\therefore \sim P \vee W$
6. $T \vee (T \equiv U)$
   $\sim(T \vee V)$
   $\therefore \sim U$

7. $L \supset N$
   $(\sim N \supset \sim L) \supset (M \supset P)$
   $\sim P$
   $\therefore \sim M$
8. $C \vee D$
   $C \supset X$
   $D \supset \sim Y$
   $(\sim Y \vee X) \supset (\sim D \supset Y)$
   $\therefore D \equiv \sim Y$
9. $A \vee \sim L$
   $\sim C \vee L$
   $\therefore A \vee \sim C$
10. $\sim E \vee (\sim W \vee \sim A)$
    $(W \& A) \vee (\sim W \& \sim A)$
    $\therefore \sim E \vee (\sim W \& \sim A)$
*11. $(Z \& M) \supset (S \vee A)$
    $Z \supset \sim S$
    $\therefore (Z \& \sim A) \supset \sim M$
*12. $K \equiv Q$
    $(K \& V) \equiv \sim(Q \vee V)$
    $\therefore K \supset \sim V$

III. Symbolize each of the following arguments; then deduce their conclusions from their premises:

1. The number 91 is prime just in case it is not composite. The number 91 is not both prime and divisible by 7. The number 91 is divisible by 7. Therefore the number 91 is composite. (P, C, D)

2. I'm damned if I do. I'm damned if I don't. Therefore I'm damned. (M, D)

3. If argument (1) is sound then it is valid. If argument (1) is sound then its premises are true. Therefore if argument (1) is sound, then it is valid and its premises are true. (S, V, P)

4. If argument (1) is sound, then it is valid and its premises are true. Thus, if argument (1) is sound then it is valid, and if argument (1) is sound then its premises are true. (*S, V, P*)

5. If the President sounds optimistic then the stock market will go up. So if the economic news is bad but the President sounds optimistic, the stock market will go up. (*O, U, B*)

6. The runner on first base will break with the pitch, but the runner on third will score only if the catcher throws to second. The catcher will not throw to second if the runner on first breaks with the pitch. Therefore the runner on third will not score. (*F, T, S*)

7. If the front of a bicycle clunks on rough roads, either its front wheel isn't tight or its headset is loose. So if its front wheel is tight, then if a bicycle clunks on rough roads, its headset is loose. (*B, F, H*)

8. If God can create a stone too heavy for Him to lift, then there is something God cannot do. If it is not the case that God can create a stone too heavy for Him to lift, then there is something God cannot create. If there is something God cannot do then God is not omnipotent, and if there is something He cannot create then He is not omnipotent. Therefore God is not omnipotent. (*S, D, C, O*)

9. If the floor alarm did not sound, then either a thief descended from the ceiling or else a trained bird flew off with the Star of India. If a thief descended from the ceiling, the skylight alarm sounded. Thus if neither the skylight nor the floor alarm sounded, a trained bird flew off with the Star of India. (*F, T, B, S*)

10. If my plans are going to work out, then I'm going to get a D on the first test but go into the final with a high average. If I study fairly hard but skip the hard problems, then I'll get a B on the second test. If I get a B on the second test but a D on the first, I won't go into the final with a high average. Therefore if I study fairly hard, then if I skip the hard problems my plans are not going to work out. (*P, D, H, F, S, B*)

11. If the price of the stock rises we'll make money on the shares we own, and if the price of the stock falls we'll make money on warrants we sold short. So if the price of the stock either rises or falls, we'll make money on either the shares we own or the warrants we sold short. (*R, S, F, W*)

12. The set of prime numbers is either denumerable or countable. If the set of prime numbers is countable, then it is either finite or denumerable. The set of prime numbers is not finite. It follows, then, that the set of prime numbers is denumerable. (*D, C, F*)

*13. White will lose his bishop unless he moves his king. Black will either check with his queen or mate with his pawn. If White loses his bishop or Black checks with his queen, White will move his king and Black will mate with his pawn. Consequently, White will move his king and Black will mate with his pawn. (*B, K, Q, P*)

IV. Show in detail how each of the rules of inference and replacement introduced in this section may be justified by the characterizations of *material conditionals* and *material biconditionals* provided in Section 1.4.

3.3 Conditional Proof[7]

Consider the argument

If the set of prime numbers is finite, then the set of prime numbers has a smallest member and a largest member.	$F \supset (S \& L)$
Therefore if the set of prime numbers is finite then it has a largest member.	$\therefore F \supset L$

And imagine how we might think this argument through to ourselves, if we were being careful to try to make each step explicit:

I'm told that if the set of prime numbers is finite then it has both a smallest and a largest member, and I'm asked to deduce from this that if the set of prime numbers is finite then it has a largest member. What if I assume for a moment that the set of prime numbers is finite; could I then show that it has a largest member? Why of course, for there is practically nothing to show. It follows from my assumption and the premise of the original argument, by **MP**, that the set of primes has both a smallest and a largest member; and from this, in turn, it follows by **Com** and **Simp** that it has a largest member, which is what I had to show! Summing up: by assuming the antecedent of the conclusion, 'The set of prime numbers is finite', I was able to deduce its consequent, 'The set of prime numbers has a largest member'. So I *have* established that *if* the set of primes is finite then it has a largest member.

Let's find a way of recording this train of thought clearly. We can begin, as usual, by writing the premise as line (1) of our deduction, noting the conclusion off to the right:

1. $F \supset (S \& L)$ Premise   DEDUCE: $F \supset L$

The first step in our argument was to assume (temporarily) that the antecedent of the conclusion is true. This is, of course, not something we can *deduce* from (1). To make it obvious that this is merely an *assumption* which we're making for the moment, we'll label it **Assumption**, indent it to the right and point an arrow at it from the left:

1. $F \supset (S \& L)$ Premise   DEDUCE: $F \supset L$

$\longmapsto$ 2. F Assumption

---

[7]We are indebted to an anonymous reviewer for the essentials of the formulation of the rule of conditional proof given at the end of this section and to Robert Neidorf, *Deductive Forms,* New York: Harper and Row, 1967, for teaching us how best to introduce and motivate this rule and its typographical trappings.

From the premise, (1), and our assumption, (2), we obtain

3. S & L 1, 2 MP

To make it clear that (3) has only been deduced with the help of our assumption, (2), we indent line (3) also, and allow the tail of the arrow marking that assumption to run down beside it:

1. $F \supset (S \, \& \, L)$ Premise   DEDUCE: $F \supset L$

┌→ 2. F Assumption

│   3. S & L 1, 2 MP

So long as we continue to make deductions in the presence of our assumption rather than from our premise alone, we indent the lines we deduce and continue the tail of the arrow alongside:

1. $F \supset (S \, \& \, L)$ Premise   DEDUCE: $F \supset L$

┌→ 2. F Assumption

│   3. S & L 1, 2 MP

│   4. L & S 3 Com

│   5. L 4 Simp

Given (1), we have now shown that *if* (2) is true *then* (5) must be true as well. We have established the conditional '$F \supset L$', by actually deducing its consequent from its antecedent (in the presence of our premise). To show that '$F \supset L$' follows from (1) alone, and that our assumption (2) has now served its purpose and can be abandoned, we *close off* the assumption by extending the tail of the arrow horizontally under (5) and writing '$F \supset L$' beneath — thus indicating that the assumption made at line (2) is no longer operative. We call the rule permitting this move the **Rule of Conditional Proof (CP)** and delay its precise formulation until the end of this section.

1. $F \supset (S \, \& \, L)$ Premise   DEDUCE: $F \supset L$

┌→ 2. F Assumption

│   3. S & L 1, 2 MP

│   4. L & S 3 Com

│   5. L 4 Simp

└──

6. $F \supset L$ 2–5 CP

Notice that the justification for line (6) refers to all steps in the subordinate deduction running from line (2) *through* line (5). It is because of this entire *sequence* that we are sure (6) follows from (1), for the lines of that sequence show that given our premise, the consequent of (6) must be true if its antecedent is.

Of course we could have deduced (6) from (1) using just the rules we already had. The deduction would not have been any longer (though in most other cases it would be):

1. $F \supset (S \& L)$	Premise	DEDUCE: $F \supset L$
2. $\sim F \vee (S \& L)$	1 Impl	
3. $(\sim F \vee S) \& (\sim F \vee L)$	2 Dist	
4. $(\sim F \vee L) \& (\sim F \vee S)$	3 Com	
5. $\sim F \vee L$	4 Simp	
6. $F \supset L$	5 Impl	

However, this deduction is more difficult to discover than the former. It also seems considerably less natural and more contorted, for the first deduction follows much more closely the steps we might actually go through in quickly thinking our way from premises to conclusion.

When the conclusion of the argument we're working with is a conditional, **CP** is often helpful. Typically, we begin by assuming the antecedent of the conclusion and try to obtain its consequent; but it can sometimes take a bit of maneuvering before the assumption proves useful:

1. $M \supset (N \supset O)$	Premise	
2. $\sim N \supset P$	Premise	
3. $(M \supset P) \supset (O \vee \sim Q)$	Premise	DEDUCE: $\sim O \supset \sim Q$
4. $\sim O$	Assumption	
5. $(M \& N) \supset O$	1 Exp	
6. $\sim (M \& N)$	5, 4 MT	

If we next use **De M** on (6), we'll be able to play the result off against (2); and this approach will provide an opportunity to use our assumption a second time:

7. ~M v ~N	6 De M
8. M ⊃ ~N	7 Impl
9. M ⊃ P	8, 2 HS
10. O v ~Q	3, 9 MP
11. ~Q	10, 4 DS
12. ~O ⊃ ~Q	4-11 CP

EXERCISES

I. Use **CP** to construct new deductions for exercises (3), (5), (9) and (11) in group (III) of the Exercises on pages 107-108.

II. Use **CP** to construct new deductions for exercises (4), (11) and (12) in group (II) on page 107.

CP is more versatile than we have shown so far. For one thing, an assumption for **CP** need not be introduced immediately after the premises. Furthermore, the conditional established by **CP** need not be the conclusion of the original argument.

1. ~A v B	Premise	
2. ~A v (~B v C)	Premise	DEDUCE: ~A v C
3. A ⊃ B	1 Impl	
4. A ⊃ (~B v C)	2 Impl	
5. A ⊃ (B ⊃ C)	4 Impl	
⌐→ 6. A	Assumption	
7. B	3, 6 MP	
8. B ⊃ C	5, 6 MP	
9. C	7, 8 MP	
10. A ⊃ C	6–9 CP	
11. ~A v C	10 Impl	

One is permitted to assume any sentence at any point in a deduction, provided only that each assumption made is later closed off. At line (6) here, for example, we *need* not have assumed 'A' — we could have assumed '~A' or 'C' or even 'M'. But there was no reason to make any of these other assumptions. Had we assumed 'M', say, then when we later gave up that assumption we would be left with an expression beginning 'M ⊃ ...', and no such expression would have helped in obtaining '~A v C'. We chose instead to assume 'A' and work for 'C', because we knew that **CP** would then give us 'A ⊃ C' and that '~A v C' would follow immediately.

It is also useful to be able to stack several conditional proofs on top of one another, using **CP** once to establish one thing we need and then using it again to establish something else, as in:

1. (G & H) ⊃ J   Premise
2. (H v I) ⊃ G   Premise   DEDUCE: (~G ⊃ ~H) & (H ⊃ J)
  3. ~G   Assumption
  4. ~(H v I)   2, 3 MT
  5. ~H & ~I   4 De M
  6. ~H   5 Simp
7. ~G ⊃ ~H   3-6 CP
  8. H   Assumption
  9. H ⊃ G   7 Trans
  10. G   9, 8 MP
  11. G & H   10, 8 Conj
  12. J   1, 11 MP
13. H ⊃ J   8-12 CP
14. (~G ⊃ ~H) & (H ⊃ J)   7, 13 Conj

Double use of **CP** is especially common when the conclusion of the argument one is concerned with happens to be a biconditional.

1. A ⊃ B   Premise
2. B ⊃ C   Premise   DEDUCE: A ≡ (A & C)
  3. A   Assumption
  4. B   1, 3 MP
  5. C   2, 4 MP
  6. A & C   3, 5 Conj

7. $A \supset (A \& C)$	3–6 CP
→ 8. $A \& C$	Assumption
9. A	8 Simp
10. $(A \& C) \supset A$	8–9 CP
11. $[A \supset (A \& C)]$ $\& [(A \& C) \supset A]$	7, 10 Conj
12. $A \equiv (A \& C)$	11 Equiv

Line (8) here does entitle us to line (9), of course, but notice that it would have been a *mistake* to have tried to justify (9) on the basis of (6). (6) was only obtained while the assumption made at line (3) was in force and that assumption is no longer operative at line (9), having been given up earlier. In general, once an assumption has been closed off no later lines of the deduction may be justified by reference to that assumption nor to any of the lines occurring between it and the point at which it was closed off. The little three-sided boxes formed by the arrows and their tails serve to keep track of such matters.

There are times, too, when we will want to nest conditional proofs inside one another, as in:

1. $T \vee [(U \vee W) \supset V]$	Premise	DEDUCE: $\sim T \supset (U \supset V)$
→ 2. $\sim T$	Assumption	
3. $(U \vee W) \supset V$	1, 2 DS	
→4. U	Assumption	
5. $U \vee W$	4 Add	
6. V	3, 5 MP	
7. $U \supset V$	4-6 CP	
8. $\sim T \supset (U \supset V)$	2-7 CP	

This sort of thing happens especially often when we use **CP** to establish *theorems,* which is frequently a wise approach. Until now we could only begin proofs of theorems with **E-M I**, but **CP** gives us a second choice: we can start with an *assumption.*

It is usually good strategy in such situations to make as many assumptions as the problem at hand suggests. For instance, to prove the theorem '$[A \supset (B \supset C)] \supset [(A \supset B) \supset (A \supset C)]$' we may begin by assuming the antecedent '$A \supset (B \supset C)$', intending to deduce from it the consequent

'$(A \supset B) \supset (A \supset C)$'. Since the latter is itself a conditional, we could next assume its antecedent '$A \supset B$' and would need only '$A \supset C$'. This also is a conditional, so we should make yet a third assumption:

1. $A \supset (B \supset C)$		Assumption
2. $A \supset B$		Assumption
3. A		Assumption
4. $B \supset C$		1, 3 MP
5. B		2, 3 MP
6. C		4, 5 MP
7. $A \supset C$		3-6 CP
8. $(A \supset B) \supset (A \supset C)$		2-7 CP
9. $[A \supset (B \supset C)] \supset [(A \supset B) \supset (A \supset C)]$		1-8 CP

In nesting conditional proofs this way, one must be sure that the order in which assumptions are closed off is the reverse of that in which they were introduced. This restriction, as well as the others mentioned above, are packed into the following more formal statement of the rules developed in this section, which serves also as a summary.

Rule of Assumptions (Assumption)

Any sentence may be introduced as a new step in a deduction, *provided that*:
  i. it is indented to the right, and a new vertical line is introduced on the left at that point, with an arrow at its top pointing to the sentence being introduced;
  ii. the vertical line in question is eventually discontinued by an application of the rule of **CP** (see below), but continued up to that point.

Rule of Conditional Proof (CP)

Let 'q' be a step in a deduction with a continuous vertical line passing *immediately* to its left, and let 'p' be the sentence heading that vertical line. Then the corresponding sentence of the form '$p \supset q$' may be introduced as the next step, *provided that*:
  i. the vertical line in question is discontinued by extending it horizontally to the right immediately below 'q' and above '$p \supset q$';

ii. all steps beside which this vertical line passes are thereby regarded as sealed off from the rest of the deduction (that is, no *later* steps in the deduction are justified by a reference to any of these earlier ones enclosed now by a three-sided box).

EXERCISES

I. Use **CP** to construct new deductions for (8), (9), and (10) in group (II) and for (4), (7), and (10) in group (III) of the Exercises at the end of Section 3.2.

II. Construct deductions for the following, using **CP** when helpful:

1. Jones has not learned how to use correctly the expression "Smith acted of his own free will" unless he has seen Smith acting freely and heard others use the expression "Smith is acting of his own free will." If all of our actions are determined according to physical laws, then no one ever acts of his own free will and Jones has never seen Smith acting freely. So if Jones has learned how to use correctly the expression "Smith acted of his own free will," not all of our actions are determined according to physical laws. (*L, S, H, D, N*)

2. If the visual field test was reliable, then loss of peripheral vision was confirmed. Either the tonometer reading was anomalous or high intraocular pressure was confirmed. If high intraocular pressure and loss of peripheral vision were both confirmed, a diagnosis of glaucoma is suggested. Therefore if the visual field test was reliable, then if the tonometer reading was not anomalous then a diagnosis of glaucoma is suggested. (*V, L, T, H, G*)

3. If argument (1) either has a false premise or is invalid, it is unsound. Consequently, if argument (1) has a false premise it is unsound and if argument (1) is invalid it is unsound. (*F, V, S*)

4. If argument (1) has a false premise, it is unsound. If argument (1) is invalid, it is unsound. So if argument (1) either has a false premise or is invalid, then it is unsound. (*F, S, V*)

5. If either Al's claim or Bill's claim is true, then Al's claim is true if and only if Bill's claim is true. It follows that Al's claim is true if and only if Bill's is. (*A, B*)

6. If Yossarian flies his missions, then if knowingly placing one's life in danger is irrational, then Yossarian is irrational. If Yossarian asks to be grounded and a request to avoid a perilous situation is rational, then Yossarian is rational. A request to avoid a perilous situation is rational if and only if knowingly placing one's life in danger is irrational. Therefore, if Yossarian flies his missions but asks to be grounded, then knowingly placing one's

life in danger is rational and a request to avoid a perilous situation is irrational. (F, K, R, G, A)

7. $(D \vee P) \supset (G \& F)$
   $(T \supset G) \supset [(R \vee B) \supset E]$
   $(\sim S \supset H) \supset [B \& \sim(E \& U)]$
   $\therefore D \supset (S \supset \sim U)$

III. Use **CP** to establish each of the following theorems:

1. $\sim A \supset \sim (A \& B)$
2. $[(A \supset A) \supset B] \supset B$
3. $[(A \supset B) \supset C] \supset$
   $[(C \supset A) \supset (D \supset A)]$
4. $(A \supset B) \vee (B \supset C)$
5. $\sim[(A \& B) \& \sim A)]$
6. $[(\sim A \vee \sim B) \& (C \supset A)] \supset$
   $[(D \supset B) \supset (\sim C \vee \sim D)]$

IV. 1. With your new computer comes the following Warranty Card:

GUARANTEED FOR 90 DAYS, PARTS AND LABOR

(i) This guarantee is void unless the machine has received proper care and there have been no attempted repairs by unauthorized persons.

(ii) Provided there have been no such attempted repairs, however, this guarantee is void only if the machine has not received proper care.

Show that (i) follows from the biconditional "The guarantee is void if and only if either the machine has not received proper care or there have been attempted repairs by unauthorized persons."

2. Show that (ii) follows from that same biconditional.
3. Show that the biconditional follows from (i) and (ii).

3.4 Indirect Proof

Consider the argument

Either the set of even numbers is infinite, or else the set of even numbers and the set of odd numbers are both infinite.	$E \lor (E \& O)$
If the set of even numbers is infinite then for each even number there exists a larger even number.	$E \supset L$
Therefore for each even number there exists a larger even number.	$\therefore L$

One natural way to argue, in English, from those premises to that conclusion would go something like this:

Given the two premises, 'L' must be true. For suppose not. Then it would follow, with premise two, that 'E' is not true, either. But then it follows from premise one, since 'E' is false, that 'E' and 'O' are both true and so, in particular, that 'E' is. Thus, if 'L' were *not* true, 'E' would be both true and false. Since 'E' cannot be both true and false, our original supposition – that 'L' is *not* true – must have been false, and we can conclude that 'L' is true.

You may be familiar with this way of arguing from a high-school course in geometry or readings in number theory, where one commonly establishes the truth of a conclusion by supposing it false and showing that such a supposition leads to an explicit contradiction, that is, a sentence of the form '$p \& \sim p$'. You may be surprised to find that the deduction informally described above translates quite directly into one of our more formal deductions:

DEDUCE: L

Given the two premises,	1. $E \lor (E \& O)$	Premise
'L' must be true.	2. $E \supset L$	Premise
For suppose not.	3. $\sim L$	Assumption
Then it would follow, with premise two, that 'E' is not true, either.	4. $\sim E$	2, 3 MT
But then it follows from premise one, since 'E' is false, that 'E' and 'O' are both true,	5. $E \& O$	1, 4 DS
and so, in particular, that 'E' is.	6. E	5 Simp
	7. $E \& \sim E$	6, 4 Conj
Thus, if 'L' were *not* true, 'E' would be both true and false.	8. $\sim L \supset (E \& \sim E)$	3–7 CP

Since 'E' cannot be both	9. $\sim E$ v $\sim\sim E$	E-M I
true and false,	10. $\sim(E$ & $\sim E)$	9 De M
our original supposition — that 'L' is not true — must have been false,	11. $\sim\sim L$	8, 10 MT
and we can conclude that 'L' is true.	12. L	11 DN

Notice that the overall strategy (called *indirect proof*) is quite natural. To establish that our conclusion must be true if the premises are, we proceed "indirectly" by showing that its negation is *inconsistent* with them, that is, that it is *impossible* for the premises to be true while the negation of the conclusion is true also. Thus we show the argument to be valid.[8]

Though we know of no case where indirect proof provides the shortest deduction possible, it nonetheless is often the simplest approach to use. Given

1. D v C	Premise	
2. $C \supset (F$ v $D)$	Premise	
3. $\sim F$	Premise	DEDUCE: D,

for example, discovery of a deduction is quite straightforward if we begin by assuming the negation of the conclusion:

| $\rightarrow$ 4. $\sim D$ | Assumption |

DS is suggested immediately,

| 5. C | 4, 1 DS |

and it sets up

| 6. F v D | 5, 2 MP |

---

[8]Since all deductions employing this strategy end the same way, the mechanical last four steps could be eliminated if we added to our system the rule

   Reductio ("Re"). From any sentence of the form '$\sim p \supset (q$ & $\sim q)$', one may infer the corresponding sentence 'p'.

Though we shall not add such a rule, some instructors may prefer to do so.

One more application of **DS** gives us our contradiction,

```
  | 7. D                    3, 6 DS
  | 8. D & ~D               7, 4 Conj
```

and once the contradiction has been obtained we finish, as always when using indirect proof, with the sequence: **CP, E-M I, De M, MT**.

```
  |_____
```

```
  9. ~D ⊃ (D & ~D)         4–8 CP
 10. ~D v ~~D              E-M I
 11. ~(D & ~D)             10 De M
 12. ~~D                   9, 11 MT
```

DN then gives us the conclusion we want:

```
 13. D                     12 DN
```

The strategy of indirect proof also comes in handy in proving theorems, though often in such cases it is best to use it in combination with some more ordinary strategy. Thus to establish '[(C v F) & (F ⊃ C)] ⊃ C' we might begin with a normal conditional proof, but then nest an indirect one inside it:

```
→ 1. (C v F) & (F ⊃ C)        Assumption
  → 2. ~C                      Assumption
    3. C v F                   1 Simp
    4. F                       2, 3 DS
    5. (F ⊃ C) & (C v F)       1 Com
    6. F ⊃ C                   5 Simp
    7. C                       4, 6 MP
    8. C & ~C                  7, 2 Conj
  9. ~C ⊃ (C & ~C)            2–8 CP
 10. ~C v ~~C                 E-M I
 11. ~(C & ~C)                10 De M
 12. ~~C                      9, 11 MT
 13. C                        12 DN
14. [(C v F) & (F ⊃ C)] ⊃ C   1–13 CP
```

It has always been annoying to have to display such lines as (5) and (12), above in deductions. So let us agree that from this point on any application of **Com** may be combined with any application of any other rule into a single line in a deduction, provided both **Com** and the other rule are cited as justification for that line; and let's agree to allow ourselves to combine **DN** with any other rule in the same way. Then, for example, the above deduction may be presented in the following more succinct way.[9]

→	1. $(C \lor F) \& (F \supset C)$	Assumption
	→2. $\sim C$	Assumption
	3. $C \lor F$	1 Simp
	4. F	2, 3 DS
	5. $F \supset C$	1 Com, Simp
	6. C	4, 5 MP
	7. $C \& \sim C$	6, 2 Conj
	8. $\sim C \supset (C \& \sim C)$	2–7 CP
	9. $\sim C \lor \sim\sim C$	E-M I
	10. $\sim (C \& \sim C)$	9 De M
	11. C	8, 10 MT, DN
	12. $[(C \lor F) \& (F \supset C)] \supset C$	1–11 CP

EXERCISES

I. See if the strategy introduced in this section can be used to construct new deductions for (2), (8), and (13) in group (III) of the Exercises at the end of Section 3.2.

II. Again using the strategy discussed in this section, deduce the conclusions of each of the following arguments from their premises:

1. $L \lor (M \& N)$
   $M \supset (\sim L \supset \sim O)$
   O
   $\therefore L$

2. $(R \supset S) \& (F \supset W)$
   $\sim (S \lor W)$
   $\therefore \sim (F \lor R)$

---

[9]When using these short-cut devices, the order in which the rules are cited should reflect the order in which they are applied. Thus, for example at step (5) of the following deduction the citation is displayed as '**Com, Simp**' since **Commutation** is employed first.

3. $A \supset E$
   $B \supset F$
   $(E \vee F) \supset \sim G$
   $(G \vee H) \supset (A \vee B)$
   $\therefore \sim G$

4. $A \vee (B \& C)$
   $[D \vee (D \supset E)] \supset$
   $(A \supset C)$
   $\therefore C$

III. Use the strategy of indirect proof to construct alternative proofs for (1) − (4) in group (III) of the Exercises at the end of Section 3.3.

IV. Using whatever methods seem most helpful, show that each of the following arguments is deductively valid:

1. $(F \supset S) \& (R \supset W)$
   $F \& R$
   $\therefore S \& W$

2. $[(A \vee B) \& (C \vee D)] \supset E$
   $\therefore (A \& D) \supset E$

3. $(A \supset B) \supset (B \supset D)$
   $(C \& O) \supset S$
   $(\sim S \supset \sim O) \supset A$
   $\therefore (A \supset B) \supset (\sim C \vee D)$

4. D
   $A \supset B$
   $E \supset C$
   $\sim A \supset (D \supset E)$
   $(B \vee C) \supset F$
   $\therefore F$

5. $\sim A \supset [\sim A \supset (B \& B)]$
   $B \supset [A \supset (C \& \sim C)]$
   $\therefore \sim A \equiv B$

6. $A \supset B$
   $B \supset C$
   $\sim A \supset \sim C$
   $(\sim A \& \sim C) \supset F$
   $(A \& C) \supset D$
   $\therefore D \vee F$

7. $(P \vee V) \supset \sim T$
   $S \supset (Q \& U)$
   $R \supset (S \vee T)$
   $\therefore (P \& \sim Q) \supset \sim R$

8. $(A \vee L) \vee O$
   $L \equiv O$
   $\therefore O \vee A$

9. $(C \vee D) \supset (C \supset B)$
   $(\sim A \supset B) \supset \sim (D \vee E)$
   $\sim C \supset D$
   $\therefore (A \vee B) \equiv C$

10. $[(D \& C) \vee (\sim D \& \sim C)] \supset$
    $(S \vee U)$
    $\sim R \supset (D \supset C)$
    $\sim D \supset \sim C$
    $S \supset T$
    $\therefore (R \supset S) \supset (T \vee U)$

11. $(E \& W) \equiv (E \& A)$
    $\therefore E \supset (W \equiv A)$

12. $(\sim \sim A \vee \sim \sim B) \supset (D \supset \sim L)$
    $(C \supset E) \supset [D \& (L \vee \sim A)]$
    $\therefore A \supset C$

13. If Watson sinks his thirty-foot putt, then if the spectators let out a spontaneous shout then Kite will miss his mere tap-in. If Watson sinks his thirty-foot putt only if Kite misses his tap-in, one of the two will be on the practice tee tonight. Therefore if the spectators let out a spontaneous shout, or if

Kite misses his tap-in, or if Watson does not sink his thirty-foot putt, then either Watson or Kite will be on the practice tee tonight.

14. Yossarian is crazy if he flies his missions. If he is crazy, he is not obligated to fly his missions. If Yossarian asks to be grounded then he is showing concern for his own safety in the face of real, immediate danger. However if he shows such concern then he is *not* crazy and *is* obligated to fly his missions. Moreover, Yossarian won't fly his missions only if he both asks to be grounded and is not obligated to fly them. Therefore Yossarian is not obligated to fly his missions but flies them nonetheless.

*15. If the usual assumptions of intuitive set theory are all true, then the set of all sets that are not members of themselves is a member of the set of all sets that are not members of themselves if and only if the set of all sets that are not members of themselves is not a member of the set of all sets that are not members of themselves. Therefore the usual assumptions of intuitive set theory are not all true.

*16. If the Administration pursues a tight-money policy, then either the leaders of the opposition party support a tight-money policy or else they denounce governmental meddling. If the leaders of the opposition party support a tight-money policy and the economy goes steadily downhill, then they'll denounce governmental meddling and claim that their party could get the economy back on its feet. So if the Administration pursues a tight-money policy or the leaders of the opposition party support a tight-money policy, then if the economy goes steadily downhill the leaders of the opposition party will denounce governmental meddling.

CHAPTER FOUR

SEMANTIC AND DEDUCTIVE METHODS COMPARED

*4.1 Two Ways of Explicating Validity

When the term 'valid' was introduced in Chapter 1, there were *two* ways of stating, provisionally, what was meant by it:

An argument is valid if and only if its conclusion follows from its premises; that is, if and only if it is impossible for all its premises to be true but its conclusion false.

The tacit assumption was that these were two different ways of saying the same thing, but it should now be apparent that we chose to explicate these suggestions independently.

Thus, in Section 3.1 we concentrated on the notion of one sentence's *following from* certain others. We suggested in effect that so long as our attention is restricted—as it will be throughout this chapter—to arguments whose premises and conclusions are built up from sentence letters by means of the connectives '&', 'v', and '~', the term 'valid' as initially characterized might better be replaced by the more precisely defined term 'deductively valid'.

An argument is **deductively valid** if and only if it is possible to construct a deduction of its conclusion from its premises

(where it is of course understood that a **deduction** is a finite sequence of sentences beginning with the premises wherein each later line is justified on the basis of earlier ones by one of the rules **Conj, Simp, Add, DS, E-M I, Com, Assoc, Dist, DN** or **DeM**). In Sections 2.1 and 2.3, on the other hand, we noticed that truth-tables could be used to determine the possibility of the premises of such an argument being true while its conclusion was false. There the suggestion was, in effect, that the term 'valid' should be replaced with 'truth-table valid'.

An argument is **truth-table valid** if and only if there is no row of its truth-table in which all its premises are assigned the truth-value **T** but its conclusion the truth-value **F**.

These accounts of validity and of their correlatives, deductive and truth-table invalidity, appear to coincide well with the informal notions of validity and invalidity introduced in Chapter 1. Arguments that (in our technical sense) are deductively valid surely are paradigm cases of those whose conclusions may properly be said to *follow from* their premises; and arguments that are truth-table valid are obviously those for which it is *impossible* for their premises to be true while their conclusions are false.

The deductive methods of Section 3.1 may also be compared directly with the semantic methods of Sections 2.1 and 2.3; and though some of the details involved in carrying out that comparison are beyond the scope of an introductory text, the results of the comparison are not. In discussing these results, it is useful to recall from Chapters 2 and 3 that we write '$\mathscr{P}_1,...,\mathscr{P}_k \models \mathscr{C}$' to abbreviate the assertion that the argument whose premises are the sentences $\mathscr{P}_1$ through $\mathscr{P}_k$ and whose conclusion is the sentence $\mathscr{C}$ is *truth-table valid* and that we write '$\mathscr{P}_1,...,\mathscr{P}_k \vdash \mathscr{C}$' to say that it is *deductively valid*. Of course we use '$\models \mathscr{T}$' as shorthand for '$\mathscr{T}$ is a tautology' and '$\vdash \mathscr{T}$' for '$\mathscr{T}$ is a theorem', as well.

First of all, it can be shown[1] that the deductive system developed in Section 3.1 is **strongly sound** in the sense that *every deductively valid argument is truth-table valid,* that is, for any sentences $\mathscr{P}_1$ through $\mathscr{P}_k$ and $\mathscr{C}$, if $\mathscr{P}_1, ..., \mathscr{P}_k \vdash \mathscr{C}$, then $\mathscr{P}_1, ..., \mathscr{P}_k \models \mathscr{C}$. Strong soundness is thus the result that guarantees that our set of rules is "safe": only arguments that are genuinely valid according to truth-tables are capable of having their conclusions deduced from their premises. Provided that the rules are used correctly, then, we can be sure that it will never be possible to construct a deduction for an argument whose truth-table reveals the existence of a row where all its premises are assigned **T**'s but the conclusion is assigned an **F**.

---

[1] For discussions of this result and those that follow, and for indications of how they may be established, see Section 4.2.

Our deductive system can also be shown to have a related property (**weak soundness** is the technical term most often used): *every theorem is a tautology* (that is, for any sentence $\mathscr{T}$, if $\vdash \mathscr{T}$ then $\models \mathscr{T}$). Again, this is a sort of "safety" result; many sentences can be proved to be theorems of our system, but each and every one of them turns out to be a sentence that is, according to truth-tables, incapable of taking the value **F** under any circumstances at all. It follows from weak soundness that our system is also **consistent**, that is, *no contradictions are provable as theorems.*

Our rules then are safe—they can only be used to show what truth-tables suggest it would be desirable to show. But do we have *enough* of them? Do we have enough rules to be certain that the conclusion of *every* truth-table valid argument can be deduced from its premises?

The answer is "Yes", for it can also be shown that our system is **strongly complete** in the sense that *every truth-table valid argument is deductively valid:* for any sentences $\mathscr{P}_1$ through $\mathscr{P}_k$ and $\mathscr{C}$, if $\mathscr{P}_1, \ldots, \mathscr{P}_k \models \mathscr{C}$ then $\mathscr{P}_1, \ldots, \mathscr{P}_k \vdash \mathscr{C}$. Any nagging doubts we may have had that there might be arguments valid on truth-table grounds which were nonetheless too complex (or whatever) to be handled using only the ten rules of Section 3.1 can now be dispelled—if it is impossible to find an assignment of truth-values according to which the premises of an argument all have the value **T** but its conclusion has the value **F**, then we can be sure that there is at least one way to deduce its conclusion from its premises using only our ten rules.

As it happens, it is very difficult to demonstrate the strong completeness of our system. So in the next section we will first establish its **weak completeness:** *every tautology is a theorem* (that is, for any sentence $\mathscr{T}$, if $\models \mathscr{T}$ then $\vdash \mathscr{T}$). So we will show that *each* of the infinitely many tautologies of our language admits of a proof—that our system "captures" all of them as theorems. Even this result is difficult, but once it has been shown it becomes somewhat easier to establish strong completeness.

Apart from its intrinsic interest, this exact agreement of the truth-table and deductive methods has practical consequences as well. If we cannot find a deduction showing an argument to be valid, it need not be that it is invalid, for it may be that we simply lack the necessary inventiveness to find a deduction that will do the job. But the use of truth-tables requires neither ingenuity nor knowledge of strategy. One who understands how truth-tables work can mechanically test any argument for validity no matter what its degree of complexity. And in contrast with deductive methods, truth-tables provide a test for *invalidity* as well—one that, though sometimes tedious, can be routinely carried out.

Since the truth-table valid arguments are precisely the deductively valid ones, then, we are free to move back and forth between the two methods.

Invalid arguments may be shown to be so by constructing truth-tables.[2] Valid arguments with six or seven letters requiring truth-tables of 64 or 128 rows may be shown valid instead by the deduction of their conclusions from their premises. Thus with arguments of a substantial complexity, the deductive method has very practical advantages, as well as a certain naturalness that reflects our own intuitive reasoning processes.

EXERCISES

1. Let us call the deductive system developed in Section 3.1 the system **D**. Explain how to use the fact that **D** is weakly sound to establish that **D** is consistent.

2. Let **D +** be the system that results from **D** by adding the new rule of inference

   From any sentence of the form

   p & $\sim p$

   one may infer any sentence

   q.

   Are any theorems of **D +** contradictions, that is, is **D +** inconsistent? (You will find it helpful to make use of the fact that **D** itself is consistent.)

3. Let **E** be a deductive system obtained by adding some rules to **D**. Explain why **E**, if not consistent, must be weakly complete.

4. Let **D $^-$** come from **D** by dropping the rule **DN**. Show that 'A' can be deduced from '$\sim \sim A$' in **D $^-$**.

5. Show that '$\sim \sim A$' can be deduced from 'A' in **D $^-$**.

6. Do the deductions you constructed for the previous two exercises suffice to show that **DN** is superfluous as a rule of replacement in **D**? Why or why not?

*7. Let **B** be the deductive system that results from **D** by dropping **D**'s rules of replacement. **B**'s only rules, then, are **Conj, Simp, Add, DS** and **E-M I**. Exercise IV on p. 86 above shows that **B** is not strongly complete. Is **B** weakly complete? Explain.

_____

[2]One of the most important unsolved problems in sentential logic today involves the question of whether or not there exist methods for determining validity and invalidity that are inherently more efficient than the truth-table method (and the related methods of Section 2.4 and Appendix B). For a precise formulation of the question and a discussion of its importance for equivalent unsolved problems in business and industry, see Michael R. Garey and David S. Johnson's *Computers and Intractability: a Guide to the Theory of NP-Completeness*, San Francisco: W.H. Freeman, 1979.

*4.2 Some Metatheorems

In the preceding section we mentioned several results that tie the deductive methods of Section 3.1 to the truth-table methods of Sections 2.1 and 2.3. They are called **metatheorems** because they are claims *about* our deductive system rather than sentences provable within it. Though we shall not attempt to establish these results in full detail, we will try to sketch the proofs required.

METATHEOREM I (Strong soundness) Every deductively valid argument is truth-table valid.

To establish this fact, we have to show that every conclusion deducible from any premises $\mathscr{P}_1, \ldots, \mathscr{P}_k$ $(k \geq 1)$ must be assigned the truth-value **T** in every row of the truth-table in which $\mathscr{P}_1, \ldots, \mathscr{P}_k$ are all assigned **T**'s. We establish this by showing that every line, regardless of its line-number, of any deduction in which $\mathscr{P}_1, \ldots, \mathscr{P}_k$ function as premises must be assigned a **T** when they all are.

For the first k lines of such a deduction this is obvious, since those lines will be $\mathscr{P}_1$ through $\mathscr{P}_k$ themselves. And subsequent lines can only be justified by **E-M I** or else be inferred from preceding steps by one of our other nine rules, rules that have already been shown by the truth-table method to lead only from truths to other truths. Thus lines obtained by **E-M I**, being of the form '$p \lor \sim p$', are tautologies (Section 2.1) and so are assigned **T**'s regardless. Those inferred from others by **Conj, Simp, Add,** or **DS** must have **T**'s when the others do (Section 2.3). And those obtained from preceding lines by **Com, Assoc, Dist, DN,** or **De M** come from them by replacement of an expression with another equivalent to it (Section 2.1); any two such sentences, then, will have the same pattern of **T**'s and **F**'s in their truth-tables so that, in particular, the later one will have a **T** in every row of the truth-table in which the earlier one does.

Thus deductions begin with premises, and proceed from them at each step only to other sentences that must be true when those already obtained are. If this point is understood, construction of a more rigorous proof is facilitated. We show:

For each positive integer n, any line of a deduction with line-number no greater than n must be assigned a **T** whenever the premises $\mathscr{P}_1, \ldots, \mathscr{P}_k$ of that deduction are assigned **T**'s.[3]

We observe, first, that any line of such a deduction whose line-number is no greater than 1 must be $\mathscr{P}_1$ and so be assigned a **T** when $\mathscr{P}_1, \ldots, \mathscr{P}_k$ all are.

---

[3] Our proof proceeds by weak mathematical induction (see Appendix C).

Next we let m be any positive integer; we assume that any line of a deduction whose line-number is no greater than m is assigned a **T** when $\mathscr{P}_1,\ldots,\mathscr{P}_k$ are; and we show that all lines of deductions with line-numbers no greater than $m+1$ are assigned **T**'s when the premises, $\mathscr{P}_1$ through $\mathscr{P}_k$, are. Our assumption takes care of lines with numbers between 1 and m, so it is only lines numbered $m+1$ that we need be concerned with, and our discussion above has shown how to handle them. Those that are themselves premises, and those justified by **E-M I**, certainly have **T**'s when the premises do. And those that come from one or two preceding lines by one of our other rules are, as we saw, taken care of by the truth-table considerations of Sections 2.1 and 2.3. Lines inferred from preceding ones by one of these rules must have **T**'s when those earlier lines do, and by our assumption those earlier lines have **T**'s when the premises do.

Now we may conclude that every line of every deduction from the premises $\mathscr{P}_1,\ldots,\mathscr{P}_k$ has a **T** when the latter have **T**'s: the early lines of such deductions automatically have **T**'s in such cases for they *are* just $\mathscr{P}_1$ through $\mathscr{P}_k$; and later lines must have **T**'s when the premises do because our rules, singly or in combination, preserve truth.

It should be obvious that this demonstration of Metatheorem I can be modified to establish

METATHEOREM II (*Weak soundness*) Every theorem is a tautology.

After all, the first lines of proofs of theorems can only be justified by **E-M I** and hence must be tautologies; and since later lines must be true when earlier ones are, all lines of all proofs — and so all theorems — must be tautologies as well. But there is an easier way.

Let $\mathscr{T}$ be any theorem. Then any argument with $\mathscr{T}$ as conclusion will be deductively valid. In particular, the argument

$\mathscr{T} \lor \sim\mathscr{T}$

$\therefore \mathscr{T}$

will be deductively valid — we may enter its premise as the first line of our deduction, and use the lines of the proof of $\mathscr{T}$ for the rest. By Metatheorem I, then, the conclusion of this argument must have a **T** in every row of the truth-table in which its premise does. But its premise, being a tautology, has a **T** in *every* row of its truth-table. So $\mathscr{T}$ also has a **T** in every row of its truth-table, that is, $\mathscr{T}$ is a tautology itself!

COROLLARY Our deductive system is consistent, that is, no contradiction is a theorem.

To establish this, suppose to the contrary that some contradiction were provable in our system. By Metatheorem II it would have to be a tautology. But of course no contradiction is tautologous.

With our next metatheorem we want to show that we have *enough* rules, that is, that our ten rules are sufficient for deducing *all* tautologies from those which may, by **E-M I**, initiate a deduction:

METATHEOREM III (*Weak Completeness*) Every tautology is a theorem.

It will be helpful first to introduce some temporary terminology. By a **literal** we shall mean a sentence letter or the negation of a sentence letter. Thus '*A*', '*B*', '*C*', and so on are literals, as are ' $\sim\!A$ ', ' $\sim\!B$ ', ' $\sim\!C$ ', and the like. But ' $\sim\!\sim\!A$ ' is not (too many curls) and neither are '*A* & *B*' and '*B* ∨ *C*'. By an *(extended) disjunction* of literals we will mean a sentence containing any number of literals, all with wedges between them: '*A* ∨ $\sim\!B$ ' qualifies, of course, but so also do things like '(*A* ∨ $\sim\!B$) ∨ *C*' and ' $\sim\!A$ ∨ [(*B* ∨ $\sim\!C$) ∨ D]'. The term *(extended) conjunction* is to be understood similarly.

We shall be especially interested in literals, extended disjunctions of literals, and extended conjunctions of these two kinds of sentences:

DEFINITION A sentence is in **conjunctive normal form (CNF)** if and only if it is either (i) a literal, or (ii) an extended disjunction of literals, or (iii) an extended conjunction each of whose conjuncts is either a literal or an extended disjunction of literals.

Thus '*A*' and ' $\sim\!B$ ' are in CNF by clause (i); so are '*A* ∨ $\sim\!B$ ' and the extended disjunction '*A* ∨ ($\sim\!B$ ∨ $\sim\!C$)' by clause (ii); and so is the extended conjunction ' $\sim\!B$ & {(*A* ∨ $\sim\!B$) & [*A* ∨ ($\sim\!B$ ∨ $\sim\!C$)]}' by clause (iii). But ' $\sim\!(A ∨ B)$ ', ' $\sim\!\sim\!A$ ∨ (*B* ∨ $\sim\!C$)', '($\sim\!A$ & *B*) ∨ $\sim\!\sim\!C$ ' and '(*A* & *B*) ∨ *C*' are not.

To establish Metatheorem III one needs five sub-results or **lemmas**. In our discussion of these we shall omit certain details that a full-fledged proof would call for.[4]

---

[4]For amplification, see D. Hilbert and W. Ackermann, *Principles of Mathematical Logic,* New York: Chelsea, 1950, pp. 11-16, whose exposition we follow fairly closely. The proof is due to Emil Post, "Introduction to a general theory of elementary propositions", *American Journal of Mathematics*, vol. 43 (1921), pp. 163-185.

Lemma 1 For each sentence 𝒯 there exists a corresponding sentence 𝒯*
in conjunctive normal form that is interdeducible with 𝒯, that is,
that can be deduced from 𝒯 using our ten rules and from which
𝒯 can be deduced.

To build such a sentence 𝒯* from any given sentence 𝒯 we can
follow a set procedure: at Stage One drive all curls inside as far as possible by repeated uses of **De M** until no curls occur outside parentheses but
only inside, in front of sentence letters; at Stage Two, use **DN** as often as
necessary to pare down to no more than *one* curl in front of any sentence
letter; then, at Stage Three, use **Com** and **Dist** repeatedly to change parts
of the sentence of the form '*p* v (*q* & *r*)' or '(*q* & *r*) v *p*' to the form
'(*p* v *q*) & (*p* v *r*)', until we wind up with a conjunction of disjunctions of
literals.

Suppose, for example, that 𝒯 is the sentence

~{[*B* v ~(*C* & ~*D*)] & *A*}.

We first use **De M** again and again to drive all the curls that come up as we
go along as far inside as possible at the first stage. .

STAGE ONE: ~ [*B* v ~(*C* & ~*D*)] v ~*A*
       [~*B* & ~~(*C* & ~*D*)] v ~*A*
       [~*B* & ~(~*C* v ~~*D*)] v ~*A*
       [~*B* & (~~*C* & ~~~*D*)] v ~*A*

Then we use **DN** again and again until no more than one curl occurs in
front of any letter.

STAGE TWO: [~*B* & (*C* & ~~~*D*)] v ~*A*
       [~*B* & (*C* & ~*D*)] v ~*A*

Finally, we commute parts of the form '(*q* & *r*) v *p*' as they arise, and then
do distributions as often as we have to in order to obtain a conjunction
of disjunctions (of literals). This is the final stage.

STAGE THREE: ~*A* v [~*B* & (*C* & ~*D*)]
        (~*A* v ~*B*) & [~*A* v (*C* & ~*D*)]
        (~*A* v ~*B*) & [(~*A* v *C*) & (~*A* v ~*D*)]

The sentence in CNF that we finish with is the sentence $\mathscr{T}^*$ that we want. Since **De M, DN, Com,** and **Dist** are all "reversible" rules, we can be sure not only that $\mathscr{T}^*$ is deducible from $\mathscr{T}$, but also that $\mathscr{T}$ is deducible from $\mathscr{T}^*$ – to construct the deduction, one need only *reverse* the above steps.

Lemma 2 If a sentence is a tautology then so is each sentence interdeducible with it.

Let $\mathscr{T}$ be any tautology. Let $\mathscr{T}^*$ be any sentence interdeducible with it so that $\mathscr{T} \vdash \mathscr{T}^*$ and $\mathscr{T}^* \vdash \mathscr{T}$. Since $\mathscr{T} \vdash \mathscr{T}^*$ and we know from Metatheorem I that every deductively valid argument is truth-table valid, it follows that $\mathscr{T} \models \mathscr{T}^*$. By the definition of truth-table validity, this means that $\mathscr{T}^*$ must be assigned the value **T** in every row of the truth-table for these two sentences in which $\mathscr{T}$ takes the value **T**. Since $\mathscr{T}$, being a tautology, takes the value **T** in every row, then, so does $\mathscr{T}^*$. So $\mathscr{T}^*$ is also a tautology.

Lemma 3 If a sentence in CNF is a tautology, then each of its conjuncts (if any) must be a tautology also.

To see that this is so, recall that conjunctions only have **T**'s in those rows of their truth-tables in which *all* their conjuncts have **T**'s. If some conjunct of a tautologous conjunction were not itself a tautology, then that conjunct would have the truth-value **F** in at least one row of its truth-table. But in that case the conjunction as a whole would have an **F** in that row, also, and not be a tautology.

Lemma 4 If an extended disjunction of literals occurring in[5] a sentence in CNF is a tautology, then it must contain for some sentence letter both that letter and its negation.

To establish this lemma, notice that if an extended disjunction of literals does *not* contain, for at least one sentence letter, *both* that letter *and* its negation, it cannot be a tautology: it will take the truth-value **F** in the row of its truth-table wherein each unnegated letter occurring in it has the value **F** and each negated one the value **T**.

---

[5]We think of a typical sentence in CNF as a conjunction of extended disjunctions of literals. But we should not forget the special case where the sentence in CNF is not a conjunction at all but simply consists of one such extended disjunction, e.g., '$(A \lor \sim B) \lor [C \lor (\sim A \lor D)]$'.

Lemma 5 Each disjunction of literals that contains for some sentence letter both it and its negation is a theorem.

In fact it is easy to see that any such disjunction can be proved by starting with **E-M I**, using **Add** to get in the other disjuncts required, and then using **Com** and **Assoc** to get the order and grouping right.

Once lemmas 1–5 have been established, we can return to Metatheorem III and show that every tautology is a theorem. For let $\mathcal{T}$ be any tautology. By lemma 1 we can find a sentence $\mathcal{T}*$ in CNF that is interdeducible with $\mathcal{T}$. Since $\mathcal{T}$ is a tautology, lemma 2 assures us that $\mathcal{T}*$ is a tautology also. But then lemma 3 tells us that each of $\mathcal{T}*$'s conjuncts must be a tautology as well. It then follows from lemma 4 that each conjunct of $\mathcal{T}*$ is a disjunction of literals which contains for some sentence letter both that letter and its negation. By lemma 5 we can conclude that each of $\mathcal{T}*$'s conjuncts is a theorem. But then $\mathcal{T}*$ must also be a theorem—we can prove each of its conjuncts (since they're all theorems) and then use **Conj** to stick them together to get $\mathcal{T}*$. Now according to lemma 1, $\mathcal{T}$ is deducible from $\mathcal{T}*$, so if we continue the proof of $\mathcal{T}$ by proceeding to deduce $\mathcal{T}$ from *it*, we finish with a proof of $\mathcal{T}$. $\mathcal{T}$ must, then, be a theorem.

Notice that our method of establishing Metatheorem III not only shows that it is possible to prove every truth-table tautology as a theorem of our deductive system; in fact, it gives us a "recipe" for actually *constructing* the proof in each case.

To see how the whole process works, let's go through one example in detail. Suppose $\mathcal{T}$ is the tautology '$[\sim(\sim B \,\&\, B) \,\&\, (D \lor \sim\sim\sim C)]$ $\lor C$'. We first find a sentence $\mathcal{T}*$ in CNF that is interdeducible with $\mathcal{T}$ by following the procedure described for lemma 1:

$\mathcal{T}$	$[\sim(\sim B \,\&\, B) \,\&\, (D \lor \sim\sim\sim C)] \lor C$	
	$[(\sim\sim B \lor \sim B) \,\&\, (D \lor \sim\sim\sim C)] \lor C$	(De M)
	$[(B \lor \sim B) \,\&\, (D \lor \sim\sim\sim C)] \lor C$	(DN)
	$[(B \lor \sim B) \,\&\, (D \lor \sim C)] \lor C$	(DN)
	$C \lor [(B \lor \sim B) \,\&\, (D \lor \sim C)]$	(Com)
$\mathcal{T}*$	$[C \lor (B \lor \sim B)] \,\&\, [C \lor (D \lor \sim C)]$	(Dist)

As lemmas 2–4 promised, $\mathcal{T}*$ is also a tautology, and since it is in conjunctive normal form, each of its conjuncts is a tautology as well, and so each is a disjunction of literals containing for some sentence letter both that letter and its negation. By lemma 5 we can consequently prove each of its conjuncts, and then conjoin them to obtain $\mathcal{T}*$:

1. $B \lor \sim B$	E-M I
2. $(B \lor \sim B) \lor C$	1 Add
3. $C \lor (B \lor \sim B)$	2 Com
4. $C \lor \sim C$	E-M I
5. $(C \lor \sim C) \lor D$	4 Add
6. $C \lor (\sim C \lor D)$	5 Assoc
7. $C \lor (D \lor \sim C)$	6 Com
8. $[C \lor (B \lor \sim B)] \& \lfloor C \lor (D \lor \sim C)]$	3, 7 Conj

To construct our proof of $\mathscr{T}$, then, we have only to extend this proof of $\mathscr{T}^*$ by reversing the steps we went through before in obtaining $\mathscr{T}^*$ from $\mathscr{T}$:

9. $C \lor [(B \lor \sim B) \& (D \lor \sim C)]$	8 Dist
10. $[(B \lor \sim B) \& (D \lor \sim C)] \lor C$	9 Com
11. $[(B \lor \sim B) \& (D \lor \sim\sim\sim C)] \lor C$	10 DN
12. $[(\sim\sim B \lor \sim B) \& (D \lor \sim\sim\sim C)] \lor C$	11 DN
13. $[\sim(\sim B \& B) \& (D \lor \sim\sim\sim C)] \lor C$	12 De M

With this work behind us, the proof of our last metatheorem is relatively simple:

METATHEOREM IV (Strong Completeness) Every truth-table valid
argument is deductively valid.

Suppose that the argument

$\mathscr{P}_1$
.
.
.
$\mathscr{P}_k$
$\therefore \mathscr{C}$

is truth-table valid. Then there can be no row of its truth-table in which $\mathscr{P}_1$ through $\mathscr{P}_k$ are all assigned **T**'s but $\mathscr{C}$ an **F**. Neglecting internal parentheses temporarily for clarity, consider

$$\sim(\mathscr{P}_1 \ \& \ ... \ \& \ \mathscr{P}_k) \lor \mathscr{C},$$

which clearly must be a tautology — any row in which it had the value **F** would have to be a row in which $\mathscr{P}_1$ through $\mathscr{P}_k$ were assigned **T**'s but $\mathscr{C}$ an **F**, and there are no such rows. Since all tautologies are theorems, then, it is a theorem. And now we can see that the argument in question must be deductively valid, for the deduction of $\mathscr{C}$ from $\mathscr{P}_1,..., \mathscr{P}_k$ could proceed as follows. First, use **Conj** repeatedly on the premises until

$$\mathscr{P}_1 \ \& \ ... \ \& \ \mathscr{P}_k$$

has been deduced. Then use **DN** to obtain

$$\sim\sim(\mathscr{P}_1 \ \& \ ... \ \& \ \mathscr{P}_k).$$

Then insert the proof guaranteed us by Metatheorem III of the theorem

$$\sim(\mathscr{P}_1 \ \& \ ... \ \& \ \mathscr{P}_k) \lor \mathscr{C}.$$

One last step, using **DS**, then delivers $\mathscr{C}$!

EXERCISES

1. Let **E** be any deductive system obtained from our system **D** by adding some rules. Assume that **E** is inconsistent. Which of the four metatheorems established for **D** hold for **E** as well? Explain.

2. People sometimes wonder why we bother to study both **D** and the truth-table methods of Chapter 2 when they turn out to be, as we have shown in this Section, just two different ways of doing one and the same thing. One reason for studying both methods is that often one of them affords the simplest way to answer a question about the other. Consider, for example, the question "If $\mathscr{P}$ and $\mathscr{Q}$ are sentences built up from letters and the three connectives '&', 'v', and '$\sim$', and $\mathscr{Q}$ is deducible (using **D**'s ten rules) from $\mathscr{P}$, must it be the case that the disjunction

$$\sim\mathscr{P} \lor \mathscr{Q}$$

is a theorem of **D**?" For this exercise, show that the answer is "Yes." (This otherwise difficult question is quite straightforward if you make use of two of our metatheorems.)

3. Let **C** be the deductive system that results from our ten-rule system **D** by *dropping* the rules **Simplification** and **Disjunctive Syllogism**. Since anything that can be done in **C** can also be done in **D**, it is clear that Metatheorems I and II both hold for **C**. Does Metatheorem III also hold for **C**, that is, is **C** weakly complete? (For a simple answer, study carefully our demonstration of Metatheorem III for **D**.)

*4. Does Metatheorem IV hold for **C**, that is, is **C** *strongly* complete?

*5. Let **D**′ be the deductive system developed in Sections 3.1 and 3.2 for treating sentences built up from sentence letters and the *five* connectives '&', 'v', '~', '⊃', and '≡'. Explain what changes would have to be made in the proofs and metatheorems of this chapter if we wished to extend our treatment of **D** to **D**′.

*6. When the rule of conditional proof was introduced in Section 3.3 we said that it was not really needed, that we could not deduce any conclusions from any premises *with* that rule that we could not have deduced *without* it. Let **D**″ be the deductive system developed in Sections 3.1-3.4, and use the work you've done in the preceding exercise to justify that earlier claim by showing that any argument whose conclusion can be deduced from its premises using the rules of **D**″ is such that its conclusion can also be deduced from its premises using just the rules available in **D**′.

CHAPTER FIVE

QUANTIFICATIONAL LOGIC
Translation

5.1 Quantifiers and Related Apparatus

While the techniques developed in Chapters 2 and 3 apply to a wide range of arguments, there are nonetheless many that fall outside their scope. Even a simple example like

Every philosopher is wise
Aristotle is a philosopher
Therefore Aristotle is wise

cannot be handled. For since the premises and conclusion are all simple sentences containing no others as parts, the argument is of the patently invalid form

p
q
$\therefore r$

Yet of course it is valid, and to show that it is, we will need to focus on the internal composition of the premises and conclusion — in particular, on the relative positions of the terms 'philosopher', 'wise', and 'Aristotle'. Only in this way shall we be able to distinguish valid arguments such as the above from invalid ones like

Every philosopher is wise
Aristotle is wise
Therefore Aristotle is a philosopher.

To develop techniques for establishing validity and invalidity in such cases, we must first examine the internal structure of uncompounded sentences such as 'Every philosopher is wise', 'Aristotle is a philosopher', and the like. This is the subject of the present chapter.

Proper names—'Aristotle', 'Fido', 'Indiana', 'Penn Station', 'Apollo 11', and the like—are typically used in a given context to refer to a single individual, and so will be called **singular terms**. While the reference of such a term may vary from one context to another (many dogs are named 'Fido' for instance), once the context has been fixed such terms serve to single out one specific thing or individual for discussion. In addition to proper names, there is another class of singular terms known as *definite descriptions*, examples of which include 'the author of *De Anima*', 'the thirty-seventh President of the United States', and 'John's mother'. These terms do not name an individual but rather purport to uniquely *describe* one. Aristotle was not *named* 'the author of *De Anima*' but 'Aristotle'; had he not actually written *De Anima* that description would not apply to him. Sentences containing definite descriptions require a rather complex analysis, however, so we shall postpone discussion of them until Section 5.3.

Unlike 'Aristotle', such terms as 'philosopher' and 'wise' do not serve to refer, either by name or description, to one particular individual. Instead, they are essentially classificatory in function; they typically represent *properties* that a number of individuals might share—in the present case the properties of being a philosopher and of being wise. We shall call such expressions **general terms** since they may characterize or describe a general class of individuals (those having the property in question) rather than referring to just one. Thus 'philosopher' is a general term in our sense—while it does not refer to any specific individual, it describes such individuals as Aristotle and Plato.[1]

General terms may occur in the predicate of a sentence in noun ('Al is a smoker'), verb ('Helen is smoking') and adjectival ('The room is smoky') form. And though they are classificatory and descriptive, it should not be thought that general terms must *in fact* describe any actual objects. 'Unicorn' is a general term, for example, though the property of unicornhood is possessed by no individual.

A sentence typically used to ascribe a property to a specifically designated individual will be called a **singular** sentence. The grammatical subject of such a sentence is a singular term and the predicate contains a general term. The second premises and the conclusions of the arguments

---

[1] Any definite description will contain a general term, for example 'author' in 'the author of *De Anima*'. But because it contains the definite article 'the' in the singular, the definite description *itself* serves not to characterize a class of individuals, but to refer to just one. It is, therefore, a singular term.

cited earlier, for example, are singular sentences. It will be convenient also to regard truth-functional compounds of singular sentences as themselves singular. So 'Aristotle or Brutus is a philosopher' and 'Cassius is a philosopher who is not wise' count as singular sentences as well. And sentences like 'Brutus is wiser than Cassius', in which a *relation* is said to hold between two or more specified individuals (instead of a property ascribed to just one), will also be considered singular. Relational sentences of this sort will be taken up in Section 5.2; first, however, we must concentrate on those in which properties are ascribed to individuals.

To abbreviate proper names, we will use the small letters '*a*' through '*t*'. They will serve to designate specific individuals just as 'Aristotle', 'Brutus' and the like do in ordinary speech. Because their designations will be fixed in any given context in which we use them, they will be called **individual constants.**

Now if we consider the singular sentences 'Aristotle is a philosopher', 'Brutus is a philosopher' and 'Cassius is a philosopher', we may abstract from them the schematic predicate expression '... is a philosopher' common to all three. And this expression in its entirety, rather than the component general term, may be taken as representing the property of being a philosopher. Our practice will thus be to treat general terms as integral parts of predicative contexts in which they occur.

To abbreviate predicate expressions we will use the capital letters '*A*', '*B*', '*C*' and so on. We shall call them **predicate letters;** in effect, they will be used to represent properties of (and later on, relations among) individuals.[2]

With these notational devices, we have the means for expressing singular sentences in compact style. With '*a*', '*b*' and '*c*' for 'Aristotle', 'Brutus' and 'Cassius' respectively, '*P*' for 'is a philosopher' and '*W*' for 'is wise', we may translate such singular sentences as:

(1) Aristotle is a philosopher Pa
(2) Aristotle or Brutus is a philosopher $Pa \vee Pb$
(3) Cassius is a philosopher who is not wise $Pc \, \& \sim Wc$

The first premise, 'Every philosopher is wise', of our sample argument, however, will not yield to this treatment. It does *not* serve to ascribe a property to a specified individual, and so is not a singular sentence. Instead it says in effect that each individual which has the property of being a philosopher also has the property of being wise, but refers to no *specific* individual whatever. Such sentences will be called **general**—their grammatical subjects and predicates each contain general terms, with the

---

[2] It is worth emphasizing that predicate contexts are often rather complex, and they may even include singular terms (for example, '... conspired with Cassius').

subject usually containing in addition a *quantity* expression such as 'every', 'all' or 'some'. Further examples include 'Some planets support life' and 'Everything is subject to the laws of physics'. Unlike the singular variety, general sentences relate together classes of individuals.

We may sharpen this somewhat rough distinction between singular and general sentences as follows. With a singular sentence like 'Aristotle is wise', there is no distinction in meaning if we negate it "externally" ('It is not the case that Aristotle is wise') or "internally", by the insertion of 'not' into the predicate ('Aristotle is not wise'). The resulting sentences are synonymous, and both might be translated, using our old notation, as '$\sim A$'. But a typical general sentence like 'Every philosopher is wise' yields quite different results depending on how 'not' is attached. Externally, we arrive at its genuine negation 'It is not the case that every philosopher is wise' (more idiomatically, 'Not every philosopher is wise'), and this is equivalent to 'Some philosophers are not wise'. But placing 'not' into the predicate gives 'Every philosopher is not wise' which, understood literally, is equivalent to 'No philosopher is wise'.[3] Clearly the latter is *not* the negation of 'Every philosopher is wise', for both of these may well be false. Only when 'not' is applied externally to a general sentence do we derive a sentence that may properly be described as its negation.

To translate general sentences, we shall employ small letters from the end of the alphabet: '*w*', '*x*', '*y*', '*z*', '*w'*' and so on.[4] We will call them **individual variables** since they will not be used to designate specific individuals but rather to range indiscriminately over all of them. Hence such expressions as

(4)   *x* is a philosopher

and

(4') P*x*

are neither true nor false, since no particular individual is designated by '*x*', and thus neither is a sentence. (4') is related to (1) much as '$x^2 = 9$' is to '$3^2 = 9$', and both (4') and its English counterpart (4) will be called **open sentences**. An open sentence is not itself either true or false, as a sentence like (1) would be. But an open sentence may be transformed *into* a singular

---

[3]Note that some sentences that are singular in form are often used like general sentences. For example, 'The whale is a mammal' may be used not to refer to any particular whale, but rather to the entire species. As such, it has the force of 'All whales are mammals'.

[4]The unused letters '*u*', '*v*', '*u''*', '*v''*', etc. are being reserved for a special use in Chapter 7.

sentence by replacing the variables in it with individual constants. Hence a singular sentence like '*Pa*' will be termed a **substitution instance** of the open sentence '*Px*', since it is derived from the latter through such replacement. An open sentence, then, must contain at least one individual variable that is subject to replacement by a constant.

Now we can turn to the task of expressing general sentences. The use of such sentences in any given context presupposes at least implicitly a definite *universe of discourse* — the range of individuals under discussion in that context. The universe may range anywhere from the actual universe to something much smaller. When an instructor looks out over his class and says "Everyone is here," for example, he clearly does not mean every person *whatever* but rather every person enrolled in his course. Were we to object that his remark is false because, say, the British Prime Minister is not there, the objection would hardly be entertained seriously. The universe of discourse is tacitly restricted to enrolled persons, and hence the sentence would be counted as false only if at least one enrolled person were not in attendance.

It will be helpful in examining general sentences to restrict ourselves initially to a small and artificial universe of discourse. We shall therefore consider a universe with three individuals *a*, *b*, and *c*, together with five properties which we shall represent by the letters '*F*', '*G*', '*H*', '*J*', and '*K*'. (We need not worry here about *which* specific individuals and properties they might be.) Now whether a sentence is true or false in this universe will depend on which individuals exhibit which properties. We shall suppose that they are distributed as follows (a '+' means that the individual has the property, a '−' that it lacks it):

	F	G	H	J	K
a	+	−	+	−	−
b	−	+	+	−	+
c	+	+	+	−	−

A glance at this chart will show that the following singular sentences (among others) are *true*:

Fa   ~Ga   Fb v Hb   Fa & ~Jc

Moreover, each of the following general sentences (stated for the moment in English phraseology) are true:

(5) Something is F   (7) Nothing is J

(6) Everything is H   (8) Something is not G

It may seem tempting at first to express (5) as '*Fs*' where '*s*' means 'something'. But a little reflection will show this to be absurd. To use '*s*' as an individual constant presupposes that it names or designates a specific individual, but there is clearly no individual *named* 'something'. An exhaustive enumeration of the individuals of our universe would include just *a*, *b*, and *c*. There is no fourth dubbed 'something', much less a fifth named 'nothing' to which the property *J* is ascribed in (7). Moreover, (5) admits of distinct "negations" that the expression '*Fs*' — admitting only the single negation ' ~*Fs*' — cannot handle. For 'Something is not *F*' is *true* in our universe (since *b* does not have *F*) but 'It is not the case that something is *F*' is *false* (*a* and *c* both have *F*).

It will be better to approach (5) from a different angle. Clearly (5) will be false just in case *none* of the individuals in our universe of discourse have *F*, and therefore will be *true* just in case *there is at least one* that does. Thus 'some' here means 'there is at least one'. And because our universe has so few individuals in it, we can reformulate the point. Since 'Something is *F*' is true just in case there is at least one thing that is *F*, that is, true if and only if either *a*, *b*, or *c* is *F*, then it is true just in case at least one substitution instance of the open sentence '*Fx*' is true.[5] And since '*Fa*' and '*Fc*' are both true, (5) is also.

(5), then, is most fully expressed as

There is at least one individual such that it is F

and, since the variable '*x*' ranges over individuals, this in turn may be reexpressed as

There is at least one x such that x is F.

Now '*x* is *F*' is rendered in symbols '*Fx*', and we saw that (5) is true just in case at least one individual in the universe of discourse has the property *F*. We shall express a sentence having such a truth-condition by attaching the expression '($\exists x$)' to '*Fx*' and enclosing the latter in parentheses. Hence (5) is expressed as

(5') ($\exists x$)(Fx).

---

[5]This alternate interpretation of (5) is known as the *substitution* interpretation. We use it only as a pedagogical device, and only here, where the universe of discourse under discussion is finite. When more realistic universes are considered, such as the set of real numbers, we shall drop this mode of speech.

The sign '($\exists x$)' may be read 'there is at least one x such that...' (or more briefly 'for some x,...') and is called an **existential quantifier**.

It should be noted that the second occurrence of 'x' in ($5'$) is no longer subject to replacement by a constant, for attaching '($\exists x$)' to 'Fx' has yielded not another open sentence but a genuine sentence, one that is the translation of 'Something is F' and so is either true or false. The role of the second 'x' in ($5'$) parallels that of the pronoun 'it' in the first of the above reformulations of (5), that is, it carries cross-reference to the first 'x' just as 'it' carries cross-reference to 'individual'. Clearly 'y' or 'z' would have done as well provided that the quantifier had been altered appropriately – '($\exists y$)(Fy)' and '($\exists z$)(Fz)' would also do as translations of (5).

Turning now to (6), we notice that it is true because every individual has the property H; and this may be put by saying that (6) is true in our universe of discourse just in case each substitution instance of the open sentence 'Hx' is true. Since all of 'Ha', 'Hb', and 'Hc' are true, then, (6) is too. So (6) may be more fully expressed as

For any individual whatever, it is H

or as

For any x, x is H.

We shall express a sentence that is true just in case each and every individual in the universe of discourse has the property H by attaching the expression '(x)' to the open sentence 'Hx'. Thus (6) is expressed by

($6'$) $(x)(Hx)$.

The sign '(x)' may alternately be read as 'given any x,...' or 'for all x,...'. We shall call it a **universal quantifier**.

By the same token, (7) is true because none of the individuals in our universe has J. That is, (7) is true in our universe just in case all substitution instances of 'Jx' are false, and this means, of course, when all substitution instances of '$\sim Jx$' are *true*. So we may express (7) by attaching the universal quantifier to the open sentence '$\sim Jx$':

($7'$) $(x)(\sim Jx)$.

Since (7) is true just in case *no* substitution instance of '*Jx*' is true, it follows that '$(\exists x)(Jx)$' is false just in case (7) is true. Hence (7) may also be expressed as the *negation* of '$(\exists x)(Jx)$',

(7″) $\sim(\exists x)(Jx),$

whose English counterpart would be 'There is no individual that has *J*'. More generally, where the schematic expression '... *x*...' represents any open sentence containing '*x*', sentences of the following forms are equivalent:[6]

$(x)(\sim(\ldots x\ldots))$ and $\sim(\exists x)(\ldots x\ldots).$

Turning now to (8), its truth is apparent from the fact that at least one individual in our universe lacks the property *G*. Hence at least one substitution instance of '$\sim Gx$' is true, and accordingly the existential quantifier may be employed:

(8′) $(\exists x)(\sim Gx).$

But this is no different from saying that (8) is true just when *not all* things are *G*. Therefore (8) may alternately be expressed using the universal quantifier as

(8″) $\sim(x)(Gx)$

which goes into English as 'Not everything is *G*'. So the following pair of forms are equivalent:

$(\exists x)(\sim(\ldots x\ldots))$ and $\sim(x)(\ldots x\ldots)$

---

[6]Parentheses are shown in the left-hand schema as a reminder that they are necessary in the case of a typical substitution for ...*x*... such as '*Fx* & *Gx*' — without them, we would have '$(x)(\sim Fx \ \& \ Gx)$' instead of the desired '$(x)(\sim(Fx \ \& \ Gx))$'. Of course when the substitution for ...*x*... is not complex (e.g. '*Fx*' or '$\sim Gx$') the inner parentheses are superfluous and should be mentally erased from the above schema. The same considerations apply to similar schemata displayed subsequently in the text.

The following sentences, each with two distinct predicate letters, are also true in our specified universe:

(9) All F are H   (11) Some F are G
(10) No F are K   (12) Some F are not G

Each of these sentences is of one of the following forms:

A: all ... are - - -   **I:** some ... are - - -
E: no ... are - - -   **O:** some ... are not - - -.

The labels 'A', 'E', 'I' and 'O' were originally applied to these forms in the traditional logic dating from Aristotle, and were regarded as basic to nearly all inferences involving general sentences (for convenience, these labels will be extended to equivalent English forms like 'every ... is ---' and to their counterparts in symbols as well). With the development of modern symbolic logic, they have come to be recognized as less fundamental than earlier logicians had supposed, but they still provide us with an excellent point of departure for the study of general sentences. It should be noted here that despite appearances sentence (12) is *not* the negation of (11) since both are true in our universe. Rather, since claiming that some things which are *F* are *not G* is the same as denying that all of them *are G*, (12) is equivalent to the negation of the corresponding **A**-sentence 'Not all *F* are *G*'. Indeed, many readers may find this a more natural reading of the **O** form than the more stilted 'Some *F* are not *G*'.

When we consider how an **A**-sentence like (9) should be expressed, it might seem natural at first to try '$(x)(Fx \& Hx)$'. But a moment's reflection will show that this will not do, for it is *false* in our universe, whereas (9) is obviously true. Since 'Fb' is false, the open sentence '$Fx \& Hx$' has the false substitution instance '$Fb \& Hb$', and this suffices to make '$(x)(Fx \& Hx)$' false as well. It in fact asserts something far stronger than (9), for it says that everything is *both F* and *H* (compare 'All fathers are husbands' with 'Everything (whatever) is both a father and a husband'). On the other hand, it is true of each individual in our universe that *if* it is *F,* then it is also *H*; accordingly, all substitution instances of '$Fx \supset Hx$' are true. For '$Fa \supset Ha$' and '$Fc \supset Hc$' have both a true antecedent and a true consequent, and '$Fb \supset Hb$' has a false antecedent. '$Fx \supset Hx$' can have a false substitution instance only when the antecedent is true and the consequent false, and this clearly matches up with the intended meaning of (9); it will be false, that is, if at least one individual has *F* but does *not* have *H*. Thus (9) is more fully expressed in the conditional form

For any x, if x is F then x is H

and hence may be translated as

(9') $(x)(Fx \supset Hx)$.

While (9') is true in our universe, '$(x)(Fx \supset Gx)$' is not, for '$Fx \supset Gx$' admits of the false substitution instance '$Fa \supset Ga$'.

Now the E-sentence (10)—'No F are K'—is true because there is no individual that has F and *also* has K. Accordingly, the open sentence '$Fx \supset {\sim}Kx$' has all true substitution instances. Both antecedent and consequent are true in '$Fa \supset {\sim}Ka$' and '$Fc \supset {\sim}Kc$', and '$Fb \supset {\sim}Kb$' has a false antecedent. So (10) says in effect that each and every F is *not* K and may be expressed as

(10') $(x)(Fx \supset {\sim} Kx)$.

But to say that no F are K is to *deny* that there are things which are both F and K. Hence (10) may also be translated by the equivalent sentence '${\sim}(\exists x)(Fx \,\&\, Kx)$'.

In view of the foregoing, it is tempting to translate the I-sentence (11)—'Some F are G'—as '$(\exists x)(Fx \supset Gx)$'. However, it is easily shown that this is incorrect. Consider the I-sentence 'Some F are K'; this is clearly false in our universe since no individual that has F also has K. Yet '$(\exists x)(Fx \supset Kx)$' is *true* (the open sentence '$Fx \supset Kx$' has at least one true substitution instance, '$Fb \supset Kb$') so it cannot be the correct rendering of 'Some F are K'. By the same token '$(\exists x)(Fx \supset Gx)$' fails to correctly translate (11). It is too weak a claim, in that it will be true whenever there is at least one individual that is *not F,* and (11) obviously requires for its truth that there be at least one individual that *is F* and moreover is G. This means that (11) will be true just in case '$Fx \,\&\, Gx$' has at least one true substitution instance—which it does, since '$Fc \,\&\, Gc$' is true. Thus (11) becomes 'There is at least one x such that x is F and x is G', which goes into symbols as

(11') $(\exists x)(Fx \,\&\, Gx)$.

Thus where 'Fx' abbreviates 'x is a father' and 'Gx' abbreviates 'x is generous', (11') translates the English sentence, 'Some fathers are generous'.

So whereas 'All F are G' is translated with '$\supset$' instead of '&', the situation is just the reverse with 'Some F are G'—we should not be misled by the occurrence of 'are' in both. More generally, the *main* connective of the open sentence governed by a universal quantifier is usually ' $\supset$ ', but it is '&' for existentially quantified cases. Note also that E- and I- sentences are contradictories—the negation of the E-sentence is equivalent to the corresponding I-sentence and vice versa. For '$(x)(Fx \supset \sim Gx)$' will be true just in case *no* individual has both F and G, whereas (11') will be true just in case at least one *does* have both F and G. Thus, the following pairs are equivalent forms:

$(\exists x)(...x... \& ---x---)$ and $\sim(x)(...x... \supset \sim(---x---))$
$(x)(...x... \supset \sim(---x---))$ and $\sim(\exists x)(...x... \& ---x---)$

The O-sentence (12) — 'Some F are not G' — may be handled on the model provided by (11). Thus it is true just in case at least one individual has F but lacks G. Accordingly, the open sentence '$Fx \& \sim Gx$' must have at least one true substitution instance—which in fact it does: '$Fa \& \sim Ga$'. We thus have

(12') $(\exists x)(Fx \& \sim Gx)$.

We saw earlier that (12) says the same as 'Not all F are G'; hence an equivalent translation would be ' $\sim(x)(Fx \supset Gx)$'. The latter makes it evident that O-sentences and A-sentences are also contradictories, for (12') will be true just in case at least one individual has F but not G, whereas '$(x)(Fx \supset Gx)$' will be true if and only if *no* individual is such that it has F but lacks G. Thus we obtain the equivalent pairs:

$(\exists x)(...x... \& \sim(---x---))$ and $\sim(x)(...x... \supset ---x---)$
$(x)(...x... \supset ---x---)$ and $\sim(\exists x)(...x... \& \sim(---x---))$

Moving beyond the artificial universe of discourse now, we will examine a variety of common English sentences each of which may be subsumed under one of the four basic forms. Provided also is a two step translation for all of them. A-sentences admit of a vast number of grammatical variations, among them:

All donations are tax-deductible
$(x)(x$ is a donation $\supset x$ is tax-deductible) $\qquad\qquad (x)(Dx \supset Tx)$

Every parking violator will be fined
$(x)(x$ is a parking violator $\supset x$ will be fined) $\qquad\qquad (x)(Px \supset Fx)$

Any metal will conduct electricity
(x)(x is a metal ⊃ x will conduct electricity)   (x)(Mx ⊃ Ex)

Each member of the board voted affirmatively
(x)(x is a member of the board ⊃ x voted affirmatively) (x)(Bx ⊃ Ax)

Whales are mammals
(x)(x is a whale ⊃ x is a mammal)   (x)(Wx ⊃ Mx)

A penny saved is a penny earned
(x)(x is a saved penny ⊃ x is an earned penny)   (x)(Sx ⊃ Ex)

If it's made in Japan, it's dependable
(x)(x is made in Japan ⊃ x is dependable)   (x)(Jx ⊃ Dx)

Two common versions of the **E** form are:

No freshmen can play on the varsity
(x)(x is a freshman ⊃ ~ (x can play on the varsity))   (x)(Fx ⊃ ~Vx)

None of the council members liked the plan
(x)(x is a council member ⊃ ~ (x liked the plan))   (x)(Cx ⊃ ~Px)

Turning now to **I**-sentences:

Some vertebrates are reptiles
   (∃x)(x is a vertebrate & x is a reptile)   (∃x)(Vx & Rx)

There exists at least one prime number that is even
   (∃x)(x is a prime number & x is even)   (∃x)(Px & Ex)

There are smiles that make you happy
   (∃x)(x is a smile & x makes you happy)   (∃x)(Sx & Hx)

A salesman stopped by today
   (∃x)(x is a salesman & x stopped by today)   (∃x)(Sx & Tx)

Whales were spotted off the port bow
   (∃x)(x is a whale & x was spotted off the port bow) (∃x)(Wx & Bx)

Note that sentences beginning with the indefinite article ('a penny saved', 'a salesman') or a plural noun ('whales') require a universal quantifier in some cases but an existential one in others. Contexts of typical use generally indicate which one is called for. The old maxim about a penny saved clearly is intended as a claim about any such coin we care to choose, whereas someone who remarks that a salesman stopped by surely does not mean to claim that *every* salesman whatever paid a visit. Finally, two common varieties for the **O**-sentence are:

Some mutual funds are not good investments
($\exists x$)(x is a mutual fund & $\sim$ (x is a good investment))   ($\exists x$)(Fx & $\sim Ix$)

There are logicians who cannot think straight
($\exists x$)(x is a logician & $\sim$ (x can think straight))   ($\exists x$)(Lx & $\sim Tx$)

EXERCISES

Put each of the following into symbols using the suggested notation.

1. Al is a man who enjoys life. (a - Al; Mx - x is a man; Lx - x enjoys life.)
2. Either Charlie will be on time or Helen will be furious. (c - Charlie; Tx - x will be on time; h - Helen; Fx - x will be furious.)
3. All contestants arrived on time. (Cx - x is a contestant; Ax - x arrived on time.)
4. Some corporations do not act responsibly. (Cx - x is a corporation; Rx - x acts responsibly.)
5. Not every politician is ambitious. (Px - x is a politician; Ax - x is ambitious.)
6. Some athletes have a drug problem. (Ax - x is an athlete; Dx - x has a drug problem.)
7. None of the products tested were highly rated. (Px - x is a tested product; Rx - x was highly rated.)
8. There's a tavern in the town. (Tx - x is a tavern; Ix - x is in the town.)
9. Politicians were observed entering the casino. (Px - x is a politician; Ox - x was observed entering the casino.)
10. You'll like all our products. (Lx - you will like x; Px - x is one of our products.)
11. There are no students who enjoy taking finals. (Sx - x is a student; Fx - x enjoys taking finals.)
12. Flight controllers have a tough job. (Fx - x is a flight controller; Jx - x has a tough job.)
13. Some of the answers will either be explained in class or posted on my door. (Ax - x is an answer; Ex - x will be explained in class; Px - x will be posted on my door.)
14. There are no doctors who do not play golf. (Dx - x is a doctor; Gx - x plays golf.)
15. Every recruit who complains will draw kitchen duty. (Rx - x is a recruit; Cx - x complains; Kx - x draws kitchen duty.)

Let us now consider sentences of somewhat greater complexity. The sentence

Any cat that howls at night is a nuisance

is basically an **A**-sentence with a conjunction as antecedent. Using 'Hx' for 'x howls at night' and other obvious notation, it becomes

$(x)[(Cx \& Hx) \supset Nx]$.

And the following is just a slightly more complex **I**-sentence with a disjunction as right conjunct:

Some Republicans are either liberal or moderate

and consequently is translated as follows:

$(\exists x)[Rx \& (Lx \lor Mx)]$

Parentheses are important here, for '$(\exists x)[(Rx \& Lx) \lor Mx]$' makes the quite different claim that something is either a Republican liberal or else is moderate.

Sentences containing 'only' can be especially troublesome. A correct translation of those beginning with this word usually requires inverting the position of the component general terms. Thus the true sentence

Only persons over 62 may receive Social Security retirement benefits

should be rendered as

$(x)(Sx \supset Px)$

where 'Sx' is 'x may receive Social Security retirement benefits' and 'Px' is 'x is a person over 62'. It should *not* be translated '$(x)(Px \supset Sx)$', which makes the much different—and false—claim that *every* person over 62 may receive the benefits; presumably a 67 year old citizen of Yugoslavia is not eligible for Social Security. The sentence 'Cats and dogs are vertebrates' may be translated in either of the equivalent forms

$(x)[(Cx \lor Dx) \supset Vx]$ $\qquad\qquad\qquad$ $(x)(Cx \supset Vx) \& (x)(Dx \supset Vx)$

but it must not be rendered '$(x)[(Cx \& Dx) \supset Vx]$', for this says that anything that is *both* a cat and a dog is a vertebrate—hardly the intended meaning. Still other sentences raise problems concerning the main connective. Consider for example

There are men who achieve greatness if they work hard

This says that there exist (individuals who are) men who (are such that) if they work hard, they achieve greatness. Accordingly, we have

$(\exists x)[Mx \& (Wx \supset Gx)]$

rather than '$(\exists x)[(Mx \& Wx) \supset Gx]$'. The latter would be true merely if no individual is a hard-working man (since then *any* substitution instance of '$(Mx \& Wx) \supset Gx$' would be true owing to a false antecedent). The English sentence does not say that there are individuals that, *if* they are men who are hard-working, achieve greatness; it says rather that there are individuals who *are* men and who, if they work hard, achieve greatness. So as is usually the case with existentially quantified sentences, it is '&', not '$\supset$', that is the main connective. Similar considerations apply to

All students are eligible if and only if they are not freshmen

This sentence is just another variation of the **A** form with a biconditional construction in the consequent. Hence, as with most universally quantified sentences we shall be considering, it is '$\supset$' and not '$\equiv$' which is the main connective:

$(x)[Sx \supset (Ex \equiv \sim Fx)]$

Sentences asserting or denying existence are easily handled in our notation, and no appeal to an open sentence like 'x exists' is either necessary or desirable (indeed it is a mystery as to what could be meant by a sentence like '$(\exists x)(\sim x$ exists)'). For the range of arguments with which we shall be concerned in the next two chapters, the existential quantifier is all that is needed to translate existence claims, as the following examples show:

Neutrinos exist (There are neutrinos) $\qquad\qquad\qquad$ $(\exists x)(Nx)$
Dragons do not exist (There are no dragons) $\qquad\qquad$ $\sim(\exists x)(Dx)$

The latter of course may also be translated '$(x)(\sim Dx)$' since it makes the equivalent claim that nothing is a dragon. Truth-functionally compound sentences are translated as they were in Chapter One, but using our present notation instead of sentence letters. The sentence

If Japan agrees to limit trade, every protectionist will cheer

is a conditional with a singular sentence as antecedent and an **A**-sentence as consequent. Thus, using obvious notation, it becomes:

$Aj \supset (x)(Px \supset Cx)$

Finally, it is important to notice that words like 'anyone', 'someone', 'everybody' and 'whoever' restrict the individuals under consideration to human beings. Consequently, unless the universe of discourse is explicitly stipulated as including only persons, the translation of sentences containing such words requires use of an open sentence like 'x is a person' ('Px' usually). Thus, 'Somebody has broken in' should be expressed as '$(\exists x)(Px \& Bx)$' and 'Whoever broke in left hurriedly' as '$(x)[(Px \& Bx) \supset Lx]$'.

EXERCISES

I. Which of the following are true and which false in the universe described earlier in this section? Briefly explain your answers.
1. $(x)[(Gx \& \sim Fx) \supset Kx]$
2. $(\exists x)[Fx \& (Gx \equiv Hx)]$
3. $(Fb \lor Gc) \supset (\exists x)(\sim Hx)$
4. $(x)[(Fx \lor Kx) \supset (Hx \& Gx)]$
5. $(x)(Jx \supset Kx)$
6. $(\exists x)(Jx \& Kx)$

II. Put each of the following into symbols using the suggested notation.
1. Mammals are all warm-blooded. (Mx - x is a mammal; Wx - x is warm-blooded.)
2. Those who ignore history are condemned to repeat it. (Ix - x ignores history; Cx - x is condemned to repeat history.)
3. He who laughs last, laughs best. (Px - x is a person; Lx - x laughs last; Bx - x laughs best.)
4. If Al or Charlie wins the race, somebody will be disappointed. (a - Al; c - Charlie; Wx - x wins the race; Px - x is a person; Dx - x will be disappointed.)

5. Everything enjoyable is either immoral, illegal, or fattening. (*Ex-x* is enjoyable; *Mx-x* is moral; *Lx-x* is legal; *Fx-x* is fattening.)

6. If George is allowed to speak, no one will be bored. (*g* - George; *Sx-x* is allowed to speak; *Px-x* is a person; *Bx-x* will be bored.)

7. Only a fanatic would advocate violence. (*Fx-x* is a fanatic; *Vx-x* advocates violence.)

8. None but the strong survive. (*Sx-x* is strong; *Vx-x* survives.)

9. Persons arrested for jaywalking will be prosecuted if—but only if—they refuse to pay a fine and have a prior history of arrest. (*Px - x* is a person; *Ax - x* is arrested for jaywalking; *Sx - x* will be prosecuted; *Rx - x* refuses to pay a fine; *Hx - x* has a prior history of arrest.)

10. Among railroad employees, only conductors and brakemen may ride the caboose. (*Rx - x* is a railroad employee; *Cx - x* is a conductor; *Bx - x* is a brakeman; *Ox - x* may ride the caboose.)

11. There are conservatives who gnash their teeth if they read the *Nation*. (*Cx-x* is a conservative; *Tx-x* gnashes *x*'s teeth; *Nx-x* reads the *Nation*.)

12. The only animals in my house are cats. (*Ax-x* is an animal in my house; *Cx-x* is a cat.)

13. Dolphins and porpoises grin and frolic in the sea. (*Dx-x* is a dolphin; *Px-x* is a porpoise; *Gx-x* grins; *Fx-x* frolics in the sea.)

14. Any man will be driven to drink who cannot tolerate children. (*Mx-x* is a man; *Dx-x* will be driven to drink; *Tx-x* can tolerate children.)

15. Some politicians will be reelected only if their Mafia connections are not revealed. (*Px-x* is a politician; *Rx-x* will be reelected; *Mx-x*'s Mafia connections are revealed.)

16. Not all adults who are willing to pay double will be admitted after 8:00 P.M. (*Ax-x* is an adult; *Ex-x* will be admitted after 8:00 P.M.; *Wx-x* is willing to pay double.)

17. Horses exist but unicorns do not. (*Hx-x* is a horse; *Ux-x* is a unicorn.)

18. No adults will be admitted after 8:00 P.M. unless they are willing to pay double. (*Ax-x* is an adult; *Ex-x* will be admitted after 8:00 P.M.; *Wx-x* is willing to pay double.)

19. Not only radicals but revolutionaries attended the convention. (*Rx-x* is a radical; *Vx-x* is a revolutionary; *Cx-x* attended the convention.)

20. No Democratic candidate is both a liberal and an advocate of trade restrictions, nor is there any Democratic candidate who is not one of the two. (*Dx-x* is a Democratic candidate; *Lx-x* is a liberal; *Ax-x* advocates trade restrictions.)

21. Except for bluegills, no fish is hooked without a sinker. (*Bx-x* is a bluegill; *Fx-x* is a fish; *Hx-x* is hooked; *Sx-* a sinker is used to hook *x*.)

*III. Since 'All *J* are *F*' is *true* in the universe specified in this section but 'Some *J* are *F*' is false, the argument

$(x)(Jx \supset Fx)$

$\therefore (\exists x)(Jx \, \& \, Fx)$

is *invalid*. Explain and discuss this rather surprising result, using what you know about the truth-conditions for **A**- and **I**-sentences. Can you suggest a natural additional premise that along with '$(x)(Jx \supset Fx)$' will validly yield '$(\exists x)(Jx \, \& \, Fx)$'?

5.2 Multiple Quantification and Relations

Thus far, our concern has been with general sentences containing a single quantifier, but there are many that contain two or more. Some are straightforwardly truth-functional. For example, if we restrict our attention to the positive integers for the moment, the disjunction

(1) Either everything is even or everything is odd

is *false* since the component sentences, 'Everything is even' and 'Everything is odd', are both false. We may express (1) in symbols in either of the following ways:

(1') $(x)(Ex) \vee (x)(Ox)$

(1") $(x)(Ex) \vee (y)(Oy)$

Although (1") is usually preferred, it does not matter whether we use distinct variables for the quantifiers of the component sentences, just as it does not matter whether an open sentence with one variable subject to replacement by a constant is written '*Fx*' or '*Fy*'. For uniformity we shall now drop the practice of using the same variable since, as we shall shortly see, there are sentences that *require* the use of distinct variables.

In contrast to (1), the single-quantifier sentence

(2) Everything is either even or odd

(2') $(x)(Ex \vee Ox)$

is *true* since each substitution instance of '*Ex* $\vee$ *Ox*' is true. (Unlike (1'), (2') is not a truth-functional compound, for though it contains a compound open sentence it does not contain any genuine, "true-or-false" sentences as parts.) Since (1') is false but (2') true, it is clear that sentences of the forms

(a) $(x)(...x...) \vee (y)(---y---)$

(b) $(x)(...x... \vee ---x---)$

are not equivalent, for we have produced a case where there is a difference in truth-value. Sentences of the form (a) *imply* those of form (b) — an argument with a sentence of the first form as premise and one of the latter as conclusion is valid—but we have just seen that the reverse does not hold; sentences of form (b) do not imply the corresponding ones of form (a).

Similar considerations apply to

(3)   Something is both C and D

(3′) $(\exists x)(Cx \; \& \; Dx)$

which is simply the I-sentence 'Some C are D' and the sentence

(4)   Something is C and something is D

(4′) $(\exists x)(Cx) \; \& \; (\exists y)(Dy).$

If we let 'Cx' represent 'x is a cat' and use 'Dy' for 'y is a dog', then while (4′) is true (3′) is false. (4′) tells us that at least one thing is a cat and at least one thing is a dog, but (3′) says that at least one thing is *both* a cat and a dog. So sentences of the forms:

(c) $(\exists x)(...x... \; \& \; \text{---}x\text{---})$

(d) $(\exists x)(...x...) \; \& \; (\exists y)(\text{---}y\text{---})$

are not equivalent, though a sentence of form (c) *implies* one of form (d). But, as the foregoing example shows, an argument with a sentence of form (d) as premise and one of form (c) as conclusion is invalid.

The foregoing considerations make use, on the intuitive level, of the concepts of equivalence and validity for sentences containing quantifiers. But we have not yet given any precise explanation or analysis of them as regards general sentences. The analysis given in Chapter 2 for truth-functional sentences clearly is not applicable here, since truth-tables cannot be used to treat uncompounded sentences like '$(x)(Fx)$' and '$(\exists x)(Fx \; \& \; Gx)$'. Therefore we must eventually provide new accounts of validity and equivalence that will accord with our insights concerning, for example, the validity of inferences from (a) to (b) and from (c) to (d). We shall deal with this in Chapter Six, but first there are further types of sentences demanding our attention.

Let us compare the expressions

(5) $(x)(Fx \supset Gx)$

(6) $(x)(Fx) \supset Gx.$

(5) differs in appearance from (6) only in the placement of parentheses, but this difference is very important. For while (5) is a complete sentence and so is true or false, (6) is an *open* sentence having no truth-value. Since '*Gx*' occurs outside the parentheses, (6) is a compound having '(*x*)(*Fx*)' and the open sentence '*Gx*' as components. As such, (6) itself is an open sentence, since the '*x*' in '*Gx*' is subject to replacement by individual constants—though of course neither '*x*' in the component '(*x*)(*Fx*)' is so subject. Thus (6) yields as substitution instances the sentences:

(x)(Fx) ⊃ *Ga*   If everything is *F* then *a* is *G*
(x)(Fx) ⊃ *Gb*   If everything is *F* then *b* is *G*
(x)(Fx) ⊃ *Gc*   If everything is *F* then *c* is *G*.

But since (5) is a sentence (and so either true or false), *no* occurrence of '*x*' in it is subject to replacement by a constant. Rather, the role of '*x*' in '*Gx*' of (5) is that of a pronoun carrying reference back to the '*x*' of the quantifier—*not* of a marker for a position where constants may be substituted. That is, '*x*' plays the same role in (5) as 'it' plays in

Each individual is such that if *it* is *F* then *it* (that same individual) is *G*.

So the parentheses in (5) show that the '*x*' of '*Gx*' refers back to the '*x*' in the quantifier; but the parentheses in (6) show that only the '*x*' of '*Fx*' refers back, leaving the '*x*' of '*Gx*' open to substitution. Whereas (5) is formed by attaching '(*x*)' to '*Fx* ⊃ *Gx*', (6) is a compound open sentence with '(*x*)(*Fx*)' and the open sentence '*Gx*' as components.

We might visualize the situation as follows—the arrows below show that the circled variables carry back reference to the indicated quantifiers, and the underlying brackets show how much of the entire expression is governed by the quantifiers:

(5) (x̂)(F ⓧ ⊃ G ⓧ)
(6) (x̂)(F ⓧ) ⊃ Gx

We shall describe the portion governed by a quantifier—the portion enclosed in subsequent parentheses—as the **scope** of that quantifier. Thus in (5) the scope of '(*x*)' extends throughout the entire remaining expression, but in (6) it extends only to the connective ' ⊃ '. A variable that is *within* the scope of a quantifier containing an occurrence of that same variable always carries back reference, but one *outside* the scope of any quantifier marks a place where a constant may be substituted—it never carries back reference. In this latter case, the entire expression is an open sentence.

We noticed at the outset that it is immaterial what variable is used in an open sentence—we could just as well write '*Fy*' as '*Fx*'. Thus the open sentence (6) may be rewritten as

(6') $(x)(Fx) \supset Gy$

where '*y*' is subject to replacement; clearly (6') admits of the very same substitution instances as (6). But let us now compare

(7) $(x)(Fx \supset Gy)$.

Here, the parentheses make it clear that the scope of '(x)' extends over the entire expression, and hence '*y*' is within its scope. Even so, '*y*' does *not* carry back reference, since the quantifier contains a *different* variable; only the '*x*' of '*Fx*' serves as a pronoun. Unlike (5), then, but like (6) and (6'), (7) is an open sentence and as substitution instances has:

$(x)(Fx \supset Ga)$
$(x)(Fx \supset Gb)$
$(x)(Fx \supset Gc)$.

Like the '*y*' in (6'), '*y*' here marks a place where an expression designating an individual may be substituted. It is only when '*y*' is within the scope of a quantifier like '$(\exists y)$' or '(y)' that it carries pronominal cross reference. The importance of quantifier scope may be brought out by noting that the substitution instances of (6') and (7) are *not* equivalent. For example, if *a* has *F* but *b* does not, and in addition neither *a* nor *b* has *G*, then '$(x)(Fx \supset Ga)$' will be *false* but '$(x)(Fx) \supset Ga$' *true* since in the former case the substitution instance '$Fa \supset Ga$' is false, and in the latter the antecedent '$(x)(Fx)$' is false, making the conditional true.

The distinction in role between serving as cross-referring pronoun and marking a place for substitution of constants is sufficiently important to warrant the introduction of some new terminology. We shall describe occurrences of variables in the former role—as well as occurrences within quantifiers themselves—as **bound**, and occurrences in the latter role as **free**. More formally, we have:

DEFINITION   An occurrence of an individual variable is **bound** if and only if it either is part of a quantifier or else lies within the scope of a quantifier containing an occurrence of that same variable.

Moreover, an occurrence of an individual variable is **free** if and only if it is not bound, that is, it is neither part of a quantifier nor within the scope of any quantifier containing an occurrence of that same variable.

From these definitions and the foregoing discussion, it follows that:

An open sentence must have an individual variable with at least one free occurrence;

In a true-or-false sentence, all occurrences of all variables (if any) are bound.

In sentences (1′) through (5) all variables occur bound; in (6) 'x' occurs free in 'Gx', and bound elsewhere. In (6′) and (7) 'y' occurs free, but all occurrences of 'x' are bound.

Both '$Fx \supset Gy$' and '$(x)(Fx \supset Gy)$' are open sentences, but in the first both 'x' and 'y' occur free whereas in the second only 'y' does. However, in the expression

(8) $(x)(y)(Fx \supset Gy)$

no variable occurs free, hence it is not an open sentence. Rather, it says that for any individuals x and y if the first is F then the second is G. The parentheses show that the scope of '(y)' extends over the remainder of the expression, thus binding the right-hand occurrence of 'y'. We shall leave it understood that when two quantifiers occur juxtaposed as in (8), the one on the left has the other within its scope—parentheses will not be used in such cases. Thus, (8) may be diagrammed:

It should be noted that whereas (1′) and (4′) can be expressed using distinct variables for the component quantifiers, a sentence like (8) *requires* distinct variables, for otherwise we should not know which variable is bound by which quantifier. Whenever the scope of one quantifier lies within the scope of another, distinct variables are a necessity.

Care must be taken to distinguish open sentences in which more than one variable occurs free—as in '$Fx \supset Gy$'—from those which, like '$Fx \supset Gx$' and '$(y)(Fx \supset Gy)$', contain just one. For while '$Fa \supset Ga$' and

'*Fb* ⊃ *Gb*' are substitution instances of '*Fx* ⊃ *Gx*', the sentence '*Fa* ⊃ *Gb*' is not; different constants cannot replace occurrences of the same variable. On the other hand, the same constant may of course replace distinct variables; thus '*Fx* ⊃ *Gy*' has not only '*Fa* ⊃ *Gb*' but '*Fa* ⊃ *Ga*' and '*Fb* ⊃ *Gb*' as legitimate substitution instances.

EXERCISES

I. Which of the following are open sentences? Specify the free occurrences in each case.
1. $Fa \& \sim Gx$
2. $Fx \supset (\exists y)(Gy)$
3. $Fa \lor Gb$
4. $(x)[Fx \supset (Gx \& \sim Hz)]$
5. $(\exists x)[Fx \& (Gx \equiv Hx)]$
6. $(x)(Fx) \supset (Ha \lor \sim Gx)$

II. Consider the open sentence
$(\exists x)(Fx) \supset [Gy \supset (Hx \& \sim Jy)]$.
Which of the following are substitution instances?
1. $(\exists x)(Fx) \supset [Ga \supset (Hb \& \sim Jc)]$
2. $(\exists x)(Fx) \supset [Ga \supset (Ha \& \sim Ja)]$
3. $(\exists x)(Fx) \supset [Gb \supset (Hc \& \sim Jb)]$
4. $(\exists x)(Fa) \supset [Gb \supset (Hc \& \sim Jb)]$

Open sentences with more than one free variable have so far all been truth-functionally compound in form, such as '*Fx* ⊃ *Gy*'. There are, however, many that are not compound but are instead *relational*, for example, '*x* is larger than *y*' (in symbols, '*Lxy*'). A relational term like 'larger than' may be regarded as another species of general term, but instead of classifying individuals singly, it classifies *pairs* of individuals. Just as a general term like 'philosopher' represents a property of individuals, so a relational term represents a *relation* holding among two or more individuals. The relation represented by 'larger than' holds of pairs of individuals and is thus termed a *dyadic* (or *binary*) relation; it holds, for example, between Boston and Albuquerque respectively, but not between Boston and Chicago or for that matter Albuquerque and Boston.

A *triadic* (or *tertiary*) relation holds among three individuals; 'between' thus holds for the ordered triad Chicago–Milwaukee–Indianapolis, but not the triad Milwaukee–Chicago–Indianapolis. There are also relations that may hold among four, five, and more individuals, but they are less commonly found in ordinary speech. An example of a four-individual relation, however, is provided by the sentence, 'France and

Britain fought Germany in World War II'. Moreover, we saw in Chapter 1 that sentences like

Al and Bill are opponents

cannot sensibly be construed as conjunctions; the above is certainly not a compound of 'Al is an opponent' and 'Bill is an opponent'. Rather, it is a relational sentence, and may be rephrased as

Al is an opponent of Bill,

which we shall translate as 'Oab'.

As with general terms signifying properties, we shall treat relational terms as supplying the form of the sentences in which they occur. Thus from the singular sentences

Boston is larger than Albuquerque	Lba
Boston is larger than Des Moines	Lbd
Boston is larger than Chicago	Lbc
Albuquerque is larger than Boston	Lab

we may abstract the schema '...is larger than---', and as before we shall use capital letters to abbreviate such expressions in our notation. A singular sentence such as 'Lba' is thus a (true) substitution instance of the open sentence 'Lxy'; on the other hand, 'Lbx' is an open sentence with just *one* free variable and has only the first three of the above sentences as substitution instances.

Let us consider some further relational sentences and see how they may be expressed in symbols:

(9)   Al struck something

(10)   Something struck Al

(11)   Nothing appeals to Bill

(12)   Whatever Alice wants, she gets

(13)   Al hit Bill with a brickbat

(14)   God helps those who help themselves

Now (9) will be true just in case at least one individual was struck by Al. So using 'Sxy' for 'x struck y' and 'a' for 'Al', we get

(9') $(\exists x)(Sax)$

or, in words, 'There is at least one x such that a struck x'. Similarly, (10) is rendered

(10') $(\exists x)(Sxa)$.

Using obvious abbreviations, (11) may be expressed

(11') $(x)(\sim Axb)$.

Now, (12) maintains that for any individual, if Alice wants it she gets it. So using 'a' for 'Alice', 'Wxy' for 'x wants y' and 'Gxy' for 'x gets y', we may express (12) as

(12') $(x)(Wax \supset Gax)$.

In (13) a triadic relation is expressed by '...hit---with——'. Therefore if we use '$Hxyz$', 'a', 'b', and 'Bx'('x is a brickbat'), we obtain

(13') $(\exists x)(Bx \ \& \ Habx)$.

(14) has the force of 'All those who help themselves are such that God helps them', *not* 'All those whom God helps help themselves'. Using 'Hxy' for 'x helps y' and 'g' for 'God', we thus derive

(14') $(x)[(Px \ \& \ Hxx) \supset Hgx]$,

where 'Hxx' conveys the notion of helping oneself.

We have seen that the words 'all' and 'any' may often be interchanged without affecting the meaning of the containing expression. Thus, 'All philosophers are wise' and 'Any philosopher is wise' are both expressed as '$(x)(Px \supset Wx)$'. However, there are other contexts where the words play different roles — if you are offered a box of candy and asked to take *any* piece, you surely will commit a social blunder by taking *all* of them. The same contrast holds between 'every' and 'any'. The sentences

Bruno can lick every man in the house
Bruno can lick any man in the house

generally carry the same meaning, and both are rendered as the **A**-sentence '$(x)(Mx \supset Lbx)$', where 'Mx' is 'x is a man in the house' and 'Lbx' is 'Bruno can lick x'. But note what happens when the important word 'not' is inserted in both:

(15) Bruno cannot lick every man in the house

(16) Bruno cannot lick any man in the house.

Here, 'every' and 'any' operate differently, for (15) says in effect that not every man in the house is one Bruno can lick, that is, there is at least one Bruno cannot lick. But (16) says that *no* man in the house is one Bruno can lick. Thus (15) may be expressed as an **O**-sentence (and the contradictory of '$(x)(Mx \supset Lbx)$'), whereas (16) may be rendered as an **E**-sentence:

(15′) $(\exists x)(Mx \;\&\; \sim Lbx)$, or alternatively $\sim(x)(Mx \supset Lbx)$

(16′) $(x)(Mx \supset \sim Lbx)$.

We must pursue this matter further, for correct translation of many sentences containing 'every', 'any', 'some', and the like can be an intricate task. Let us consider the sentences:

(17)   If everything has mass then something has velocity

(18)   If something has mass then something has velocity

(19)   If anything has mass then something has velocity

(20)   If anything has mass then it has velocity

(21)   If something has mass then it has velocity.

(17) is quite straightforward; we simply put 'Everything has mass' and 'Something has velocity' into symbols and insert a horseshoe between them:

(17′) $(x)(Mx) \supset (\exists y)(Vy)$.

Likewise, (18) becomes

(18′) $(\exists x)(Mx) \supset (\exists y)(Vy)$.

In both (17′) and (18′), the scope of the left-hand quantifier extends only to the connective '$\supset$', not over the entire expression. Moreover, (19) is *also* correctly translated as (18′), since the force of 'anything' in (19) is that of 'at least one thing'.

But when we compare (19) with (20), a striking contrast emerges. Since 'it' in (20) carries reference back to the antecedent, we must use the same variable in both antecedent and consequent. But if we now attempt to translate (20) using (19) as our model, we get into trouble. For in '$(\exists x)(Mx) \supset Vx$', we do not have a genuine sentence, but an *open* sentence with a free occurrence of 'x' in the consequent. Suppose, then, we bind this occurrence by extending the scope of the quantifier across the entire expression, thus getting '$(\exists x)(Mx \supset Vx)$'. Now at least we have a sentence that is true or false. However, it is an *incorrect* translation of (20). For it says that there is some individual such that if it has mass then it has velocity, and this lacks the generality of (20). (20) clearly says that *no matter which* individual we pick, it has velocity if it has mass. Therefore, (20) must be translated both using the universal quantifier and with scope extending across the entire expression. Thus, it comes out as the **A**-sentence

(20') $(x)(Mx \supset Vx)$

and means, the same as 'Everything (anything) that has mass has velocity.' Moreover, (21) is *also* translated as (20'), since 'something' in such a context clearly carries the same force as 'anything' in (20). Thus 'something' is not always to be translated by means of '$(\exists x)$'. Just as there are English contexts where 'every' and 'any' play the same role, so there are contexts where 'some' and 'any' do. Each of the following is thus translated as (20'):

$\begin{Bmatrix} \text{Everything} \\ \text{Anything} \end{Bmatrix}$ which has mass has velocity

If $\begin{Bmatrix} \text{anything} \\ \text{something} \end{Bmatrix}$ has mass, then it has velocity.

Moreover, (18) and (19) admit of an alternate translation which, like that of (20) and (21), contains a universal quantifier with scope extending over the entire expression. That is, (18) and (19) may be translated not just as (18') but also as

(18") $(x)[Mx \supset (\exists y)(Vy)]$.

(18′) and (18″) are equivalent,[7] but two important facts about the latter should be kept in mind: first, the use of distinct variables is essential in (18″), since the scope of the existential quantifier lies within that of the universal; and second, the scope of the universal quantifier *must* extend across the entire expression even though there is no occurrence of '*x*' in the consequent. If it did not, we would have '(*x*)(*Mx*) ⊃ (∃*y*)(*Vy*)' — (17′) — which makes the quite different claim that if *everything* has mass, something has velocity. More fully, the contrast between (17′) and (18″) is best appreciated in terms of their English paraphrases:

(17′) *If*, for every *x*, *x* has mass, *then* there is a *y* such that *y* has velocity
(18″) For every *x*, *if x* has mass, *then* there is a *y* such that *y* has velocity.

The equivalence just noted may be stated more generally as follows. Let '…*x*…' represent any open sentence with a free occurrence of '*x*' (for example, '*Fx*', '*Fx* ⊃ ~*Gx*', '*Fx* ∨ (∃*y*)(*Gy*)') and let '*p*' represent any expression whatever as long as it does not contain '*x*'; then sentences of the following forms are equivalent:

(∃*x*)(…*x*…) ⊃ *p*
(*x*)(…*x*… ⊃ *p*),

but these are *not* equivalent to sentences of the form

(*x*)(…*x*…) ⊃ *p*.

Of course the same holds using variables '*y*', '*z*', and so on. Clearly (18′) is derivable from the first schema substituting '*Mx*' for '…*x*…' and '(∃*y*)(*Vy*)' for '*p*', and (18″) is derivable from the second using the same substitutions. The nonequivalent (17′) is derivable from the third schema but not, of course, from the second.

Relying on what we have just learned, let us tackle three somewhat more complicated examples:

(22) If all stockholders were in attendance, some of the executives were upset

---

[7]Readers who have difficulty seeing the equivalence of (18′) and (18″) should consider, first, whether (18′) can be *true* when the open sentence '*Mx* ⊃ (∃*y*)(*Vy*)' has a *false* substitution instance, and second, whether (18′) can be *false* when all substitution instances of '*Mx* ⊃ (∃*y*)(*Vy*)' are *true*.

(23) If any (some) stockholders were in attendance, some of the executives were upset
(24) If any (some) stockholders were in attendance, they were tipped off by at least one insider.

(22) is easily expressed as a conditional with '$(x)(Sx \supset Ax)$' as antecedent and '$(\exists y)(Ey \ \& \ Uy)$' as consequent:

(22′) $(x)(Sx \supset Ax) \supset (\exists y)(Ey \ \& \ Uy)$.

(23) is also a conditional, but with '$(\exists x)(Sx \ \& \ Ax)$' as antecedent:

(23′) $(\exists x)(Sx \ \& \ Ax) \supset (\exists y)(Ey \ \& \ Uy)$.

But since (23′) is of the form '$(\exists x)(...x...) \supset p$' with '$Sx \ \& \ Ax$' substituted for '$...x...$' and '$(\exists y)(Ey \ \& \ Uy)$' for '$p$', we may, according to the equivalence principle just noted, reexpress (23) in the form '$(x)(...x... \supset p)$',

(23‴) $(x)[(Sx \ \& \ Ax) \supset (\exists y)(Ey \ \& \ Uy)]$.

Again, the outer brackets are essential, since the scope of '(x)' must extend across the entire expression. Otherwise we would have '$(x)(Sx \ \& \ Ax) \supset (\exists y)(Ey \ \& \ Uy)$', which makes the quite different claim that if *everything whatever* is both a stockholder and in attendance, some executives were upset. Finally, (24) must be handled along the lines of (23″), since 'they' obviously refers back to 'stockholders' in the antecedent. Hence the variable used in the antecedent must also occur in the consequent, and the scope of the universal quantifier must accordingly extend over the entire expression so as to bind that occurrence. Using 'Txy' for 'x was tipped off by y', and 'Ix' for 'x is an insider', we have:

(24′) $(x)[(Sx \ \& \ Ax) \supset (\exists y)(Iy \ \& \ Tyx)]$

EXERCISES

Put each of the following into symbols, using the suggested notation.
1. Somebody loves Alice. (*Px-x* is a person; *Lxy-x* loves *y; a* - Alice.

2. Al hates everything that Charlie likes. (*a* - Al; *Hxy-x* hates *y*; *c* - Charlie; *Lxy-x* likes *y*.)
3. If everything is in motion, then Aristotle was wrong. (*Mx-x* is in motion; *a* - Aristotle; *Wx-x* was wrong.)
4. If anything is in motion, Zeno was wrong. (*Mx-x* is in motion; *e* - Zeno; *Wx-x* was wrong.)
5. If anything is in motion, it is subject to time-warp. (*Mx-x* is in motion; *Tx-x* is subject to time-warp.)
6. If something moves, Clint will shoot it. (*Mx* - *x* moves; *c* - Clint; *Sxy* - *x* will shoot *y*.)
7. If every suspect is arrested, some libertarians will protest. (*Sx-x* is a suspect; *Ax-x* is arrested; *Lx-x* is a libertarian; *Px-x* will protest.)
8. If any suspect is arrested, at least one bail bondsman will be pleased. (*Sx-x* is a suspect; *Ax-x* is arrested; *Bx-x* is a bail bondsman; *Px-x* will be pleased.)
9. If any suspect is arrested, he will be provided with a lawyer. (*Sx-x* is a suspect; *Ax-x* is arrested; *Pxy-x* will be provided with *y*; *Lx-x* is a lawyer.
10. Burt is jealous of any man liked by Sally. (*b* - Burt; *Jxy-x* is jealous of *y*; *Mx-x* is a man; *Lxy-x* likes *y*; *s* - Sally.)
11. If any circuits are defective, then if no electrician is available, some operations will have to be curtailed. (*Cx-x* is a circuit; *Dx-x* is defective; *Ex-x* is an electrician; *Ax-x* is available; *Ox-x* is an operation; *Tx-x* will have to be curtailed.)
12. If any circuits are defective, they won't be repaired unless an electrician is available. (*Cx-x* is a circuit; *Dx-x* is defective; *Rx-x* will be repaired; *Ex-x* is an electrician; *Ax-x* is available.)

Techniques for Translating Complex Sentences

(24′) leads us to relational sentences containing more than one quantifier. Beginning with some elementary cases, it should be intuitively clear that

$(x)(y)(Rxy)$
$(y)(x)(Rxy)$

are equivalent; both mean 'Everything bears R to everything'. At best, the only difference between the two expressions would be an insignificant one of phrasing—the latter might more naturally be expressed in the passive voice: 'Everything has R borne to it by everything'. But just as the two English sentences come to the same thing, so do their symbolic counterparts. Moreover, the same holds for the existential quantifier; '$(\exists x)(\exists y)(Rxy)$' and '$(\exists y)(\exists x)(Rxy)$' are equivalent and correspond to 'Something bears R to something'.

On the other hand, order of quantifiers becomes important when one is universal and the other existential, for the following are *not* equivalent:

(25) $(x)(\exists y)(Cxy)$

(26) $(\exists y)(x)(Cxy)$.

To see this, let 'Cxy' mean 'x causes y'; then (25) says

Everything causes something (or other)

or more fully

For any x, there is at least one y such that x causes y.

Notice that this does not say that there is a y which is caused by each and every x. Rather, it makes the weaker claim that for any such x, a y can be found that x causes, but it need not be the *same y* for all the x's—hence the parenthetical phrase 'or other'. However, (26) *does* make the stronger claim—it says that at least one y has every x for a cause, and so corresponds to the English

Something is caused by everything

or, eliminating the passive voice:

Something is such that everything causes it.
There is at least one y such that for any x, x causes y.

Unlike (25), then, (26) will be true only if at least one individual can be found that *all* individuals cause. But though (25) and (26) are not equivalent, it should now be apparent that the stronger claim (26) *implies* (25), for if there is at least one individual that everything causes, then everything causes something (*that* individual if nothing else). The same contrast occurs if we reverse the variables 'x' and 'y' to obtain '$(x)(\exists y)(Cyx)$' and '$(\exists y)(x)(Cyx)$'. The first says that everything has a cause while the second says that there is something that causes everything.

To amplify a bit, let us consider the more familiar cases:

(27)   Every boy likes some girl

(28)   Some girl is liked by every boy.

(27) would most naturally be understood as meaning 'some girl or other', while (28) would have the force of 'Some girl is such that she is liked...'. So interpreted, (28) asserts the existence of a girl who is universally liked by the younger male population, whereas (27) makes the more plausible claim that for any boy, a girl can be found whom he likes. (27) says that every boy is such that there is a girl he likes, and (28) says that there is a girl such that every boy likes her. (27) thus may be regarded as a complex A-sentence with an existentially quantified clause following the main connective '$\supset$'. On the other hand, (28) is basically an I-sentence with a universally quantified clause to the right of '&'. More fully, then, we have:

(27') For any x, if x is a boy, then there is a y such that y is a girl and x likes y
       $(x)[x$ is a boy $\supset (\exists y)(y$ is a girl & x likes $y)]$
(28') There is a y such that y is a girl, and for any x if x is a boy then x likes y.
       $(\exists y)[y$ is a girl & $(x)(x$ is a boy $\supset x$ likes $y)]$

In symbols, then, we derive:

(27") $(x)[Bx \supset (\exists y)(Gy \& Lxy)]$
(28") $(\exists y)[Gy \& (x)(Bx \supset Lxy)]$.

Turning now to further cases of multiple quantification, we consider:

(29)   Every suspect who carries a gun will be denied bail

(30)   No athletes win any contests unless they train for them

The technique of *paraphrasing inward* is helpful in translating such sentences. The first objective is to find the *main* quantifier — that with widest scope — and the main connective. Secondly, put both into symbols, leaving the rest of the sentence in English for the moment. Then symbolize as many of the subordinate connectives and open sentences as make themselves apparent. Finally, find the subordinate quantifiers and, in successive steps, put each into symbols along with the remaining English expressions within their scope. (29) provides an elementary example of the technique, for the first step rather obviously yields

$(x)(x$ is a suspect who carries a gun $\supset x$ will be denied bail).

However, it is clear that a further quantifier is embedded in the antecedent, for the indefinite article 'a' preceding 'gun' has the force of 'any', 'some', and 'at least one'. Thus, using 'Sx' for 'x is a suspect' and 'Bx' for 'x will be denied bail', our next step will be

$(x)[(Sx$ & there is a gun which x carries) $\supset Bx]$.

Finally, putting the subordinate quantifier into symbols along with the remaining bit of English within its scope, we have

(29') $(x)\{[Sx$ & $(\exists y)(Gy$ & $Cxy)] \supset Bx\}$

where 'Cxy', of course, means 'x carries y'.

Some readers may have felt inclined to read (29) as sayng that *any* suspect who carries *any* gun will be denied bail. Understood this way, the following translation naturally suggests itself:

(29″) $(x)(y)\{[Sx$ & $(Gy$ & $Cxy)] \supset Bx\}$.

(29″) is also a perfectly acceptable translation and thus is equivalent to (29'). But here the "y-quantifier" is universal and has its scope extending over both the antecedent and consequent of the component open sentence. That they are equivalent may be gleaned from a look at their English counterparts:

For any x, if x is a suspect and there is a gun y which x carries, then x will be denied bail
For any x and y, if x is a suspect and y a gun x carries, then x will be denied bail.

It is essential in (29″) that the y-quantifier have scope extending across the main connective, for the sentence

$(x)\{[Sx$ & $(y)(Gy$ & $Cxy)] \supset Bx\}$

is definitely *not* equivalent to (29') and (29″)—it says that if x is a suspect and *everything* whatever is a gun he carries then..., which is hardly the intended meaning.

(30), however, demands a translation along the lines of (29″) rather than (29′). As a first step, we have

(x)(x is an athlete who wins any contest ⊃ x trains for it),

since to say no athlete wins *unless* he/she trains is to say that every athlete who wins *does* train. Were we now to proceed as we did for (29), we could first get

(x)[(Ax & x wins any contest) ⊃ x trains for it],

and since 'it' in the consequent refers back to the antecedent, the same variable must be used in both places, thus giving

(x){[Ax & (∃y)(Cy & Wxy)] ⊃ Txy}.

But this cannot be right, since the occurrence of 'y' in the consequent is *free*—the scope of the y-quantifier extends to, but not past, the main connective. To bind this occurrence, the scope of the y-quantifier must of necessity extend over both antecedent and consequent; hence a proper translation must take the form of (29″) rather than (29′):

(30′) (x)(y){[Ax & (Cy & Wxy)] ⊃ Txy}.

Alternately, (30) may be equivalently read as 'For any athlete, any contest he/she wins is one for which he/she trains'. This most naturally goes into symbols as

(30″) (x){Ax ⊃ (y)[(Cy & Wxy) ⊃ Txy]}.

Let us now consider:

(31) A rolling stone gathers no moss

(32) Where there's smoke there's fire.

The translation of (31) is straightforward. Paraphrasing inward, we first get

(x)(x is a rolling stone ⊃ x gathers no moss).

Clearly the consequent contains a further quantifier signaled by 'no', and may be more clearly brought into focus by

$(x)(Rx \supset$ no moss is such that x gathers it$)$,

where 'Rx' is 'x is a rolling stone'. The rather stilted (but clearer) English in the consequent may now be replaced using 'Mx' and 'Gxy':

(31′) $(x)[Rx \supset (y)(My \supset {\sim}Gxy)]$.

(32) is also easily handled by the technique of paraphrasing inward. First, let us find a more perspicuous (though less snappy) English version. Presumably the claim is that any *place* where there is smoke is one at which there is fire as well, so our first step is

$(x)(x$ is a place where there is smoke $\supset$ there is fire at $x)$.

Plainly there is a quantifier in both the antecedent and consequent to be expressed, as is better seen by

$(x)[(Px$ & there is smoke at $x) \supset$ there is fire at $x]$.

So, using 'Axy' for 'x is at y' and other obvious notation, we have

(32′) $(x)\{[Px$ & $(\exists y)(Sy$ & $Ayx)] \supset (\exists z)(Fz$ & $Azx)\}$.

Note that two existential quantifiers are used here; smoke is one thing, fire another.

English idioms often conceal relational terms, and a correct translation often depends on exposing the hidden relational structure. For example, the sentence

(33) It's a sin to tell a lie

seems unproblematic at first blush, but telling a lie presupposes a teller and a hearer, neither of whom is explicitly mentioned. Thus (33) must first be reformulated to bring out the implicit triadic relation:

Anyone who tells a lie to someone (anyone) has sinned.

Thus we would have successively:

$(x)(x$ is a person who tells a lie to someone $\supset x$ has sinned)

$(x)[(Px$ & x tells a lie to someone) $\supset Sx]$

$(x)\{[Px$ & $(\exists y)(Ly$ & x tells y to someone)] $\supset Sx\}$

(33') $(x)\{\{Px$ & $(\exists y)[Ly$ & $(\exists z)(Pz$ & $Txyz)]\} \supset Sx\}$

where '$Txyz$' means 'x tells y to z'.

Sentences containing words like 'always', 'sometimes', 'never', 'whenever' usually have temporal significance and hence require an open sentence like 'x is a time'. For example,

(34) Barry always goes into a rage when George delivers a speech

means roughly 'Anytime George delivers a speech is one at which Barry goes into a rage'. So using 'Tx' for 'x is a time', '$Dxyz$' for 'x delivers y at z', 'Rxy' for 'x goes into a rage at y', and other obvious notation, (34) becomes

(34') $(x)\{[Tx$ & $(\exists y)(Sy$ & $Dgyx)] \supset Rbx\}$.

On the other hand, there are contexts where 'always' and the like do not carry temporal significance, for example:

(35) The cube of a positive integer is always a positive integer.

This really says no more than that every cube of a positive integer is itself a positive integer; hence no open sentence 'x is a time' is needed. Employing some obvious arithmetic notation, we have:

(35') $(x)(y)[(Ix$ & $y = x^3) \supset Iy]$.

The paraphrasing-inward technique is shown to its best advantage in translating highly complex sentences with several quantifiers:

(36) Any politician who votes against every reform bill will be defeated by a candidate who offers something to every minority group.

Digging out the main quantifier and connective, we have

(*x*)(*x* is a politician who votes against every reform bill ⊃ *x* will be defeated by a candidate who offers something to every minority group).

In the consequent, we have presumably the claim that *any* candidate who offers something to every minority group will defeat *x*; so our next step is:

(*x*){(*Px* & *x* votes against every reform bill) ⊃ (*y*)[(*Cy* & *y* offers something to every minority group) ⊃ *Dyx*]}.

Next, we put the remainder of the antecedent into symbols; it says that every reform bill is such that *x* votes against it.

(*x*){[*Px* & (*z*)(*Rz* ⊃ *Vxz*)] ⊃ (*y*)[(*Cy* & *y* offers something to every minority group) ⊃ *Dyx*]}.

Now the remaining English clause in the consequent contains an ambiguity. It might mean either of the following:

Something is such that *y* offers it to every minority group
Every minority group is such that *y* offers it something (or other).

Clearly the latter is the intended meaning; the sentence does not claim that the candidate will offer the same plum to minorities with quite different interests and needs. So where '*Oxyz*' means '*x* offers *y* to *z*', we finally have (resorting now to the extra variable '*x'*'):

(*x*){[*Px* & (*z*)(*Rz* ⊃ *Vxz*)] ⊃ (*y*){{*Cy* & (*w*)[*Mw* ⊃ (∃*x'*)(*Oyx'w*)]} ⊃ *Dyx*}}.

For all its complexity, however, what we have here is just a very long **A**-sentence, '(*x*)[...*x*... ⊃ ---*x*---]', where both antecedent and consequent are themselves quite intricate.

EXERCISES

I. For each sentence on the left, specify the sentence on the right that correctly translates it.

Px - x is a person Cx- x is a car Dxy - x drives y

1. Someone drives a car
2. Someone drives every car
3. Everyone drives some car (or other)
4. No one drives every car
5. No one drives any car
6. Not everyone drives a car

a. $(\exists x)[Px \,\&\, (y)(Cy \supset Dxy)]$
b. $(x)[Px \supset (\exists y)(Cy \,\&\, Dxy)]$
c. $(x)[Px \supset (\exists y)(Cy \,\&\, {\sim}Dxy)]$
d. $(\exists x)[Px \,\&\, (\exists y)(Cy \,\&\, Dxy)]$
e. $(\exists x)[Px \,\&\, (y)(Cy \supset {\sim}Dxy)]$
f. $(x)[Px \supset {\sim}(\exists y)(Cy \,\&\, Dxy)]$

II. Put each of the following into symbols, using the suggested notation.
1. For every problem, there's a solution. (Px-x is a problem; Sxy-x is solution to y.
2. Everybody likes something. (Px-x is a person; Lxy-x likes y.)
3. Somebody likes everything. (same as (2))
4. Each person pursues at least one goal in life. (Px-x is a person; Pxy-x pursues y; Gx-x is a goal in life.)
5. There is a goal in life that everyone pursues. (same as (4))
6. Nobody likes everything. (same as (2))
7. Nobody likes anything. (same as (2))
8. All of the applicants received an offer. (Ax-x is an applicant; Rxy-x received y; Ox-x is an offer.)
9. Some politicians can outtalk every salesman. (Px-x is a politician; Oxy-x can outtalk y; Sx-x is a salesman.)
10. Anyone who likes Betty likes everyone. (Px-x is a person; Lxy-x likes y; b - Betty.)
11. Anyone who listens to a prohibitionist is either mad or naive (Px-x is a person; Bx-x is a prohibitionist; Lxy-x listens to y; Mx-x is mad; Nx-x is naive.)
12. Some people don't care about anybody. (Px-x is a person; Cxy-x cares about y.)
13. Nothing ever changes. (Tx-x is a time; Cxy-x changes at y.)
14. Charlie likes anyone who flatters him, (c - Charlie; Lxy-x likes y; Px-x is a person; Fxy-x flatters y.)
15. Every student learns something from each of his teachers. (Sx-x is a student; Txy-x teaches y; $Lxyz$-x learns y from z.)
16. Anything not at rest perseveres in motion unless acted upon by an external force. (Rx-x is at rest; Px-x perseveres in motion; Axy-x acts upon y; Fx-x is an external force.)
17. People who keep everything they earn for themselves are neither generous nor prudent. (Px-x is a person; $Kxyz$-x keeps y for z; Gx-x is generous; Rx-x is prudent; Exy-x earns y.)

18. Delilah had a ring on every finger and had a finger in every pie. (*d* - Delilah; *Oxy-x* is on *y*; *Rx-x* is a ring; *Fxy-x* is a finger of *y*; *Px-x* is a pie; *Ixy-x* is in *y*.) .

19. Any man who hates dogs has W.C. Fields for a friend. (*Mx-x* is a man; *Hxy-x* hates *y*; *Dx-x* is a dog; *f*-Fields; *Fxy-x* has *y* for a friend.)

20. Any watch sold to anyone by Crumley is a stolen item. (*Wx-x* is a watch; *Sxyz-x* sells *y* to *z*; *c*-Crumley; *Px-x* is a person; *Lxyz-x* stole *y* from *z*..)

III. Now do the same for these:

1. If any system malfunctions, nobody will be able to repair it unless some new parts are ordered. (*Sx-x* is a system; *Mx-x* malfunctions; *Px-x* is a person; *Rxy-x* will be able to repair *y*; *Nx-x* is a new part; *Ox-x* is ordered.)

2. If every system malfunctions, neither technicians nor trouble-shooters can make any repairs without full authorization. (*Sx-x* is a system; *Mx-x* malfunctions; *Tx-x* is a technician; *Bx-x* is a trouble-shooter; *Rx-x* is a repair; *Cxy-x* can make *y*; *Ax-x* has full authori-·zation.)

3. Valid arguments are derivable only from valid argument forms. (*Vx-x* is a valid argument; *Dxy-x* is derivable from *y*; *Fx-x* is a valid argument form.)

4. Only valid arguments are derivable from valid argument forms. (Same notation as (3).)

5. Nothing ventured, nothing gained. (*Vxy-x* ventures *y*; *Gxy-x* gains *y*.)

6. There is a politician who is despised by all citizens who despise any politician at all. (*Px-x* is a politician; *Dxy-x* despises *y*; *Cx-x* is a citizen.)

7. A watched pot never boils. (*Px-x* is a pot; *Wxyz-x* watches *y* at *z*; *Bxy-x* boils at *y*; *Tx-x* is a time.)

8. There is a prime number between 14 and 18 greater than any number that is a square root of a number less than 300. (*Px-x* is prime; *Nx-x* is a number; *f*-14; *e*-18; $x > y$-*x* is greater than *y*; $x = \sqrt{y}$-*x* is a square root of *y*; $x < y$-*x* is less than *y*; *Bxyz-x* is between *y* and *z*; *a* - 300.) By the way, is (8) true?

9. A politician who appeases every pressure group evades every issue with a platitude. (*Px-x* is a politician; *Axy-x* appeases *y*; *Gx-x* is a pressure group; *Ix-x* is an issue; *Exyz-x* evades *y* with *z*; *Tx-x* is a platitude.)

10. It's an ill wind that blows no good to anyone. (*Ix-x* is ill; *Wx-x* is a wind; *Bxyz-x* blows *y* to *z*; *Gx-x* is a good; *Px-x* is a person.)

IV. Let us add to our notation some standard mathematical symbols, and suppose our universe of discourse to be restricted to the series of positive integers: 1, 2, 3, *ad infinitum*. Which of the following are true and which false? Explain briefly in each case.

1. $(x)(\exists y)(y > x)$
2. $(x)(\exists y)(x > y)$
3. $(\exists y)(x)(y > x)$
4. $(\exists y)(x)(y \neq x \supset y > x)$
5. $(\exists x)(y)(y \neq x \supset y > x)$

In each of the following '*Bxyz*' means '*x* is between *y* and *z*'.

6. $(\exists x)(\exists y)(\exists z)(Bxyz)$
7. $(x)(\exists y)(\exists z)(Bxyz)$
8. $(y)(\exists x)(\exists z)(y \neq 1 \supset Byxz)$
9. $(\exists y)(x)(\exists z)(x \neq y \supset Bxyz)$
10. $(x)(y)(z)[Bxyz \equiv (z < x \ \& \ x < y)]$

V. The following remark, often attributed to Lincoln, is ambiguous:
    You can fool some of the people all of the time.
    State first in English and then in symbols the two different meanings. Be sure to specify your notation.

VI. Let the universe of discourse again be the set of positive integers. Match the following English sentences with their correct translations.

1. There is a greatest positive integer.
2. There is no greatest positive integer.   a. $(x)(\exists y)(y > x)$
3. There is a positive integer than which none is greater.   b. $(\exists x)(y)(y \not> x)$
       c. $\sim(\exists x)(y)(x > y)$
       d. $(\exists x)(y)(y \neq x \supset x > y)$
4. For each positive integer, there is a positive integer than which it is not greater.

*5.3 Identity and Definite Descriptions

Among dyadic relations, the most important for our purposes is that of **identity**. We shall abbreviate the expression 'x is identical with y' not by 'Ixy' but in the more customary fashion '$x = y$' (we shall also write its negation as '$x \neq y$' instead of '$\sim x = y$'). Identity explicates one of the three important uses of the verb 'to be'. The other two have already been accounted for and are mentioned here briefly. First we have the *existential* sense of 'is', as in

There is a man whom all women admire.

Clearly 'is' here has the force of 'exists' and is handled by means of the existential quantifier. Second, there is the *predicative* (or *copulative*) sense of 'is' — the role it plays, for example, in a singular sentence where a property is ascribed to an individual:

Aristotle is wise.

The verb 'is' here plays no essential role as far as logic is concerned — it is an integral part of the predicate schema '... is wise', and no special sign is needed for it in our notation. Hence the above is simply translated 'Wa'.

The third sense is that of identity; 'is' here means 'is identical with'. In ordinary discourse, the 'is' of identity commonly has two singular terms flanking it. Singular sentences of this form will thus be true just in case the very same individual is designated or referred to by both terms. Examples would be:

(1) Eric Blair is George Orwell

(2) Muhammad Ali is Cassius Clay

(3) Taiwan is Formosa

(4) Bonn is the capital of West Germany

(5) The author of *Profiles in Courage* is the 35th President of the U.S.

(6) The greatest heavyweight champion is Joe Louis

(7) The product of 4 and 16 is the cube root of 262,144

Each of these is true because its component singular terms refer to one and the same individual — which is *not* to say that the sentence is about the component *terms*. (1) is no more *about* the terms 'Eric Blair' and 'George Orwell' than 'Boston is a city' is about the term 'Boston'; rather, the subject matter of (1) is the co-designated man who, for example, wrote the novel *1984*. Moreover, though each sentence in the above list is true, it should not be inferred that the component terms have the same informational content. Clearly we understand different things by the two definite descriptions in (7), though with some minimal arithmetic knowledge it is easily determined that those descriptions have the same number as their reference and hence that (7) is true. It is in this way that an identity sentence may be true—an individual thus being designated twice over—without thereby being trivial, as is the case with the uninformative '64 = 64'. Though one and the same number is referred to by 'the product of 4 and 16' and by 'the cube root of 262,144', it is specified or singled out differently owing to the lack of synonymy between the descriptions. Or again, while every fluent English speaker understands the terms 'the Morning Star' and 'the Evening Star', those with little or no knowledge of astronomy are unaware that they refer to the same celestial body, and hence that the sentence

The Morning Star is the Evening Star

is true.

The identity relation exhibits three important properties worth noting. First, a dyadic relation—here represented by 'Rxy'—is said to be **symmetric** if and only if the following holds of it:

$$(x)(y)(Rxy \supset Ryx).$$

For example, '...is married to---' expresses a symmetric relation but '...is husband of---' does not. Identity is obviously symmetric, for given any x and y, if $x = y$ then $y = x$. A relation is said to be **transitive** if and only if:

$$(x)(y)(z)[(Rxy \ \& \ Ryz) \supset Rxz].$$

The relation represented by '...is larger than---' is transitive, but that by '...is mother of---' is not. Identity is also transitive, since if $x = y$ and $y = z$ then $x = z$. Finally, a relation is said to be **reflexive** if and only if

$$(x)(Rxx),$$

that is, every individual bears R to itself. Though identity is reflexive (every substitution instance of '$x = x$' is true and necessarily so), the relation expressed by '. . . is taller than ---' clearly is not.[8]

Sentences (1), (2) and (3) all contain proper names flanking 'is', and using obvious notation are expressed:

(1') $e = g$

(2') $m = c$

(3') $t = f$.

Sentences (4) through (7), however, each contain at least one definite description. We have yet to discuss the translation of such terms, and shall temporarily put off further discussion of them.

Identity enables us to express many arithmetical sentences that heretofore could not be handled, for example

(8) There are at least two spies in the room.

Now, we know that 'There is at least one spy in the room' is expressed simply as '$(\exists x)(Sx)$', so it might be thought that (8) is adequately rendered as

$(\exists x)(\exists y)(Sx \ \& \ Sy)$.

---

[8]Among relations which are *not* symmetric some are **asymmetric,** others not. A relation is asymmetric if and only if:

$(x)(y)(Rxy \supset \sim Ryx)$.

Thus '. . . is husband of ---' expresses an asymmetric relation, but '. . . loves ---' expresses one which is neither symmetric nor asymmetric. Similarly, there are relations that are merely *not* transitive and those that are **intransitive**:

$(x)(y)(z)[(Rxy \ \& \ Ryz) \supset \sim Rxz]$

as may be seen by contrasting '. . . loves ---' with '. . . is mother of---'. Likewise, we may contrast relations that are merely not reflexive (again '. . . loves ---') with those which are **irreflexive** (for example, '. . . is more important than ---'):

$(x)(\sim Rxx)$.

However, this will not do, for suppose that in fact only one spy, whom we shall designate as 'a', is in the room; then (8) is *false* but the above is *true* since the substitution instance '$Sa \ \& \ Sa$' is true. This sentence of our notation, then, does not provide the desired translation, since it allows that x and y might be the same individual. To remove this possibility, it is necessary to add the further conjunct '$x \neq y$', so that (8) becomes

(8') $(\exists x)(\exists y)(Sx \ \& \ Sy \ \& \ x \neq y).$[9]

By the same token, the sentence

(9)   There are at least three spies in the room,

or in other words 'There are three or more spies in the room', may be expressed in symbols as

(9') $(\exists x)(\exists y)(\exists z)(Sx \ \& \ Sy \ \& \ Sz \ \& \ x \neq y \ \& \ y \neq z \ \& \ x \neq z).$

Sentences asserting the existence of at least four, five, and more spies may be handled in similar fashion.

Now consider the sentence

(10)   There is a senator who can outtalk everyone.

A moment's reflection will show that

$(\exists x)[Sx \ \& \ (y)(Py \supset Oxy)]$

does not do (10) full justice as it is normally understood, for if *every* person is outtalked by the senator then he outtalks himself as well, which is hardly plausible even for senators. The force of 'everyone' in (10) is that of 'everyone *else*', and hence the appropriate translation would be

(10') $(\exists x)\{Sx \ \& \ (y)[(Py \ \& \ y \neq x) \supset Oxy]\}.$

We have already considered many types of sentence containing the

---

[9]For brevity, we sometimes drop interior parentheses when no ambiguity results.

troublesome word 'only'; yet a further example is

(11) Al is the only spy in the room.

(11) is best understood as claiming, first, that Al is a spy in the room, and second, nothing else is. Thus we have the conjunction

(11′) Sa & $(x)(x \neq a \supset {\sim}Sx)$.

Sentences containing the expressions 'at most' and 'no more than' may also be expressed using the identity sign. Consider:

(12) No more than one spy is in the room

(13) At most two spies are in the room.

Both sentences would generally be understood as *not* claiming the existence of at least one spy despite the fact that they may be reformulated to read, 'There is (are) at most…'. In (12), the claim amounts to: *if* there is a spy in the room, he's the only one. So (12) becomes

(12′) $(x)(y)[(Sx \& Sy) \supset x = y]$.

Similarly, (13) is expressed

(13′) $(x)(y)(z)[(Sx \& Sy \& Sz) \supset (x = y \lor y = z \lor x = z)]$.

In other words, if x, y, and z are all spies, then at least "two" of them are the same.

Now that we have obtained translations of (8) and (13), we are able to handle:

(14) There are two spies in the room
      Exactly two spies are in the room.

To say there are *exactly* two is to say there are at least two *and* at most two. So (14) may be translated using the tools provided by (8) and (13):

(14′) $(\exists x)(\exists y)\{Sx \& Sy \& x \neq y \& (z)[Sz \supset (z = x \lor z = y)]\}$.

The last conjunct in effect ensures that the distinct spies x and y are the *only* ones. A less complicated example would be

(15) There is exactly one deity,

which accordingly becomes

(15′) $(\exists x)[Dx \ \& \ (y)(Dy \supset y = x)]$.

EXERCISES

Put each of the following into symbols using the identity sign.
1. At least one god exists. (Gx-x is a god.)
2. There are two gods at most. (Same as (1).)
3. No more than three gods exist. (Same as (1).)
4. There are at least two satellites of Mars. (Sxy-x is a satellite of y; m-Mars.)
5. No more than two suspects were indicted. (Sx-x is a suspect; Ix-x was indicted.)
6. New York is the only American city larger than Los Angeles. (n-New York; Ax-x is an American city; Lxy-x is larger than y; l-Los Angeles.)
7. New York is larger than all (other) American cities. (Same as (6).)
8. Gerry, Dave and at least two more suspects were indicted. (g-Gerry; d-Dave; Sx-x is a suspect; Ix-x was indicted.)
9. Cal and Bill were the only suspects who confessed. (c-Cal; b-Bill; Sx-x is a suspect; Cx-x confessed.)
10. There is at least one but no more than two inferior planets. (Px-x is an inferior planet.)
11. There are exactly two senators representing each state. (Sx-x is a senator; Rxy-x represents y; Tx-x is a state.)
12. *Elementary Symbolic Logic* is co-authored. (e-*Elementary Symbolic Logic*; Axy-x authored y.)

Definite descriptions were briefly introduced in Section 5.1 as a variety of singular term in addition to that of proper names. Instead of naming the designated individual, such a term uniquely describes it. Thus in (4) above, 'the capital of West Germany' is a definite description, and the use of 'the' in the singular is characteristic of such expressions (though they are sometimes abbreviated in idiomatic speech to the point where 'the' disappears, for example 'Al's father'). When 'the' is not used in the singular, the descriptive phrase is not definite — it does not purport to refer descriptively to just *one* individual. Thus, 'the men in the next room' is not a definite description, but in 'the capital of West Germany' we have an expression designating the one and only one individual truly described as 'capital of West Germany'.[10]

Since it has a descriptive dimension, any definite description must have at least one general term and hence may be analyzed in terms of an open sentence. Thus, for 'the capital of West Germany', we have 'x is capital of West Germany', and the use of 'the' carries with it the presumption of exactly one individual so described. We may thus reformulate the description as

the (individual) x such that x is capital of West Germany

and we shall abbreviate the phrase 'the x such that ...' by attaching an inverted iota to 'x' and enclosing both in parentheses, thus giving '$(\imath x)$'. We shall call this sign the **iota operator**, and attaching it to an open sentence thus provides us with a means of translating definite descriptions into our notation. So using 'Cx' for the above open sentence, the description is rendered

$(\imath x)(Cx)$.

The iota operator is *not* a quantifier — when attached to an open sentence the result is not a sentence true or false, but a descriptive singular *term*.

We are now in a position to translate (4)-(7) at the beginning of this section.

(4) Bonn is the capital of West Germany

(5) The author of *Profiles in Courage* is the 35th president of the U.S.

(6) The greatest heavyweight champion is Muhammad Ali

(7) The product of 4 and 16 is the cube root of 262,144

---

[10]It should be noted that there are expressions that grammatically resemble definite descriptions but are best construed as proper names, such as 'the Windy City', 'the Sunshine State', 'the Empire State Building'. Here, the use of capital letters indicates that the designated individuals are not being uniquely described as such, but are named — albeit through the use of terms felt to be especially characteristic of their bearers ('windy' of Chicago, 'sunshine' of Florida). In fact, such expressions often serve as nicknames.

Employing obvious abbreviations, (4) becomes:

(4') $b = (\imath x)(Cx)$.

Using 'Wx' for 'x · wrote *Profiles in Courage*' and 'Px' for 'x is 35th President of the U.S.', (5) becomes:

(5'.) $(\imath x)(Wx) = (\imath y)(Py)$

Now (6) contains the superlative term 'greatest' and such terms may be handled as follows: in speaking of a greatest heavyweight champion, x, we mean that x is a heavyweight champion greater than any *other* such champion, i.e. '$Hx \,\&\, (y)[(Hy \,\&\, y \neq x) \supset Gxy]$'. So we may express (6) as

(6') $(\imath x)\{Hx \,\&\, (y)[(Hy \,\&\, y \neq x) \supset Gxy]\} = j$.

Importing now some arithmetic vocabulary, (7) may be expressed

(7') $(\imath x)(x = 4 \times 16) = (\imath y)(y = \sqrt[3]{262{,}144})$.

A description, like a constant, may replace a free variable in an open sentence to yield a true-or-false sentence as substitution instance. Thus, where 'a' is 'Aristotle' and '$(\imath x)(Axd)$' is 'the author of *De Anima*', both terms may replace the 'x' of 'Px' ('x is a philosopher') thus giving

Aristotle is a philosopher Pa
The author of *De Anima* is a philosopher $P(\imath x)(Axd)$

Thus, singular descriptions of the form '$(\imath x)(\ldots x \ldots)$' would seem to behave no differently from the individual constants of our symbolic notation. Yet there is an important disparity that requires us to push further our analysis of sentences containing descriptions. Our use of individual constants thus far has proceeded on the supposition that in any context where it occurs, a constant always designates a particular individual of the presumed universe of discourse. However, this supposition cannot be extended to the singular descriptions of our notation, for it should be apparent that we can form at will such descriptions that fail to designate or refer to any individual at all. For example:

($\imath x$)(x hit 65 home runs in one major league season)
($\imath x$)(x hit 55 home runs in one major league season)

The first fails to refer because *no* individual has hit 65 home runs, the second because *more than one* has hit 55 home runs. We can easily multiply such examples:

($\imath x$)(x is a man who has run a two-minute mile)
($\imath x$)(x is monarch of France)
($\imath x$)(x has squared the circle)
($\imath x$)(x is a senator from Indiana)

We are thus left with a problem concerning the analysis of sentences with singular descriptions, for we must now face the possibility of reference failure. Keeping this in mind, how are we to analyze such sentences? In particular, under what conditions will they be true and under what false? Let us consider a sentence with a nonreferring description:

(16)　The man who has run a two-minute mile is bald,

or in other words, where 'Tx' is 'x is a man who has run a two-minute mile' and 'Bx' is 'x is bald':

(16')　$B(\imath x)(Tx)$.

One answer to the above questions is to maintain that such a sentence has no truth-value, since the presupposition of unique reference is not satisfied. This accords well with our intuitions, for if someone were (seriously) to assert (16) our response would very likely take the form of denying that there is such a man, *not* of claiming that (16) is false — which would carry the suggestion that there *is* such a man but that he is not bald. To say that (16) has no truth-value is not, according to this view, to say that it has no meaning. Rather, it makes perfectly good sense in that it is easily understood by any fluent speaker of English. The claim is that for (16) — or (16') — to be considered a candidate for *either* truth *or* falsity a certain presupposition must be satisfied:

(17)　There is exactly one man who has run a two-minute mile.

(17')　$(\exists x)[Tx \mathbin{\&} (y)(Ty \supset y = x)]$.

Since, however, (17) is false, the question of the truth-value of (16) does not arise. Only if (17) is itself true will (16) be considered true or false— true if the designated individual *is* bald, false if he is not.[11]

However tempting this approach may be, we shall not pursue it any further here. Natural though it is to regard sentences with nonreferring descriptions as lacking a truth-value, the practice has serious drawbacks with respect to the evaluation of arguments containing such sentences in our notation. Our accounts of validity, equivalence, and the like clearly hinge on sentences having a truth-value on *all* interpretations, and were we now to remove this condition, new and highly complex definitions would be required—definitions the adequacy of which is still a matter of intense philosophic controversy. We will do better, then, if we can find an account of sentences with definite descriptions that, first, ascribes to them a truth-value on all interpretations, and second, accords reasonably well with our intuitions concerning their use in ordinary discourse and their behavior in deductive inference.

Such an approach has been offered by Bertrand Russell and has long had wide acceptance.[12] On this view, (17) is not a presupposition for (16) having a truth-*value*, but simply for its being true—(17) in effect gives us part of the truth-conditions for (16). On Russell's view, then, (16) *asserts* the existence of exactly one man who has run a two-minute mile instead of presupposing it as a condition for having a truth-value. More fully, Russell claims (16) will be true if and only if:

i. at least one man has run a two-minute mile,
ii. at most one man has run a two-minute mile, and
iii. that man is bald.

Hence, according to Russell, (16) is *false*, since it has the same truth-conditions as the existentially quantified sentence

---

[11]For more on this approach, see Gottlob Frege, "On Sense and Reference," in P. Geach and M. Black (Eds.), *Philosophical Writings of Gottlob Frege*, Oxford: Blackwell, 1952. In his formal language, however, Frege simply assigns a reference to what would otherwise be nonreferring descriptions. A different though related approach may be found in P. F. Strawson, "On Referring," in A. Flew (Ed.), *Essays in Conceptual Analysis*, London: Macmillan, 1956. Both essays are classics and have been reprinted in many other anthologies on logic and language.
[12]Bertrand Russell, "On Denoting," *Mind*, vol. 14(1905), pp. 479-493. Reprinted in many places, among them: H. Feigl and W. Sellars (Eds.), *Readings in Philosophical Analysis*, New York: Appleton-Century-Crofts, 1949; and I. Copi and J. Gould (Eds.), *Contemporary Readings in Logical Theory*, New York: Macmillan, 1967 (the latter contains the Frege and Strawson essays as well). See also Russell's *Introduction to Mathematical Philosophy*, London: Allen and Unwin, 1920, Chapter 16, and his *Principia Mathematica* (with A. N. Whitehead), vol. I, Cambridge: Cambridge Univ. Press, 1910.

(16″) $(\exists x)[Tx$ & $(y)(Ty \supset y = x)$ & $Bx]$.

More generally, a sentence containing a definite description and therefore one that has been provisionally expressed in our notation as being of the form

$--- (\imath x)(...x...) ---$

will hereafter be regarded as equivalent to one of the form

$(\exists x)[...x... $ & $(y)(...y... \supset y = x)$ & $---x---]$.

On this analysis, a sentence purporting to uniquely describe an individual and ascribe a property to it will be false when either:

(a) the description fails to refer (either because there is no individual so described or because there are two or more), or
(b) the description does refer but the individual lacks the property in question.

Thus (16) would also be false, according to (b), if there were a man who has run a two-minute mile but who is not bald. Following Russell, then, we shall translate (16) *and* (16′) as (16″), thus dispensing with the iota operator and expressions of the form '$(\imath x)(...x...)$' for the balance of our discussion. Notice that upon adopting this course, singular descriptions will no longer occur in our notation. Sentences like (16) and (16′) are now reexpressed in such a way that there is no *one* expression or phrase in their translation—as in (16″)—that replaces the description; instead, the entire sentence does the job. Russell describes this as the *contextual elimination* of definite descriptions; they are not defined away in the sense in which, say, 'bachelor' is by 'unmarried man', but rather defined within the sentential context as a whole. In going from (16) or (16′) to (16″), the description 'the man who has run a two-minute mile' ('$(\imath x)(Tx)$') simply disappears, its role taken over by the entirety of (16″).

Let us now consider the ostensible negation of (16) and of (16′):

(18)   The man who has run a two-minute mile is not bald.

(18′) $\sim B(\imath x)(Tx)$.

Since (16) and (16') can each be *false* in two different ways, clearly (18) and (18') can be *true* in two different ways—either if (a) holds or if (b) holds. Since (a) and (b) are distinct conditions for truth, it follows that (18) and (18') on Russell's theory can both be understood in two different ways. Understood one way, they will be true if and only if there *is* just one man who has run a two-minute mile but who is *not* bald, and on this reading the descriptions 'the man who has run a two-minute mile' and '$(\imath x)(Tx)$' are said to have **primary occurrence** in (18) and (18') respectively. Understood the other way, they will be true if no man or more than one man has run a two-minute mile, and on this second reading the descriptions are said to have **secondary occurrence** in (18) and (18'). The expressions (18) and (18') must therefore, according to Russell, be regarded as ambiguous. If the component descriptions are taken to have primary occurrence, (18) and (18') are both symbolized

(18p) $(\exists x)[Tx \; \& \; (y)(Ty \supset y = x) \; \& \sim Bx]$.

When the descriptions in (18) and (18') are taken as having primary occurrence, then a condition for the sentence's being true is that there *be* an individual referred to. When the descriptions are regarded as having secondary occurrence, however, the sentences may be true even if there is no individual designated by those descriptions. Hence, a quite different translation is required when the descriptions have secondary occurrence. Here, (18) and (18') must be understood to mean

(18s) $\sim(\exists x)[Tx \; \& \; (y)(Ty \supset y = x) \; \& \; Bx]$.

We may now translate (4), (5), and (6) in accordance with Russell's technique. (4) says that there is exactly one x which is capital of West Germany and is (identical with) Bonn. Thus we have:

(4") $(\exists x)[Cx \; \& \; (y)(Cy \supset y = x) \; \& \; x = b]$

(5") $(\exists x)(\exists y)[Wx \; \& \; (z)(Wz \supset z = x) \; \& \; Py \; \& \; (w)(Pw \supset w = y) \; \& \; x = y]$

In (5"), the first two conjuncts say in effect that exactly one individual wrote *Profiles in Courage* and the second pair say that exactly one was president of the U.S. The fifth conjunct ensures their identity. (6) now becomes

(6") $(\exists x)\{Hx \; \& \; (y)[(Hy \; \& \; y \neq x) \supset Gxy] \; \& \; x = a\}$.

(7) is left as an exercise for the patient reader.

EXERCISES

I. Put each of the following in symbols using the suggested notation. Use Russell's method on those containing definite descriptions. The identity sign is required for all.

1. The author of *Huckleberry Finn* is a humorist. (*Axy-x* authored *y*; *f-Huckleberry Finn*; *Hx-x* is a humorist.)

2. The author of *Huckleberry Finn* is Mark Twain. (*Axy-x* authored *y*; *f*-Huckleberry Finn; *m*-Mark Twain.)

3. Reagan's greatest triumph is *Bedtime for Bonzo*. (*r*-Reagan; *Gxy-x* is greater than *y*; *Txy-x* is a triumph for *y*; *b-Bedtime for Bonzo*.)

4. The man who defeated Dempsey fought only heavyweights. (*Mx-x* is a man; *Dxy-x* defeated *y*; *d*-Dempsey; *Fxy-x* fought *y*; *Hx-x* is a heavyweight.)

5. Nelson's most recent speech was untelevised. (*n*-Nelson; *Sxy-x* is a speech given by *y*; *Rxy-x* is more recent than *y*; *Tx-x* was televised.)

6. The only man who hated every child is W. C. Fields. (*Mx-x* is a man; *Hxy-x* hates *y*; *Cx-x* is a child; *f*-W. C. Fields.)

7. Only the best applicant will be hired. (*Ax-x* is an applicant; *Bxy-x* is better than *y*; *Hx-x* will be hired.)

8. Everybody loves my baby. (*Px-x* is a person; *Lxy-x* loves *y*; *Bx-x* is a baby; *Lx-x* belongs to me.)

9. Only Audrey hit Frank on his arm. (*a*-Audrey; *f*-Frank; *Axy-x* is an arm of *y*'s; *Hxyz-x* hit *y* on *z*.)

10. Audrey hit only Frank on his arm. (Same notation as (9).)

11. Audrey hit Frank on his only arm. (Same notation as (9).)

*12. For each number *x*, $\log_2(x)$ = the number *y* such that $2^y = x$. (*Nx* − *x* is a number; *Lxy* − *x* = $\log_2(y)$; *Rxy* − 2^x = *y*.)

*II. Express

I want a girl just like the girl that married dear old Dad

such that, first, the description 'the girl that married dear old Dad' has primary occurrence, and second, has secondary occurrence. (*Gx-x* is a girl; *Wx*-I want *x*; *Lxy-x* is just like *y*; *Mxy-x* married *y*; *d*-dear old Dad.)

CHAPTER SIX

QUANTIFICATIONAL LOGIC
Semantics

6.1 Interpretations and Invalidity

In the course of our discussion we have discovered some valid argument forms and several equivalence principles. While we did not *prove* validity or equivalence in these cases, they were sufficiently obvious to be grasped intuitively. We know, for example, that each of the following arguments are valid, although when the premise and conclusion are interchanged, the resulting arguments are invalid in each case:

$(\exists x)(Fx \ \& \ Gx)$	$(x)(Fx) \lor (y)(Gy)$	$(\exists y)(x)(Cyx)$
$\therefore (\exists x)(Fx) \ \& \ (\exists y)(Gy)$	$\therefore (x)(Fx \lor Gx)$	$\therefore (x)(\exists y)(Cyx)$

Moreover, each of the sentence pairs below were seen to be equivalent in Chapter 5:

$\sim (\exists x)(Fx)$	and	$(x)(\sim Fx)$
$(\exists x)(\sim Fx)$	and	$\sim (x)(Fx)$
$(\exists x)(Fx) \supset (\exists y)(Gy)$	and	$(x)[Fx \supset (\exists y)(Gy)]$

Further, it is natural to suppose that there are sentences of quantificational logic that are analogues of the tautologies of truth-functional logic, that is, sentences that are true under all conditions (though we have yet to say just what this phrase comes to for sentences containing quantifiers). Thus, consider the sentence

(1) $(x)(Fx \lor \sim Fx)$.

Clearly this will be true no matter what property we take the predicate letter to represent and no matter what individuals make up our universe of discourse. Thus (1) contrasts sharply with, say, '$(x)(Fx \supset Gx)$', which comes out *true* when 'F' is taken to represent the property of being a man, 'G' represents the property of being mortal, and our universe of discourse is the set of human beings. But it comes out *false* for that same universe when 'F' represents being mortal and 'G' being a man (if 'a' designates a specific woman, then '$Fa \supset Ga$' is false in this latter case). We shall call sentences that, like (1), are true under all conditions **logical truths**. Further examples are:

(2) $(x)[(y)(Fy) \supset Fx]$

(3) $(x)[(\exists y)(\sim Gxy) \supset \sim (z)(Gxz)]$

(4) $\sim (\exists x)(Fx) \equiv (y)(\sim Fy)$

(5) $(\exists x)(Fx \ \& \ Gx) \supset [(\exists y)(Fy) \ \& \ (\exists z)(Gz)]$.

Again, no matter what universe of discourse is assumed nor which properties the predicate letters represent, (2) through (5) come out true.[1]

Now two main tasks lie before us. First, we must precisely *define* the concepts of logical truth, validity, and equivalence for quantificational logic, just as we gave parallel definitions in Chapter 2 of similar notions for truth-functional logic. Second, we must develop a method for *establishing* logical truth, validity, and equivalence — one that will serve the same purpose as the methods of deduction and truth-tables served for truth-functional sentences. The second job is left for Chapter 7; the first is handled in the present section.

These are quite separate tasks. A proper characterization of validity does not automatically yield a method for establishing it. However, as we shall soon see, it easily yields methods for establishing *invalidity* (as well as *nonequivalence*, and so on). This is important, for the method developed in Chapter 7 is an extension of the method of deduction introduced in Chapter 3, and it is therefore a method for establishing validity, equiv-

---

[1] It should be plain from the above that what has been called a "sentence" of our symbolic notation, such as '$(x)(Fx \supset Gx)$', differs in an important respect from a sentence of English like 'All men are mortal'. For while the latter has a fixed meaning, the former does not. The letters 'F' and 'G' can represent *different* properties in different contexts, as the foregoing example shows, so that in *one* context '$(x)(Fx \supset Gx)$' means in our notation what is (permanently) meant in English by 'All men are mortal', and in another what is meant by 'All mortals are men'. Whether a sentence of our notation is true or false in a given context depends upon which properties the predicate letters have been taken to represent — only then do we know which English sentence it abbreviates in that context. In calling a sentence like (1) a logical truth, we mean that every English sentence it can abbreviate is true in any universe of discourse, for example, 'Everything is either red or not red'.

alence, and logical truth *only*—not for their respective opposites.[2] Unfortunately, in quantificational logic with relations and multiple quantifiers, there is no *mechanical* routine analogous to the truth-tables of sentential logic for determining *both* validity and invalidity, equivalence and nonequivalence, and so on. This is not to say merely that no such method has thus far been discovered; rather, it has been demonstrated that there is no such method to be found.[3] Methods such as those developed in Chapter 7, then, are all we have for showing validity, equivalence, and logical truth, and the decidedly nonmechanical methods to be presented here must suffice for their opposite numbers.

We can perhaps best develop the desired accounts of validity, equivalence, and logical truth by first considering how invalidity of quantificational arguments may be established. Let us return to an invalid argument noted in the preceding chapter:

(I) $(x)(\exists y)(Cyx)$
   $\therefore (\exists y)(x)(Cyx)$.

We interpreted 'Cyx' as 'y causes x', and on this interpretation the argument becomes:

(A) Everything has a cause
   Therefore something causes everything.

But of course there are many arguments in English that are of exactly the same form and hence may also be represented in our notation by (I). For example, if 'Cyx' is now interpreted as 'y creates x', we have:

(B) Every being was created by some being
   Therefore some being created every being.

Now if we restrict our universe of discourse to persons—thus eliminating the need for 'Px'—and interpret 'Cyx' as 'y is an acquaintance of x', (I) symbolizes the argument:

---

[2]More precisely, the method allows one to establish a sentence as a logical truth by establishing it as a theorem (deducible from no premises) and to establish two sentences as equivalent by showing them interdeducible with one another.
[3]See Alonzo Church, "A Note on the Entscheidungsproblem," *Journal of Symbolic Logic,* vol. 1 (1936), pp. 40-41.

(C) Everyone has someone for an acquaintance
    Therefore someone is an acquaintance of everyone.

Now suppose someone stubbornly refuses to acknowledge the invalidity of arguments (A) and (B). We might then offer (C) to show their invalidity, for as their mutual translation by means of (I) reveals, (C) is of the same form as the other two and it is far more obvious that the premise is true and the conclusion false. Surely everyone is such that someone or other is an acquaintance of his, but it is, to say the least, implausible that someone is an acquaintance of every single person. This technique is usually called *refutation by logical analogy* and is commonly employed in practical debate; one attempts to refute his opponent's argument by citing an argument with the same form having obviously true premise(s) and false conclusion. Since the latter argument is invalid, then of course any argument of exactly similar form is invalid as well.

Suppose, however, that our antagonist deftly points out that as 'acquaintance' is ordinarily understood, newborn babies can hardly be said to have acquaintances at all. They therefore provide the needed counterexample for showing the premise of (C) to be false. We must then respond by producing an argument whose invalidity is more blatantly obvious even than (C), and it is not hard to find. Let us this time restrict the universe of discourse to the natural numbers, 0, 1, 2, 3,... and interpret 'Cyx' as 'y is greater than x'. Then (I) translates the argument:

(D) For every number there is a number greater than it
    Therefore some number is greater than every number.

Clearly the premise is true, for no matter what number, n, we pick, there is another—its successor $n + 1$—that is greater. But the conclusion is false—no number is greater than *all* numbers, since among other things no number can be greater than *itself.*

We can achieve the same end by considering a much smaller universe of discourse, one with just two individuals: Mount Everest and Mount Rushmore. Interpreting 'Cyx' as 'y differs in size from x', (I) now symbolizes:

(E) Everything differs in size from something or other
    Therefore something differs in size from everything.

Since Everest and Rushmore do in fact differ, the premise is true, and since neither differs in size from itself, the conclusion is false.

Let us now take stock of what we have done here. We showed (A) invalid by logical analogy with arguments (D) and (E), for they have the same form as (A) and in addition an obviously true premise with obviously false conclusion. More importantly, we showed argument (I) of our symbolic notation to be invalid by first specifying a universe of discourse and then assigning a dyadic relation to the letter '*C*' such that when (I) was interpreted in this way, we obtained a true premise and false conclusion. Each of (A) through (E) may be regarded as the result of *interpreting* (I) in different ways; hence to show (I) invalid, we need only find at least one interpretation yielding a true premise and false conclusion. To show an argument *in English* invalid, we first put it in symbols. It then remains to produce an interpretation — in effect, another English argument of that form — wherein all premises' are clearly true and the conclusion is just as clearly false.

This notion of "interpreting" one or more sentences of our notation is essential for explicating the concepts of validity, equivalence, and logical truth as they pertain to quantificational logic. Let $\mathscr{T}$ be any sentence of our notation. To give an **interpretation** of $\mathscr{T}$, one must first *specify a universe of discourse over which the variables of $\mathscr{T}$ are to range* (if any individual constant occurs in $\mathscr{T}$ it is necessary also to specify which individual of that universe the constant designates), and second, *assign a property or relation* (as the case may be) *to each distinct predicate letter occurring in $\mathscr{T}$*.[4] Note that an interpretation as defined here is the analogue in quantificational logic of a *truth-table row* in the sentential logic of

---

[4]We shall for convenience extend the use of 'predicate letter' to include those letters that represent *relations* in a given context. We are now in a position to define recursively a quantificational sentence of our notation:

  i. Each predicate letter followed by a sequence of constants is a sentence (for example, '*Fa*', '*Bab*', '*Ljct*');
  ii. The result of prefixing '~' to a sentence is a sentence;
  iii. The result of inserting '&', 'v', '⊃', or '≡' between two sentences, enclosing the result in parentheses, is a sentence;
  iv. The result of replacing one or more constants in a sentence with some variable μ ('*x*', '*y*', '*z*', and so on) not already occurring in it, enclosing the result in parentheses unless it is already so enclosed, and prefixing either $(\exists\mu)$ or (μ) to the whole, is a sentence;

nothing is a sentence unless it can be shown to be so on the basis of the above clauses.

  In practice we sometimes replace parentheses with brackets and braces in the interest of readability, and often omit parentheses when doing so causes no ambiguity — for example, we write '$(x)(y)(Fxy)$' instead of '$(x)((y)(Fxy))$'. Some authors prefer adding a "clause 0" at the beginning:

Each sentence letter ('*A*', '*B*', '*C*', and so on) is a sentence.

If this addition is made, the semantics presented here must be expanded slightly.

Chapter 2. Just as the truth-value of a compound sentence is uniquely determined by an assignment of truth-values to its component sentences, so here
the truth-value of a quantificational sentence is uniquely determined once a
universe of discourse is specified and properties (and objects) are assigned
to its component predicate letters (and constants).

We may thus give an interpretation of the sentence

$(x)[Px \supset (\exists y)(Lxy)]$

as follows:

universe of discourse: 1, 2, 3, . . .
interpret 'Px' as 'x is a prime'
interpret 'Lxy' as 'x is greater than y'

Clearly the sentence comes out true on this interpretation; for each positive integer, x, if it is a prime number, then there is in this universe a y such
that x is greater than it (the number 1 will always do for 'y', since it is not
prime). Another interpretation is:

universe of discourse: human beings
interpret 'Px' as 'x is a man'
interpret 'Lxy' as 'x is father of y'.

Here the sentence comes out false, since not every man is a father.

Now, an argument consists of two or more sentences one of which is
taken to follow from the others. So we need to introduce the notion of
interpreting *several* sentences jointly. To this end, let us consider the
argument:

(6) $(x)(Hx) \lor (y)(Jy)$

(7) $(\exists x)(\sim Jx) \& (\exists y)(Gy)$

(8) $(\exists x)(Kx \lor Gx)$

(9) $\therefore Fa \supset (\exists x)(Hx \& Kx)$

The following provides an interpretation on which all four sentences are
true:

universe of discourse: the integers 7, 6, 2 designated respectively by '*a*', '*b*', and '*c*'
interpret '*Fx*' as '*x* is prime'
interpret '*Gx*' as '*x* is even'
interpret '*Hx*' as '*x* is positive'
interpret '*Jx*' as '*x* is a square root of a negative integer'
interpret '*Kx*' as '*x* is divisible by 3'.

Back in Section 5.1 an artificial universe was employed to assist us in understanding the truth-conditions of general sentences. That universe was specified by this chart:

	F	G	H	J	K
a	+	−	+	−	−
b	−	+	+	−	+
c	+	+	+	−	−

Note that the plus-minus arrangement coincides exactly with the above interpretation of sentences (6)-(9), and hence the chart gives an alternate means of specifying that interpretation. The only difference between the two is that the chart does not tell us just which properties are represented by the five predicate letters nor which three individuals comprise the universe of discourse. The interpretation of course makes all of that explicit. For example, (6) on this interpretation says that either everything (in the specified universe) is positive or else everything is a square root of a negative integer; (9) says that if 7 is a prime number then something is both positive and divisible by 3. Both are true on the above interpretation, as are (7) and (8); hence if this argument is invalid, the above interpretation will not show it so, since the conclusion is not false. That the argument *is* in fact invalid is shown by the following interpretation on which the conclusion is false but the premises are true:

universe of discourse: Lincoln, Washington, and Franklin D. Roosevelt, designated respectively by '*a*', '*b*', and '*c*'
interpret '*Fx*' as '*x* is a Republican'
interpret '*Gx*' as '*x* is a Democrat'
interpret '*Hx*' as '*x* is a former U.S. President'
interpret '*Jx*' as '*x* is a former commanding general'
interpret '*Kx*' as '*x* is a woman'.

This new interpretation gives us:

	F	G	H	J	K
a	+	−	+	−	−
b	−	−	+	+	−
c	−	+	+	−	−

And from this it is easily seen that the conclusion is false, since it now has a true antecedent and false consequent. Moreover, all premises come out true on this interpretation, so the argument has been shown invalid. Just as an invalid truth-functional argument must have at least one row with T's for all premises and F for the conclusion, so an invalid quantificational argument must have at least one interpretation of its sentences yielding all true premises and false conclusion. More fully:

A quantificational argument is **invalid** if and only if there is at least one universe of discourse and at least one way of interpreting all predicate letters (and constants) occurring in its sentences such that all premises are rendered true and the conclusion false.

The foregoing suggests that we should simply define 'valid' in terms of there being no interpretation in which all premises are true and the conclusion is false. This is in fact what we shall do, but first a qualification must be made. By 'no interpretation' we must mean 'no *non-empty* universe of discourse...'. That is, the case of the empty universe – one with no individuals at all – must be excluded from consideration. This universe exhibits counterintuitive characteristics; because of its unique and curious nature it is best that it be excluded from consideration. Otherwise, the plainly valid argument

$(x)(Fx)$
$\therefore (\exists x)(Fx)$

would have an interpretation with a true premise and false conclusion and thus be *invalid*. For its conclusion is false in the empty universe since it asserts the existence of at least one individual; but its premise is *true*, since—as a little reflection shows—it is equivalent to '$\sim(\exists x)(\sim Fx)$', which *denies* the existence of something that is not F. Indeed, a great many intuitively valid arguments have true premises and false conclusion when interpreted with respect to the empty universe. Hence we shall restrict ourselves to universes with at least one individual. This restriction in practice comes to very little, since the universes presupposed in actual contexts of argument generally are known to be non-empty.

Our definitions proceed as follows:

DEFINITION   A quantificational sentence is a **logical truth** if and only if it is true on every interpretation.

DEFINITION   A quantificational sentence is **satisfiable** if and only if it is true on at least one interpretation.

DEFINITION   A quantificational sentence is a **logical falsehood** if and only if it is false on every interpretation.

Thus, '$(x)(Fx \supset Fx)$' is a logical truth, '$(x)(Fx \supset Gx)$' is satisfiable, and '$(\exists x)(Fx \, \& \sim Fx)$' is a logical falsehood. Note that every logical truth is *also* satisfiable, though of course the reverse does not hold.

DEFINITION   A quantificational argument with premises $\mathscr{P}_1, \ldots, \mathscr{P}_n$ and conclusion $\mathscr{C}$ is **valid** if and only if there is no interpretation upon which all of $\mathscr{P}_1, \ldots, \mathscr{P}_n$ are true and $\mathscr{C}$ false.

DEFINITION   Two quantificational sentences are **equivalent** if and only if there is no interpretation upon which one is true and the other false.

Validity may also be defined in terms of the corresponding conditional being a logical truth, and equivalence in terms of the logical truth of the corresponding biconditional. To show two sentences *not* to be equivalent, it suffices to exhibit any interpretation upon which they differ in truth-value.

Finding an interpretation that shows invalidity is not always an easy task. As with constructing deductions back in Chapter Three, it often involves trial and error and a little ingenuity (as a general method, the use of plus and minus charts like those displayed earlier should be avoided since they are useless for arguments involving relations). Almost any subject matter can serve as the basis for deriving an interpretation that gives clearly true premises and false conclusion—that is, where the truth-values are a matter of "common knowledge". Readers should feel free to choose interpretations from those areas of knowledge with which they are most familiar. Hence with a simple case like

$(x)(Fx \supset Gx)$

$(x)(Hx \supset Gx)$

$\therefore \ (x)(Fx \supset Hx)$

those whose interests lie in politics or in physics or in literature should quickly arrive at examples like

universe of discourse: politicians
interpret '*Fx*' as '*x* is a Republican senator'
interpret '*Gx*' as '*x* is an elected official'
interpret '*Hx*' as '*x* is a Democratic senator'

universe of discourse: metal wires
interpret '*Fx*' as '*x* is made of copper'
interpret '*Gx* as '*x* conducts electricity'
interpret '*Hx*' as '*x* is made of silver'

universe of discourse: writers
interpret '*Fx*' as '*x* is a novelist'
interpret '*Gx*' as '*x* writes prose'
interpret '*Hx*' as '*x* is a journalist'

However, certain areas of interest are especially useful in this regard, and one of them is biology (in particular zoological examples), for here there exists a widely known structure of species, genera and the like available for our purpose. To illustrate, let us consider

$(x)[Ax \lor Wx) \supset Gx]$

$(\exists x)(Ax \ \& \ Fx)$

$(x)(Fx \supset \sim Wx)$

$(\exists x)(Ax \ \& \ Tx)$

$\therefore (\exists x)(Fx \ \& \ Tx)$

The first premise tells us in effect that anything which is an *A* or a *W* is a *G*, so we must find a property for '*Gx*' that "includes" all those things that qualify either as *A*'s or as *W*'s. Using the set of animals as our universe of discourse, then, we can pick for '*Ax*' and for '*Wx*' classes that are subsumable under the genus (or more accurately, "subphylum") assigned to '*Gx*'; for example, if for '*Gx*' we use '*x* is a vertebrate', then '*x* is a mammal' and '*x* is a reptile' will fill the bill. We now have a true first premise:

Mammals and reptiles are vertebrates.

Moreover, for '*Fx*' we need a family of mammals (to make true the second premise) that are *not* reptiles (to make true the third premise). So if we pick the property of being a dog for '*Fx*' we get respectively:

Some mammals are dogs

No dogs are reptiles

To render the fourth premise true but the conclusion false, we need a property for '*Tx*' such that some mammals *have* that property but no dogs do. Choosing a family of mammals incompatible with being a dog will do the job, so let us pick that of being a cat (though picking a property that cats have but dogs do not would do just as well),

Some mammals are cats

Therefore, some dogs are cats

Our interpretation is now complete:

universe of discourse: animals

interpret '*Ax*' as '*x* is a mammal'

interpret '*Wx*' as '*x* is a reptile'

interpret '*Gx*' as '*x* is a vertebrate'

interpret '*Fx*' as '*x* is a dog'

interpret '*Tx*' as '*x* is a cat'

Elementary mathematics, and especially arithmetic, provides us with another subject matter that often yields quick results, as may be seen with the following case:

$(x)[(y)(Rxy) \supset Fx]$

$\therefore (x)[(\exists y)(Rxy) \supset Fx]$

Let us take the set of positive integers as our universe of discourse, and cast about for an appropriate relation for '*Rxy*' and property for '*Fx*'. The premise tells us that for each positive integer, if it bears *R* to *all* positive integers, then it has *F*. However, the conclusion tells us that if it bears *R* to *at least one* of them, then it has *F*. Suppose, then, we choose the "less-than-or-equal" relation for '*Rxy*' (that is, the relation commonly expressed by the open sentence '$x \leq y$'), and the property of being odd for '*Fx*'. The premise now says that for any positive integer *x*, if *x* is less than or equal to all such integers then it is odd—which is true since only the number one qualifies for x and it is indeed odd. The conclusion, however, states that if x is less than or equal to at least one positive integer, it is odd, and there are of course many counterexamples to this, among them the number 6. So the following simple interpretation works:

universe of discourse: positive integers

interpret 'Rxy' as '$x \leqslant y$'

interpret 'Fx' as 'x is odd'

It must be kept in mind that if an argument contains individual constants, they must be assigned objects from the universe of discourse. To show the argument

$(x)(Rxx)$

∴ Rab

invalid we must interpret 'a' and 'b' as well as the predicate letter. So we can use an interpretation like

universe of discourse: the positive integers

interpret 'Rxy' as '$x = y$'

interpret 'a' as '2'

interpret 'b' as '3'

However, mathematical interpretations with infinite universes also have a theoretical significance that is worth noting. If an argument can be shown invalid on an interpretation with a universe of discourse containing n individuals (where n is some positive integer), its invalidity might not be showable for any interpretation having a universe of *less than* n individuals. The above argument can be shown invalid on certain interpretations having just two objects in the specified universe, but not on any having just a single individual. For example, we could have shown invalidity using a universe whose *only* members are the numbers 2 and 3, with 'Rxy' again being assigned the equality relation. But a one-object universe cannot work, for the constants 'a' and 'b' would have to be two different names for the *same* individual. Hence whenever the premise '$(x)\ (Rxx)$' is true, the conclusion 'Rab' (in the above example '$a = b$') would be true instead of false as we require.

On the other hand, when invalidity can be shown using an n-element universe of discourse, it can in addition always be shown using an interpretation that has a universe with *more than n* objects. Our sample argument is easily shown invalid not just on an interpretation with a universe of two objects, but also on interpretations with universes of three, four, or *any* greater number of objects. And as our above interpretation shows, the universe may even contain an infinite number of objects ("the positive integers").

A little reflection on this fact leads to the supposition that if we rely solely on interpretations with *infinite* universes, we will always be able to show invalidity no matter what invalid argument we encounter (though of course some insight and trial-and-error may still be needed). This is indeed the case; it was proven in the early part of this century and is known as the *Skolem-Löwenheim Theorem* (though its usual formulation is given in much more general and abstract terms than will be done here):

For any quantificational argument, if there is *any* interpretation on which it has all true premises and false conclusion, then there is an interpretation of the same kind whose universe of discourse is the set of positive integers.

So if it is invalid, then it can always be shown invalid using an interpretation with an infinite universe of discourse. In other words, if it can be done at all, it can always be done using a universe like the positive integers (or the integers generally or the rational numbers, etc.), and employing appropriate arithmetic relations and properties for the predicate letters. Herein lies the importance of mathematical interpretations: a given argument might not be demonstrably invalid on some interpretations with finite universes, but one with an infinite universe will work if anything will. In fact, there are even arguments whose invalidity cannot be shown on any interpretation with a finite universe no matter how large (an example of such an argument is given at the end of the next section). However, for the balance of this chapter our focus will be on arguments whose invalidity can be established using finite universes as well as infinite ones.

EXERCISES

I. For each of the following arguments, find an interpretation that shows it invalid.

1. $(x) (Fx \supset Gx)$
   $(x) (Fx \supset Hx)$
   $\therefore (x) (Gx \supset Hx)$

2. $(x) (Ax \supset Bx)$
   $(\exists x)(Cx \ \& \sim Bx)$
   $\therefore (\exists x)(Ax \ \& \sim Cx)$

3. $(x)(Ex \supset Fx)$
   Fa
   $\therefore Ea$

4. $(x)(Jx \supset Kx)$
   $(\exists x)(Jx \& Hx)$
   $\therefore (x)(Hx \supset \sim Kx)$

5. $(x)[(Fx \vee Gx) \supset Hx]$
   $(\exists x)(Hx \& \sim Fx)$
   $\therefore (x)[\sim Fx \supset (Gx \& Hx)]$

6. Rab
   $\therefore (\exists x)(Rxx)$

7. $(x)(Fx \supset Gx)$
   $(\exists x)(Fx \& Hx)$
   $(\exists x)(Gx \& \sim Fx)$
   $\therefore (\exists x)(Gx \& \sim Hx)$

8. $(x)[Fx \supset (\exists y)(Gy \& Rxy)]$
   $\therefore (x)[Fx \supset (y)(Gy \supset Rxy)]$

9. $(x)[(Ax \& Bx) \supset Cx]$
   $\therefore (x)(Ax \supset Cx) \& (x)(Bx \supset Cx)$

10. $(x)(Fx \supset Gx) \supset (y)(Hz \supset Jy)$
    $\therefore (\exists x)(Fx \& Gx) \supset (y)(Hy \supset Jy)$

11. $(x)[Ax \supset (\exists y)(Kxy)]$
    $\therefore (\exists y)(x)(Ax \supset Kxy)$

12. $(x)[Fx \supset (\exists y)(Gy \& \sim Rxy)]$
    $\therefore (\exists y)[Gy \& (x)(Fx \supset \sim Rxy)]$

II. Find an interpretation showing the *converse* of logical truth (5) from this section to be satisfiable. Then find another showing it *not* to be a logical truth.

III. Show each pair to be nonequivalent by finding an interpretation on which the two sentences differ in truth-value:
   1. $(x)(Fx \supset Gx) \vee (y)(Hy \supset Gy)$ and $(x)[(Fx \vee Hx) \supset Gx]$
   2. $(x)(Fx \supset Gx)$ and $(x)(Fx) \supset (y)(Gy)$

IV. Where $\mathcal{T}$ is any sentence, an interpretation is often described as a **model** of $\mathcal{T}$ (and $\mathcal{T}$ is thus said to **have a model**) if and only if $\mathcal{T}$ is true on that interpretation. Where $\mathcal{P}_1, \mathcal{P}_2, \ldots, \mathcal{P}_n$ are any sentences, an interpretation is said to be a **model** of $\mathcal{P}_1, \mathcal{P}_2, \ldots, \mathcal{P}_n$ if and only if *all* are true on that interpretation. Using this new terminology, complete the following:

The sentence $\mathcal{T}$ is a logical truth if and only if each _____ of $\mathcal{T}$ is a _____ of $\mathcal{T}$.

The sentence $\mathcal{T}$ is satisfiable if and only if $\mathcal{T}$ has a _____.

The sentences $\mathcal{T}$ and $\mathcal{U}$ are equivalent if and only if each _____ of _____ is a _____ of _____, and each _____ of _____ is a _____ of _____.

The argument with premises $\mathcal{P}_1$, $\mathcal{P}_2$,..., $\mathcal{P}_n$ and conclusion $\mathcal{C}$ is valid if and only if each interpretation of $\mathcal{P}_1$, $\mathcal{P}_2$,..., $\mathcal{P}_n$ which is a _____ of _____ is a _____ of _____.

* 6.2 More on Showing Invalidity

Often, by deliberately focusing on universes containing a very small number of individuals, invalidity can be established quickly and in more systematic fashion. Let us examine the argument:

(I) $(x)(Fx \supset Gx)$
$(\exists x)(Gx \, \& \, Hx)$
$\therefore (\exists x)(Fx \, \& \, Hx)$.

Now consider a universe of discourse with just one individual, *a*. Then in *that* universe, the first premise will be true under the same conditions as '$Fa \supset Ga$', since the latter is the only substitution instance of '$Fx \supset Gx$'. By the same token, the second premise will be true in this universe just in case '$Ga \, \& \, Ha$' is true, and the conclusion just in case '$Fa \, \& \, Ha$' is true. By assigning truth-values to the simple components 'Fa', 'Ga', and 'Ha', we can make the premises true and conclusion false of the truth-functional argument:

(II) $Fa \supset Ga$
$Ga \, \& \, Ha$
$\therefore Fa \, \& \, Ha$.

But if the assignment shows *this* argument invalid, then it shows (I) invalid as well. Since the premises and conclusion of (I) have the same truth-conditions for a one-individual universe as the corresponding premises and conclusion of (II), we have in effect shown that in any universe with just one individual there is an interpretation of the predicate letters of (I) yielding true premises and false conclusion. In the present case, such a truth-value assignment is easily come by:

Fa	Ga	Ha
F	T	T

Now consider the invalid inference:

(III) $(\exists x)(Fx)$
$(\exists x)(Gx)$
$\therefore (\exists x)(Fx \, \& \, Gx)$.

Here, for a one-individual universe we would have the corresponding truth-functional argument:

Fa
Ga
∴ Fa & Ga

which is valid by the rule of **Conj**. Yet we know that there is *some* universe of discourse in which it will have true premises and false conclusion. So let us try a universe with two individuals, *a* and *b*. Under what conditions will the first premise of (IV) be true in such a universe? Since 'Fx' has two substitution instances, 'Fa' and 'Fb', clearly '$(\exists x)(Fx)$' will be true just in case at least one of them is true, that is, just in case the disjunction '$Fa \lor Fb$' is true. Similar considerations hold for the second premise and conclusion; the latter, for example, has the same truth-conditions in a two-individual universe as '$(Fa \& Ga) \lor (Fb \& Gb)$'. So for such a universe, the corresponding truth-functional argument would be:

(IV)   Fa v Fb
       Ga v Gb
       ∴ (Fa & Ga) v (Fb & Gb).

We may now produce a truth-value assignment yielding true premises and false conclusion, thus showing this argument – and hence (IV) – invalid:

Fa	Ga	Fb	Gb
T	F	F	T

Generally, for a universe with *n* individuals, the sentence '$(\exists x)(Fx)$' will have the same truth-conditions as the disjunction

$Fa_1 \lor Fa_2 \lor \ldots \lor Fa_n$.[5]

Similarly, a universally quantified sentence such as '$(x)(Fx)$' will be true just in case *all* substitution instances of 'Fx' are true and therefore just in case the conjunction of those instances is true. So for an *n*-individual universe, '$(x)(Fx)$' will be true under the same conditions as

---

[5]For brevity, we shall here drop parentheses in extended disjunctions and conjunctions.

Fa_1 & Fa_2 & ... & Fa_n.

We shall call such disjunctions and conjunctions **truth-functional expansions** of the corresponding existentially and universally quantified sentences. Further examples of such expansions for a three-individual universe are:

$(x)(Fx \vee Gx)$: $(Fa \vee Ga)$ & $(Fb \vee Gb)$ & $(Fc \vee Gc)$
$(x)(Fx) \vee (y)(Gy)$: $(Fa$ & Fb & $Fc) \vee (Ga$ & Gb & $Gc)$
$(\exists x)(Fx)$ & $(\exists y)(Gy)$: $(Fa \vee Fb \vee Fc)$ & $(Ga \vee Gb \vee Gc)$

Multiply-quantified sentences with relational terms are a bit more complicated, since each quantifier must be broken down separately. Consider again

$(x)(\exists y)(Cyx)$.

Breaking down the main quantifier first, we derive for a two-individual universe

$(\exists y)(Cya)$ & $(\exists y)(Cyb)$.

Breaking down the existential quantifiers then gives us

$(Caa \vee Cba)$ & $(Cab \vee Cbb)$.

By considering several interpretations, we saw earlier in this chapter that the argument

$(x)(\exists y)(Cyx)$
$\therefore (\exists y)(x)(Cyx)$

is invalid, and now we can show its invalidity by obtaining the truth-functional expansion for the conclusion. Thus, '$(\exists y)(x)(Cyx)$' becomes successively

$(x)(Cax) \vee (x)(Cbx)$
$(Caa$ & $Cab) \vee (Cba$ & $Cbb)$.

We then show that the argument

$(Caa \lor Cba) \& (Cab \lor Cbb)$
$\therefore (Caa \& Cab) \lor (Cba \& Cbb)$

is truth-functionally invalid. The following assignment of truth-values renders the premise true and the conclusion false:

Caa	Cab	Cba	Cbb
F	T	T	F

If a is Mt. Everest and b is Mt. Rushmore, the interpretation presented earlier leads directly to this truth-value assignment.

However, not every invalid quantificational argument can be *proved* invalid through the use of truth-functional expansions. Consider, for example,

$(x)(y)(z)[(Ryx \& Rzy) \supset Rzx]$
$(x)(\exists y)(Ryx)$
$\therefore (\exists x)(Rxx)$.

This argument is invalid, as the following interpretation shows:

universe of discourse: the positive integers
interpret 'Rxy' as '$x > y$'

However, the method presented in this section is of no use here. The reason is simple: the above argument exhibits true premises and false conclusion only for interpretations where the universe of discourse contains *infinitely* many objects. Obviously, then, writing out truth-functional expansions and finding truth-value assignments is out of the question, and there is thus a theoretical limitation on the method used in this section. The method of Section 6.1, then, remains the primary one. The Skolem-Löwenheim Theorem told us that an infinite universe will work if anything will; the above example shows as that there are arguments where *only* an infinite universe will do the job.

However, invalid arguments containing no *relational* expressions are unproblematic in this respect. There is always some finite universe adequate to the task, and moreover a purely mechanical method is, at least in theory, available.

More fully, for quantificational arguments containing no relational expressions, the following result[6] holds: where the number of predicate letters is n, if there is no truth-value assignment yielding true premises and false conclusion for a universe of 2^n individuals, then the argument is valid—no assignment in *any* non-empty universe will yield such a combination. So, in theory one could mechanically test such arguments for validity *and* invalidity by going directly to the 2^n-individual case. In practice, however, this is usually an unwise strategy. Most invalid arguments may be so established using a much smaller universe, and even for a relatively simple argument with only four predicate letters, the strategy would call for a universe of no less than 16 individuals—along with the correspondingly huge truth-functional expansions.

The 2^n-individual strategy nonetheless constitutes a mechanical routine for determining validity, equivalence, and logical truth as regards non-relational sentences and the arguments in which they occur. However, as noted earlier, there is no such routine for quantificational logic with relations and multiple quantifiers.

EXERCISES

I. Go back and do the examples in (I) of the preceding exercise set using truth-functional expansions.

II. Using truth-functional expansions, show each of the following invalid:

1. $(x)[Fx \vee (Gx \& Hx)]$
   $(\exists x)[Ex \& (Gx \& Hx)]$
   $(\exists x)[Ex \& (Gx \supset \sim Hx)]$
   $\therefore (x)(Ex \vee Fx)$

2. $(x)(Ax) \supset (\exists y)(Py)$
   $(x)(Qx) \supset (\exists y)(Ay)$
   $\therefore (x)[Qx \supset (y)(Py)]$

3. $(x)[\sim Cx \supset (y)(Gy)]$
   $(x)(Gx) \supset (\exists y)(Jy \& \sim Ky)$
   $(x)(Gx) \supset (\exists y)(Jy \& Ky)$
   $\therefore (x)(\sim Jx \supset Cx)$

---

[6]P. Bernays and M. Schönfinkel, "Zum Entscheidungsproblem der mathematischen Logik," *Mathematische Annalen,* vol. 99 (1928), pp. 342-372.

CHAPTER SEVEN

QUANTIFICATIONAL LOGIC
Deductive Methods

7.1 The Rules EI, EG, UI, UG, and QN

Certain arguments whose premises and conclusions involve quantifiers can be shown valid with no more deductive apparatus than is available from Chapter 3:

1. $(\exists x)(Px) \supset (Gae \supset Laf)$ Premise

2. $\sim Laf$ Premise   DEDUCE: $(\exists x)(Px) \supset \sim Gae$

3. $(\exists x)(Px)$ Assumption

  4. $Gae \supset Laf$ 1, 3 MP

  5. $\sim Gae$ 2, 4 MT

6. $(\exists x)(Px) \supset \sim Gae$ 3–5 CP

But if we restrict ourselves to just those earlier rules we will not be able to construct a deduction even for the obviously valid argument:

1a. $(\exists x)[Px \& (Gxe \& Lxf)]$ There exists at least one prime number greater than 8,000,000 and less than $8,000,000! + 2$

∴ $(\exists x)(Px \& Gxe)$ Therefore there exists at least one prime number greater than 8,000,000.

The scope of the quantifier in (1a), unlike that of the quantifier in (1), extends across the whole sentence. For that reason (1a) is clearly not a conjunction, and so neither **Simp** nor any of the other rules of Chapter 3 apply.

Of course if we were entitled to suppose that 8,000,001 ('a' for ease of reference) is a prime number greater than 8,000,000 and less than 8,000,000! + 2 — that is if we were entitled to

2a. *Pa & (Gae & Laf)*,

we could go on to argue

3a. *(Pa & Gae) & Laf*   2a Assoc

4a. *Pa & Gae*   3a Simp

5. *(∃x)(Px & Gxe)*

intending to justify the move from (4a) to (5) by noting that any argument of the form

...a...
∴ (∃x)(...x...)

must be valid (if a has a certain property, then obviously *something* does). Or if we could infer that 8,000,002 ('b' here) is a prime number between 8,000,000 and 8,000,000! + 2, that is, if we had

2b. *Pb & (Gbe & Lbf)*,

we could argue

3b. *(Pb & Gbe) & Lbf*   2b Assoc

4b. *Pb & Gbe*   3b Simp

5. *(∃x)(Px & Gxe)*

Indeed, if we knew of any particular individual c that is a prime number between 8,000,000 and 8,000,000! + 2, we could argue from

2c. *Pc & (Gce & Lcf)*

to our conclusion:

3c. $(Pc$ & $Gce)$ & Lcf 2c Assoc

4c. Pc & Gce 3c Simp

5. $(\exists x)(Px$ & $Gxe)$

But none of this, of course, provides us with a *deduction* of (5) from (1a), for neither (2a) nor (2b) nor anything like (2c) follows from (1a). Our premise does not tell us *which* thing or things (there may be several) are prime numbers between 8,000,000 and 8,000,000! + 2 and so does not entitle us to any of these steps. The difficulty seems to be that (1a) tells us that there exists at least one number of the sort in question but gives us no name for it, and so leaves us with no way to refer to it or talk about it.

In ordinary intuitive reasoning we often meet this sort of situation in a direct and straightforward way: given no name for a thing we are assured exists and need to refer to, we provide a temporary one by means of a locution like "let v be one such" or "pick one, and call him 'John Doe'." To handle the argument considered here, for example, it would be natural to argue along the following lines:

1. Our premise assures us that there exists at least one prime number greater than 8,000,000 and less than 8,000,000! + 2.

2. Let v be one such; that is, pick any prime number greater than 8,000,000 and less than 8,000,000! + 2 and call it 'v' (if there are several such numbers, it won't matter which one is chosen).

3. Then of course v is a prime number greater than 8,000,000 as well as being less than 8,000,000! + 2.

4. So v is (among other things) a prime number greater than 8,000,000.

5. Consequently there does exist at least one prime number greater than 8,000,000 (v, for example, whichever number it happens to be).

To make use of this device in our deductive system, we will agree to keep the hitherto unused letters 'u', 'v', 'u''', 'v''' and so forth (which we will call **pseudo-names**) available for use only in situations like this one, where we want to refer to a thing of a certain sort but do not know or care *which* thing of that sort it is. (In the example above we are told that there are prime numbers between certain limits, but are not told which numbers between those limits are prime.) Thus our pseudo-names 'u', 'v', and so forth, contrast sharply with individual constants such as 'a', 'b', and 'c', which are used to designate specific individuals just as genuine proper names ('Aristotle', for example) are. Having introduced pseudo-names, it must now be required that no deduction may terminate with a line containing one of them.

Then our deduction can be obtained directly from the informal argument (1)–(5) above:

1. $(\exists x)[Px \,\&\, (Gxe \,\&\, Lxf)]$ Premise   DEDUCE: $(\exists x)(Px \,\&\, Gxe)$

2. $Pv \,\&\, (Gve \,\&\, Lvf)$

3. $(Pv \,\&\, Gve) \,\&\, Lvf$ 2 Assoc

4. $Pv \,\&\, Gve$ 3 Simp

5. $(\exists x)(Px \,\&\, Gxe)$

The rule permitting us to move from *existentially* quantified open sentences like (1) to such *instances*[1] of them as (2) will be called **Existential Instantiation** (**EI**, for short) and it can be formulated most generally as follows:

Existential Instantiation (EI)

Where β is a pseudo-name (*not* a constant), one may in a deduction infer from a previously obtained existentially quantified sentence

$(\exists \mu)(\ldots \mu \ldots)$

any corresponding sentence

$\ldots \beta \ldots$

provided that: i. β does not occur in any *earlier* line of the deduction;
               ii. $\ldots \beta \ldots$ results from $\ldots \mu \ldots$ by replacing each occurrence of μ with β (making *no* other changes).[2]

---

[1] We call expressions which involve occurrences of pseudo-names (for example, '*Pv*' and '*Gve & Lvf*') *sentences* because of their similarity to such English sentences as 'It hit me on the head'. Often the context in which the sentence 'It hit me on the head' is used is all that makes clear the antecedent reference of the pronoun 'it'. Imagine that we rush into a room shortly after a number of cans have fallen from a shelf and find a friend surrounded with cans and clutching his head. "Did one of them hit you?" we might ask, and he could reply truly "Yes, *it* hit me on the head" even though none of us may have any way of finding out exactly *which* can it was that hit him there.

To footnote 4 of the preceding chapter, we now add:

     v. The result of replacing one or more constants in a sentence with a pseudo-name ('*u*', '*u'*', '*v*', '*v'*', etc.) is a sentence.

[2] We shall follow the practice throughout of employing only members of the *v*-family ('*v*', '*v'*', '*v''*', and so on) in applications of EI. Though not required, this practice will pay off later when we come to the rule called **UG** for which we are reserving pseudo-names in the *u*-family.

The reason for the restriction to pseudo-names should be clear from our preceding discussion. With proviso (i) we forestall unjustifiable attempts to treat several presumably different things as if they were one and the same by giving the same name to all of them, as in:

1. $(\exists x)(Cx)$	Premise	There are some cats
2. $(\exists x)(Dx)$	Premise	There are some dogs
3. Cv	1 EI	Let v be a cat
4. Dv	2 EI (mistake)	Let v be a dog (too?)
5. Cv & Dv	3, 4 Conj	Then v is *both* a cat *and* a dog
6. $(\exists x)(Cx$ & $Dx)$		So there is something that is both a cat and a dog.

The premises are both true but the conclusion reached is obviously false, so there must be a mistake somewhere and it's not hard to find. If we're sure there are cats (as premise one tells us) we may call one of them 'v'. Or if we're also sure there are dogs (as premise two tells us) we may prefer to call one of *them* 'v' instead. But we cannot justifiably use 'v' as a name for a cat at one point and then later, in the same context, decide to use 'v' as a name for a dog, too — doing so would be tantamount to supposing that there was some one thing that was both, and our premises give us no right to *that* supposition. The best we can legitimately do at line (4) is get 'Dv'' or 'Dv'''', using a *different* name for the dog than we used for the cat; and of course neither of these will lead to (6).

We need proviso (ii) as well, to rule out attempts to infer things like 'Pv & Gxe' and 'Pv & Gvv' from '$(\exists x)(Px$ & $Gxe)$'.

EI, then, gives us a way to get existential quantifiers *out* of deductions. But after we've gotten them out we often later want to put them back *in*, as we did earlier when we wanted to move from

Pv & Gve

to

$(\exists x)(Px$ & $Gxe)$.

The rule needed here is:

Existential Generalization (EG)

Where α is a pseudo-name *or* constant, one may in a deduction infer from a previously obtained sentence

$\ldots\alpha\ldots$

any corresponding existentially quantified sentence

$$(\exists\mu)(\ldots\mu\ldots)$$

provided that: i. μ is a variable not already occurring in $\ldots\alpha\ldots$;
            ii. $\ldots\mu\ldots$ results from $\ldots\alpha\ldots$ by replacing *at least one* occurrence of α with μ (making *no* other changes).

The inference from

1. $Fba \vee (\exists x)(Fxb)$ Either Brown will fire Adams or someone[3] will fire Brown

to

2. $(\exists y)[Fya \vee (\exists x)(Fxy)]$ Someone is such that either he will fire Adams or someone will fire him

is clearly valid so we permit **EG** on individual constants as well as on pseudo-names. Of course the move from (1) to

3. $(\exists y)[Fya \vee (\exists x)(Fxb)]$ Someone is such that either he will fire Adams or someone will fire Brown

is also valid, and that's why proviso (ii) permits replacement of only *some* of the occurrences of the pseudo-name or constant involved. *No* other changes are permitted, naturally — we can no more justifiably move from (1) to

4. $(\exists y)[Fyy \vee (\exists x)(Fxy)]$ Someone is such that either he will fire himself or someone will fire him

than we could move to '$(\exists y)[Qya \vee (\exists x)(Zxy)]$'. And of course the restriction in (i) is necessary to prevent such undesirable moves as that from (1) to

5. $(\exists x)[Fxa \vee (\exists x)(Fxx)]$,

for the latter would be totally ambiguous. (Does it say that there's some-

---

[3]For this example, we restrict our universe of discourse to persons.

one who will either fire Adams or be himself fired by someone? Or that there's someone who will either fire Adams or fire someone? Or something else? There is no way to tell which of the two rightmost occurrences of '*x*' are bound by which quantifiers.) To apply **EG** to (1) one has to pick a "new" variable not already occurring in it, as was done, for example, by choosing '*y*' in passing from (1) to (2).

Undoubtedly

6. $(\exists y)(Fya) \lor (\exists x)(Fxb)$ Either someone will fire Adams or someone will fire Brown

also follows from (1) but it ought to be noticed at the outset that **EG** does *not* permit us to go from (1) to (6) *directly* — **EG** only permits the existential quantifier to be inserted in front of a whole line of a deduction and does not allow us to insert quantifiers into the middle. To get (6) from (1) we would need something like the following deduction:

1. $Fba \lor (\exists x)(Fxb)$	Premise	DEDUCE:
		$(\exists y)(Fya) \lor (\exists x)(Fxb)$
2. $\sim(\exists x)(Fxb)$	Assumption	
3. $(\exists x)(Fxb) \lor Fba$	1 Com	
4. Fba	2, 3 DS	
5. $(\exists y)(Fya)$	4 EG	
6. $\sim(\exists x)(Fxb) \supset (\exists y)(Fya)$	2–5 CP	
7. $\sim\sim(\exists x)(Fxb) \lor (\exists y)(Fya)$	6 Impl	
8. $(\exists y)(Fya) \lor \sim\sim(\exists x)(Fxb)$	7 Com[4]	
9. $(\exists y)(Fya) \lor (\exists x)(Fxb)$	8 DN	

EXERCISES

Using the rules **EI** and **EG** in addition to those of Chapter 3, deduce the conclusions of the following arguments from their premises:

1. $(\exists x)(Fx \& Gx)$
     $\therefore (\exists x)(Gx)$

---

[4]In accordance with our agreement in Section 3.4, this application of **Com** can be combined with the ensuing application of **DN** into a single step. In our examples in this section and the next, we will continue listing each application of each rule on a separate line for the sake of clarity; after we've acquired some experience with the quantification rules, however, we will drop the separate listings even in our examples.

2. $(\exists x)(Fx \& Gx)$
    $\therefore (\exists y)(Fy)$
3. $(\exists x)(\exists y)(Rxy)$
    $\therefore (\exists y)(\exists x)(Rxy)$
4. $(\exists x)(Fx \& Gx)$
    $(\exists x)(Fx \& Hx)$
    $(\exists y)(Gy \& Fy) \supset (\exists x)(Gx \& Hx)$
    $\therefore (\exists y)(Gy \& Hy)$
5. $(\exists x)(Lxx)$
    $\therefore (\exists x)(\exists y)(Lxy)$
6. $(\exists x)(Bx) \supset Be$
    $\therefore (\exists x)(Bx) \equiv Be$
7. $(\exists x)[Sx \& \sim(Sx \& Tx)]$
    $\therefore (\exists x)(Sx) \& (\exists y)(\sim Ty)$
8. $(\exists x)[(Ax \& Bx) \& (\exists y)(Cy \& Dxy)]$
    $\therefore (\exists x)(\exists y)(Ax \& Dxy)$
9. $(x)(Fx \supset Gx) \supset [\sim(\exists y)(Gy) \supset \sim(\exists z)(Fz)]$
    $(\exists z)(Fz)$
    $\therefore (x)(Fx \supset Gx) \supset (\exists y)(Gy \lor Hy)$
*10. $\sim(\exists x)(Px \& \sim Wx)$
    Pa
    $\therefore Wa$

EG and **EI** enable us to get existential quantifiers into and out of deductions, and if we were willing to forego universal quantifiers in our translations entirely, translating the 'All philosophers are wise' argument as in Exercise 10 above, we would need no other rules. (This could always be done, since sentences of the form 'All P's are W' may invariably be translated as ' $\sim(\exists x)(Px \& \sim Wx)$ '.) But instead of course we want to be able to translate that argument in the following more natural way:

1. $(x)(Px \supset Wx)$ Premise   All philosophers are wise

2. Pa Premise   Aristotle is a philosopher

  $\therefore Wa$ Therefore Aristotle is wise.

So it will be desirable to have a rule of **Universal Instantiation (UI)** permitting us to move from *universally* quantified sentences to particular substitution instances of the open sentences within them.

Premise (1) here tells us, for instance, that *all* things that are philosophers are things that are wise and so tells us something about each thing in the universe. It clearly follows from (1), for example, that if Descartes is a philosopher then Descartes is wise, that if the Columbia River is a philosopher then the Columbia River is wise, that if Bozo the Clown is a philosopher then Bozo the Clown is wise, and so on through a variety of other such conditionals including one of special interest to us here:

3. *Pa* ⊃ *Wa*   If Aristotle is a philosopher then Aristotle is wise.

Once we have (3), of course, we can complete our deduction:

4. *Wa*   2, 3 MP

What is true of every individual naturally is true of any particular individual we happen to be interested in, so the move from (1) to (3) is clearly valid, as are each of the following arguments:

(*x*)(*Ix* & *Wx*)	(*x*)(*Lxx*)	(*x*)(∃*y*)(*Ax* ⊃ *By*)	(*x*)(*y*)(*Rxy*)	(*y*)(*Rhy*)
∴ *Ib* & *Wb*	∴ *Ldd*	∴ (∃*y*)(*Ae* ⊃ *By*)	∴ (*y*)(*Rhy*)	∴ *Rhh*

So we want to formulate a rule like this:

Universal Instantiation (UI)
Where α is a pseudo-name *or* constant, one may in a deduction infer from a previously obtained universally quantified sentence

(*μ*)(...*μ*...)

any corresponding sentence

...α...

provided that ...α... results from ...*μ*... by replacing each occurrence of *μ* with α (making *no* other changes).

Here, we allow instantiation to pseudo-names as well as to constants so that an argument like

(*x*)(*Px* ⊃ *Wx*)	All philosophers are wise
(∃*x*)(*Px*)	There exists at least one philosopher
∴ (∃*x*)(*Px* & *Wx*)	Therefore there exists at least one philosopher who is wise

can be handled along the following natural lines:

1. (*x*)(*Px* ⊃ *Wx*)   Premise

2. (∃*x*)(*Px*)   Premise   DEDUCE: (∃*x*)(*Px* & *Wx*)

3. *Pv*   2 EI

4. *Pv ⊃ Wv*   1 UI

5. *Wv*   3, 4 MP

6. *Pv & Wv*   3, 5 Conj

7. (∃*x*)(*Px & Wx*)   6 EG

(2) tells us that there are some philosophers and we choose, with (3), to use '*v*' to refer to one of them. Now along with telling us things about Aristotle, Bozo the Clown, and the Columbia River, (1) also tells us something about *v*, whatever it may be, so the move from (1) to (4) is clearly acceptable. We did the **EI** before the **UI** because of proviso (i) on **EI**: had we tried to do the **UI** first, that provision on **EI** would have prevented us from later instantiating (2) to '*v*'.

It is extremely important to remember that the quantification rules introduced here are rules of *inference*, not rules of replacement, and so can be applied only to quantifiers whose scope extends across the entire line. For example, the argument

(*x*)[*Bx* ⊃ (∃*y*)(~*Gy*)]   Everything is such that if it's blue then something is not green

∴ *Ba* ⊃ (∃*y*)(~*Gy*)   Therefore if the Atlantic Ocean is blue then something is not green

is valid, and **UI** permits passing from its premise to its conclusion. However, the argument

(*x*)(*Bx*) ⊃ ~(∃*y*)(*Gy*)   If everything is blue then nothing is green

∴ *Ba* ⊃ ~(∃*y*)(*Gy*)   Therefore if the Atlantic Ocean is blue then nothing is green.

is clearly invalid, and **UI** does *not* license such inferences.

Our last rule, **Universal Generalization (UG)**, will be designed to permit us, under certain very special conditions, to introduce universal quantifiers into deductions. Sometimes one can establish a sentence of the form

...*β*...

(*β* not a constant but a pseudo-name) in such a way as to make it clear that *all* similar sentences,

...*a*...
...*b*...
...*c*...

and so on, could have been established as well. **UG** will permit us, in such special cases, to infer the general conclusion we would clearly be entitled to:

$(\mu)(\ldots \mu \ldots)$.[5]

Consider, for example, the argument:

1. $(x)(Ix)$	Premise	Everything is interesting
2. $(x)(Wx)$	Premise	Everything is worthy of study
$\therefore (x)(Ix\ \&\ Wx)$		Therefore everything is both interesting and worthy of study.

With the rules we already have we could begin in any of several ways, including:

1. $(x)(Ix)$	Premise	1. $(x)(Ix)$	Premise	1. $(x)(Ix)$	Premise
2. $(x)(Wx)$	Premise	2. $(x)(Wx)$	Premise	2. $(x)(Wx)$	Premise
3. Ia	1 UI	3. Ib	1 UI	3. Ic	1 UI
4. Wa	2 UI	4. Wb	2 UI	4. Wc	2 UI
5. $Ia\ \&\ Wa$	3, 4 Conj	5. $Ib\ \&\ Wb$	3, 4 Conj	5. $Ic\ \&\ Wc$	3, 4 Conj

But it is clear that none of these fine beginnings is going to justify our concluding '$(x)(Ix\ \&\ Wx)$'. It does not follow from '$Ia\ \&\ Wa$' that everything

---

[5]The reader may find it helpful to think of a geometer who has established that

If *ABC* is an equiangular triangle
then *ABC* is an equilateral triangle

but wants to conclude not just that the particular triangle shown in the accompanying diagram is equilateral if equiangular but that

All equiangular triangles are equilateral.

Of course there are many quite general methods he might have used to establish the former that would justify his concluding the latter, but obviously a lot depends on the generality of the manner in which he did establish the former. If he did it by first measuring the angles of the particular triangle shown in the diagram, for example, and then measuring its sides, he is clearly *not* justified in leaping to a conclusion concerning *all* triangles.

is interesting and worthy of study, nor does it follow from 'Ib & Wb' nor from 'Ic & Wc'. To be justified in concluding that *everything* is both interesting and worthy of study we must show more than that a number of particular individuals are. What we want to show is that no matter which individual we consider, it is both interesting and worthy of study. We want to say, in effect:

Pick anything you wish, anything at all. Then it follows from (1) that it's interesting and from (2) that it's worthy of study. So of course it—that thing you picked, whatever it was—is both interesting and worthy of study. And since you were free to choose *any* individual whatever, we are justified in concluding that *everything* is both interesting and worthy of study.

When we consider how this prose might be written up as a deduction, we find a small problem similar to one encountered earlier. We are in effect inviting the readers of our deduction to pick anything they want, intending to show them that *it* is both interesting and worthy of study. But to do this we need a way of specifying whatever it is that was picked; once again, we must be able to refer to a thing whose identity may be unknown to us, and once again pseudo-names provide us with the device we need. Of course we must be careful to show this without making use of any special facts about the thing picked; only in that case can we be sure that the *particular* choice of individuals is irrelevant to our deduction—that we could have carried it out as well no matter what was picked and so be justified in concluding that all things are both interesting and worthy of study. So we argue:

1. $(x)(Ix)$	Premise	Everything is interesting
2. $(x)(Wx)$	Premise	Everything is worthy of study
3. Iu	1 UI	Then u is interesting, where u is anything whatever that you care to pick
4. Wu	2 UI	And u—that thing you picked, *whatever* it was—is worthy of study, too
5. Iu & Wu	3, 4 Conj	So of course it's both interesting and worthy of study
6. $(x)(Ix$ & $Wx)$		Since you might have picked anything at all and we've been able to show that no matter what it was it's both interesting and worthy of study, we are justified in concluding that *everything* is both. (If you think we might have *missed* something, go back and imagine you'd picked *it* from the start—our deduction works no matter *which* individual you began with.)

The rule **UG** requires a fairly complicated formulation—after all, it can permit us to pass with justification from expressions like (5) to those like (6) only when the former was obtained in so totally general a manner as to make it evident that the pseudo-name occurring in it may be taken to stand for anything whatsoever. So we must include provisions in our rule sufficient to ensure that our deduction of the expression like (5) really is obtained in such a totally general way.

Universal Generalization (UG)

Where β is a pseudo-name (*not* a constant), one may in a deduction infer from a previously obtained sentence

$$\ldots \beta \ldots$$

any corresponding universally quantified sentence

$$(\mu)(\ldots \mu \ldots)$$

provided that: i. β does not occur in any **CP** assumption that has not been closed off;
ii. β does not occur in any preceding line obtained by **EI**;
iii. μ is a variable not already occurring in $\ldots \beta \ldots$;
iv. $\ldots \mu \ldots$ results from $\ldots \beta \ldots$ by replacing each occurrence of β with μ (making *no* other changes).[6]

Our discussion should have made it clear why applying **UG** to a constant is impermissible; from the fact that a particular individual satisfies a certain condition it surely does not follow that everything does. Provisos (i) and (ii) spell out the conditions under which our method of proof will be general enough to permit passage from a sentence in which a pseudo-name occurs to a corresponding universally quantified conclusion. Without (i) we could argue

1. Fu	Assumption	
2. $(y)(Fy)$	1 UG (mistake, violating proviso (i))	
3. $Fu \supset (y)(Fy)$	1–2 CP	
4. $(x)[Fx \supset (y)(Fy)]$	3 UG	

thus "proving" something that clearly cannot legitimately be established: that each thing is such that if it's a father then everything is. The **UG** at line (2) is clearly indefensible; 'u' cannot, at that point, be construed as standing for any old thing anyone might have picked—it is, at that point,

---

[6]As indicated in footnote 2 above, we will in practice only apply **UG** to the pseudo-names 'u', 'u''', 'u'''', and so on. Since only 'v', 'v''', and so on, are being used in applications of **EI**, then, proviso (ii) is less susceptible to violation than it would otherwise be.

being assumed to stand for one of those things that's a father, and obviously *assuming* that u is a father does not justify *concluding* that everything is like u in that respect.

Without (ii) we would be permitted to argue from the true premise 'There are fathers' to the false conclusion 'Everything is a father':

1. $(\exists x)(Fx)$ Premise

2. Fv 1 EI

3. $(x)(Fx)$ 2 UG (mistake, violating proviso (ii))

The **UG** at line (3) is obviously in error. The 'v' in line (2) cannot be construed as standing for anything anyone might have cared to pick; rather, it there stands for some *particular father* we chose at line (2) to consider.

Now if that were the whole story (ii) could be recast merely to block **UG** when the pseudo-name involved came into the deduction by an earlier appication of **EI**. But that would not be sufficient, since we could then deduce from the presumably true premise 'Everything resembles something' the false conclusion 'There is something that resembles everything':

1. $(x)(\exists y)(Rxy)$ Premise

2. $(\exists y)(Ruy)$ 1 UI

3. Ruv 2 EI

4. $(x)(Rxv)$ 3 UG (mistake, violating proviso (ii))

5. $(\exists y)(x)(Rxy)$ 4 EG

Even though it was 'v' that was obtained by **EI** at line (3), 'u' *occurs* in (3) and that is sufficient, according to proviso (ii), to block the **UG** at (4). To see intuitively why the **UG** at (4) cannot be defended but must be blocked, think of the argument as taking something like the following course:[7]

1. For anything x you pick, I can find a thing y that x resembles.

2. Oh, you pick u. OK, I can find a thing y that u resembles.

3. Here it is, v—the thing u you picked resembles this thing v.

4. So *whatever* you might have picked, it resembles v and so consequently—

YOU: Wait a minute, here! Sure, after you found I'd picked u you were able to find that thing v that u resembles. But suppose I'd picked a dif-

---

[7]See J. L. Mackie, "The Rules of Natural Deduction," *Analysis*, vol. 19 (1958), pp. 27–35.

ferent thing u' instead of u at the beginning. Of course you could have found *something* that u' resembles, premise one assures us of that; but there's no reason to suppose that the thing u' resembles is going to be the *same* thing (v) that you found out u resembles.

And of course, I have no reply.

The reason for the restriction in proviso (iii) should be both obvious and familiar; we must not allow such inferences as

$Fua \supset (\exists x)(Fxx)$
$\therefore (x)[Fxa \supset (\exists x)(Fxx)]$

for the conclusion would be ambiguous and it would be impossible to tell which of the two quantifiers bind which occurrences of 'x' in its consequent. Proviso (iv), however, involves a subtlety that is not troublesome in the other rules. If we were permitted to replace just *some* occurrences of the pseudo-name with variables when applying **UG** then we could "deduce" from the truism that everything is identical with itself, '$(x)(Ixx)$', the false conclusion that only one thing exists, '$(x)(y)(Ixy)$':

1. $(x)(Ixx)$ Premise
2. Iuu 1 UI
3. $(y)(Iuy)$ 2 UG (mistake, violating proviso (iv))
4. $(x)(y)(Ixy)$ 3 UG

The best we can legitimately get at line (3) would be

3'. $(y)(Iyy)$ 2 UG,

replacing *all* occurrences of 'u' in (2) with 'y' when doing the **UG**. The situation at line (2) may be described as follows: at that point our reader has picked something arbitrarily—we've used 'u' to refer to it—and we have succeeded so far in showing that no matter what was picked, it is identical with itself. Now from that it follows that '$(y)(Iyy)$', that is, (3') follows; but it does *not* follow, and we clearly have not shown, that no matter what was picked it is identical with everything!

Deductions for arguments whose conclusions contain universal quantifiers usually involve **CP**. Of course in such cases it is important to handle one's assumptions carefully. To deal with

1. $(\exists x)(Fx) \supset (y)(Fy \supset Iy)$ Premise   DEDUCE: $(x)(Fx \supset Ix)$,

for example, we should realize that our conclusion could be obtained by **UG** if we first got '$Fu \supset Iu$' and so discover the following deduction:

2. Fu	Assumption	
3. $(\exists x)(Fx)$	2 EG	
4. $(y)(Fy \supset Iy)$	1, 3 MP	
5. $Fu \supset Iu$	4 UI	
6. Iu	5, 2 MP	
7. $Fu \supset Iu$	2-6 CP	
8. $(x)(Fx \supset Ix)$	7 UG	

Since the **UG** occurs at step (8), *after* the assumption at line (2) containing 'u' has been closed off, proviso (i) on **UG** is satisfied (as are the others). Notice also how steps (3) and (4) are required to play (2) off against (1). Since ' $(\exists x)(Fx)$' is merely the antecedent of our premise, it is not possible to apply **EI** to it: **EI** (as well as our other quantification rules) is a rule of inference, not a rule of replacement, and so cannot be applied to parts of lines.

If we had been given instead the argument

1. $(x)Fx \supset Ix)$ Premise   DEDUCE: $(x)(Fx) \supset (\exists y)(Iy)$

then **CP** is again useful, but in a different way. Here our conclusion is itself a conditional, so we should assume its antecedent, quantifier and all, and deduce its consequent:

2. $(x)(Fx)$	Assumption	
3. Fu	2 UI	
4. $Fu \supset Iu$	1UI	
5. Iu	4, 3 MP	
6. $(\exists y)(Iy)$	5 EG	
7. $(x)(Fx) \supset (\exists y)(Iy)$	2-6 CP	

Only if we assume (2), with its quantifier, and do the **EG** at step (6) before closing off that assumption will the quantifiers '(x)' and '$(\exists y)$' appear in the antecedent and consequent of (7) where we want them.

EXERCISES

Using all four of the quantification rules when necessary, deduce the conclusions of the following arguments from their premises:

1. $(x)(Ax \supset Bx)$
   $(x)(Bx \supset Cx)$
   $\therefore (x)(Ax \supset Cx)$
2. $(x)[Ax \supset (Bx \& Cx)]$
   $\sim Bd$
   $\therefore \sim Ad$
3. $(x)[Dx \supset (Ex \& Fx)]$
   $(\exists x)(\sim Fx)$
   $\therefore (\exists x)(\sim Dx)$
4. $(\exists x)(Gx)$
   $(x)[Gx \supset (Ix \& Jx)]$
   $\therefore (\exists x)(Gx \& Ix)$
5. $(x)(Kx \supset Yx)$
   $(x)[(Kx \& Yx) \supset Zx]$
   $\therefore (x)(Kx \supset Zx)$
6. $(x)[(Bx \lor Cx) \supset Dx]$
   $(\exists y)(\sim Cy \lor \sim By)$
   $(\exists z)[\sim (Ez \lor \sim Cz)]$
   $\therefore (\exists x)(Dx)$
7. $(x)(Mx \supset \sim Lx)$
   $(\exists x)(Nx \& Lx)$
   $\therefore (\exists x)(Nx \& \sim Mx)$
8. $(\exists x)(y)(Oxy)$
   $\therefore (x)(\exists y)(Oyx)$
9. $(x)[(Tx \lor Ux) \supset Vx]$
   $(x)(Vx \supset \sim Wx)$
   $\therefore (x)[Tx \supset (Vx \& \sim Wx)]$
10. $(x)(Fx \supset Ix)$
    $\therefore (x)(Fx) \supset (y)(Iy)$
11. $(\exists x)(Px)$
    $(\exists x)(Ax)$
    $(y)(z)[(Py \& Az) \supset Lyz]$
    $\therefore (\exists x)(\exists y)(Py \& Lyx)$
12. $(\exists x)[Ax \& (y)(By \supset Rxy)]$
    $(x)[Ax \supset (y)(Sy \supset \sim Rxy)]$
    $\therefore (x)(Bx \supset \sim Sx)$
13. All gods are immortal. All gods are moral. Zeus is immoral and Socrates is mortal. Therefore neither Socrates nor Zeus is a god. (Gx-x is a god; Mx-x is mortal; Rx-x is moral; d-Zeus; s-Socrates.)
14. All multiples of six are multiples of two and also multiples of three. All multiples of two are even. Therefore all multiples of six are even.

(Mxy-x is a multiple of y; f-six; b-two; c-three; Ex-x is even.)

15. All even numbers are multiples of either six or two. All multiples of six are multiples of two. Therefore all even numbers are multiples of two. (Same notation as (9).)

16. Every triangle is either scalene or isosceles. Isosceles and equilateral triangles are symmetrical. Therefore all triangles that are not scalene are symmetrical. (Tx-x is a triangle; Sx-x is scalene; Ix-x is isosceles; Ex-x is equilateral; Rx-x is symmetrical.)

17. All moths are Lepidoptera. All insects are arthropods. So if all Lepidoptera are insects then all moths are arthropods. (Mx-x is a moth; Lx-x is a Lepidoptera; Ix-x is an insect; Ax-x is an arthropod.)

In Section 5.1 we noticed that where μ is any variable any two sentences of the forms

$$\sim(\exists\mu)(\ldots\mu\ldots) \qquad \text{and} \qquad (\mu)(\sim(\ldots\mu\ldots))$$

will be equivalent, as will any two sentences of the forms

$$\sim(\mu)(\ldots\mu\ldots) \qquad \text{and} \qquad (\exists\mu)(\sim(\ldots\mu\ldots)).$$

Accordingly, we add to our system the associated rule of replacement:

Quantifier Negation (QN)
$$\sim(\exists\mu)(\ldots\mu\ldots) :: (\mu)(\sim(\ldots\mu\ldots))$$
$$\sim(\mu)(\ldots\mu\ldots) :: (\exists\mu)(\sim(\ldots\mu\ldots))$$

We *need* not add this rule, for it is redundant and we won't be able to do anything with it that we could not have done more laboriously without it. But it will prove immensely *convenient*.

Without **QN**, for example, the deduction of '$(x)(\sim Fx)$' from '$\sim(\exists x)(Fx)$' requires a sneak attack. Our premise begins with a negation sign, *not* with a quantifier, so of course it is not possible to instantiate it: **EI** (as well as **EG**, **UI** and **UG**) is a rule of inference, *not* a rule of replacement, so it cannot be applied to mere parts of sentences. Instead, we have to try something like this:

1. $\sim(\exists x)(Fx)$ Premise   DEDUCE: $(x)(\sim Fx)$

┌→ 2. Fu Assumption

│   3. $(\exists x)(Fx)$ 2 EG

4. $Fu \supset (\exists x)(Fx)$ 2–3 CP

5. $\sim Fu$ 1, 4 MT

6. $(x)(\sim Fx)$ 5 UG

(notice that the **UG** at line (6) is legitimate — 'u' does not occur in any line of the deduction obtained by **EI**, and although 'u' does occur in the assumption made at line (2) that assumption was closed off at line (4) before **UG** was applied). But *with* **QN**, the deduction is a one-step affair:

1. $\sim(\exists x)(Fx)$ Premise   DEDUCE: $(x)(\sim Fx)$

2. $(x)(\sim Fx)$ 1 QN

The usefulness of **QN** will be even more obvious if readers compare their work on an earlier exercise with the deduction we can now construct:

1. $\sim(\exists x)(Px \ \& \ \sim Wx)$ Premise

2. Pa Premise   DEDUCE: Wa

3. $(x)[\sim(Px \ \& \ \sim Wx)]$ 1 QN

4. $\sim(Pa \ \& \ \sim Wa)$ 3 UI

5. $\sim Pa \ v \sim\sim Wa$ 4 De M

6. $\sim\sim Pa$ 2 DN

7. $\sim\sim Wa$ 5, 6 DS

8. Wa 7 DN

Just as it was important to realize that our other quantification rules were mere rules of *inference,* it is worth noticing that **QN** is a rule of *replacement.* **EI, EG, UI** and **UG** only permit us to operate on entire lines of deductions, but with **QN** we can go inside them and tinker with their parts:

1. $(x)(Fx) \supset (\exists x)(Fx)$ Premise   DEDUCE: $(x)(\sim Fx) \supset (\exists x)(\sim Fx)$

2. $\sim(\exists x)(Fx) \supset \sim(x)(Fx)$ 1 Trans

3. $(x)(\sim Fx) \supset \sim(x)(Fx)$ 2 QN
4. $(x)(\sim Fx) \supset (\exists x)(\sim Fx)$ 3 QN

EXERCISES

Deduce the conclusions of each of the following arguments from their premises:

1. $\sim(x)(Fx \vee Gx)$
   $\therefore (\exists y)(\sim Fy)$

2. $(x)(\sim Fx)$
   $(\exists y)(Gy) \supset (\exists x)(Fx)$
   $\therefore (y)(\sim Gy)$

3. $\sim(\exists x)(Fx \vee Gx)$
   $\therefore (x)(\sim Fx)\ \&\ (y)(\sim Gy)$

4. $(\exists x)[\sim(Fx\ \&\ Gx)]$
   $\sim(x)(Fx\ \&\ Gx) \supset \sim(z)(Hz)$
   $(\exists z)(\sim Hz) \supset (y)(Gy)$
   $\therefore \sim(\exists y)(\sim Gy)$

7.2 Stratagems

Skill in discovering deductions goes hand in hand with a thorough familiarity with the workings of our rules, and both come with practice. Skill in developing deductions *in a reasonable amount of time* calls for more: ingenuity sometimes, along with a store of strategies and devices to draw on and a firm resolve to avoid beginnings that seem unlikely to lead anywhere.

Some of the arguments we look at have existentially quantified conclusions. In dealing with such arguments the following general rule of thumb is often helpful:

When the conclusion of an argument is existentially quantified, instantiate everything you can; then sort through the resulting lines for something from which the conclusion can be obtained by **EG**.

For example, suppose we're given the argument:

1. $(\exists x)(Hx)$ Premise
2. $(\exists x)\{Qx \,\&\, (y)[Hy \supset (Rx \,\&\, Hx)]\}$ Premise   DEDUCE: $(\exists x)(Hx \,\&\, Qx)$

How are we to obtain '$(\exists x)(Hx \,\&\, Qx)$'? It is barely conceivable that we will somehow deduce something like '$Ab \supset (\exists x)(Hx \,\&\, Qx)$' at line (22) and '$Ab$' at line (37) and so get '$(\exists x)(Hx \,\&\, Qx)$' by **MP**. But it seems far more likely that we will, instead, deduce something like '$Hv \,\&\, Qv$' or '$Hv' \,\&\, Qv'$', and then infer our conclusion by **EG**. There seems no better way to begin moving in that direction than to instantiate our two premises by **EI**, being careful to use different letters for each so as to avoid running afoul of proviso (i) on **EI**:

3. Hv 1 EI
4. $Qv' \,\&\, (y)[Hy \supset (Rv' \,\&\, Hv')]$ 2 EI

These moves suggest at least a few more:

5. $(y)[Hy \supset (Rv' \,\&\, Hv')] \,\&\, Qv'$ 4 Com
6. $(y)[Hy \supset (Rv' \,\&\, Hv')]$ 5 Simp

And obviously (6) can be used to accumulate even more information about v:

7. $Hv \supset (Rv' \,\&\, Hv')$ 6 UI

Glancing again at our conclusion, we see that obtaining it is just a matter of sorting through (3)–(7) until we get 'Hv' & Qv'':

8. Rv' & Hv'	3, 7 MP	
9. Hv' & Rv'	8 Com	
10. Hv'	9 Simp	
11. Qv'	4 Simp	
12. Hv' & Qv'	10, 11 Conj	
13. $(\exists x)(Hx$ & $Qx)$	12 EG	

Another timesaving rule of thumb suggests itself so readily that readers have perhaps already discovered it for themselves:

*When instantiating premises, do all the **EI**'s first.*

When, for example, we approach the argument

1. $(x)[(Bx \lor Cx) \supset Dx]$	Premise	
2. $(\exists z)[\sim(Ez \lor \sim Cz)]$	Premise	DEDUCE: $(\exists x)(Dx)$

we first notice the existentially quantified conclusion and see that 'Dv' would give it to us in one more step. Turning to the business of instantiating premises, we should begin with the **EI** on (2)

3. $\sim(Ev \lor \sim Cv)$	2 EI

intending to use **UI** next on (1) in order to accumulate more information about v:

4. $(Bv \lor Cv) \supset Dv$	1 UI

We can then find 'Dv' and finish:

5. $\sim Ev$ & $\sim\sim Cv$	3 De M
6. $\sim Ev$ & Cv	5 DN

7. *Cv* & ~*Ev*	6 Com
8. *Cv*	7 Simp
9. *Cv* v *Bv*	8 Add
10. *Bv* v *Cv*	9 Com
11. *Dv*	10, 4 MP
12. (∃x)(*Dx*)	11 EG

Had we made the *strategic* mistake of beginning with **UI**

3. (*Bv* v *Cv*) ⊃ *Dv*   1 UI

what would we do? We could *not* obtain '~(*Ev* v ~*Cv*)' from (2) by **EI** now because proviso (i) on **EI** blocks use of the letter '*v*' (which here already occurs in (3)). So we have to pick a *new* letter for the **EI**:

4. ~(*Ev'* v ~*Cv'*)   2 EI

and what good has the **UI** at (3) done us? To be able to get anywhere with (4) it is *v'* we need to know about, not *v*, and so we must **UI** (1) again, to '*v''* this time:

5. (*Bv'* v *Cv'*) ⊃ *Dv'*   1 UI

Notice that (3) is totally useless to us.

When the conclusion of an argument is existentially quantified, then, we begin by instantiating everything we can, being careful to do the **EI**'s first; we then sort through the resulting lines, looking for something from which our conclusion can be inferred by **EG**. But what do we do when our conclusion is *universally* quantified? Most often, in such cases, the conclusion will be of the form

(µ)(...µ... ⊃ ---µ---)

and so conditional proof suggests itself; if we assume ...*u*... (say) and deduce ---*u*---, **CP** will give us ...*u*... ⊃ ---*u*--- and our conclusion will follow by **UG**:

┌→ ...*u*...	Assumption
.	
.	
.	
---*u*---	

...*u*... ⊃ ---*u*---   CP

$(\mu)(...\mu... \supset ---\mu---)$ UG

Consider, for example, the argument

1. (*x*)[*Fx* ⊃ (*Gx* & *Hx*)]   Premise

2. (*x*)(*Gx* ⊃ *Ix*)   Premise   DEDUCE: (*x*)(*Fx* ⊃ *Ix*)

To end with '(*x*)(*Fx* ⊃ *Ix*)' we aim first for '*Fu* ⊃ *Iu*', intending to obtain the former from the latter by **UG**; and since '*Fu* ⊃ *Iu*' is a conditional we try to establish it by **CP**, assuming its antecedent

┌→ 3. *Fu*   Assumption

and trying to deduce '*Iu*'. To make some use of our assumption it is obvious that we should next instantiate our first premise:

4. *Fu* ⊃ (*Gu* & *Hu*)	1 UI
5. *Gu* & *Hu*	3, 4 MP

And now it is clear that **UI** on (2) will quickly lead us to our goal:

6. *Gu* ⊃ *Iu*	2 UI
7. *Gu*	5 Simp
8. *Iu*	6, 7 MP

Coming out of the conditional proof we have

└_____

9. *Fu* ⊃ *Iu*   3–8 CP

and **UG** now gives us our conclusion:

10. $(x)(Fx \supset Ix)$ 9 UG

Of course **UG** cannot be applied to a pseudo-name occurring in an assumption *still operative*; but the assumption (3) in which 'u' occurs here has been closed off at line (9) so the **UG** at (10) is perfectly all right.

Notice how important it is to begin with the assumption 'Fu' rather than

┌→ 3. $(x)(Fx)$ Assumption

Had we begun with this assumption, we could have proceeded

| 4. Fu 3 UI
| 5. $Fu \supset (Gu \& Hu)$ 1 UI
| 6. $Gu \& Hu$ 4, 5 MP
| 7. $Gu \supset Iu$ 2 UI
| 8. Gu 6 Simp
| 9. Iu 7, 8 MP

but now we would be stymied. If we give up the assumption at this point, to get

└────────────

10. $(x)(Fx) \supset Iu$ 3–9 CP

we have *no* way of obtaining the desired '$(x)(Fx \supset Ix)$' from (10). The best we can do is apply **UG** to get

11. $(y)[(x)(Fx) \supset Iy]$ 10 UG

And if we try the **UG** on (9) before applying **CP** we can only finish:

| 10. $(x)(Ix)$ 9 UG

11. $(x)(Fx) \supset (x)(Ix)$ 3–10 CP

We might formulate the general rule of thumb involved this way:

When the conclusion of an argument is of the form $(\mu)(\ldots\mu\ldots \supset ---\mu---)$, try assuming ... β ... (where β is a pseudo-name) and deducing --- β ---, intending to use **CP** *followed by* **UG**.

At the same time, we must notice that there are many arguments which will *not* yield to this *precise* line of attack but only to a simple but easily overlooked modification of it.

Let us confront the innocent-looking argument:

1. $(x)(y)(Exy \supset Eyx)$ Premise

2. $(x)(y)(Exy \supset Exx)$ Premise   DEDUCE: $(x)[(\exists y)(Eyx) \supset Exx]$

Since the conclusion is universally quantified, we naturally think of establishing it by first deducing '$(\exists y)(Eyu) \supset Euu$' and then using **UG**. And since '$(\exists y)(Eyu) \supset Euu$' is a conditional, our first thought is to try to establish it by assuming its antecedent and going for the consequent (this is the step in our thinking that will later have to be modified):

3. $(\exists y)(Eyu)$ Assumption

The obvious thing to do next is to use **EI** on (3), being careful to use a new pseudo-name, and then look for some help from (1):

4. Evu 3 EI

5. $(y)(Evy \supset Eyv)$ 1 UI

6. $Evu \supset Euv$ 5 UI

7. Euv 4, 6 MP

Things seem to be going well as we turn to (2).

8. $(y)(Euy \supset Euu)$ 2 UI

9. $Euv \supset Euu$ 8 UI

10. Euu 7, 9 MP

11. $(\exists y)(Eyu) \supset Euu$ 3–10 CP

All we need is that final **UG**...

12. $(x)[(\exists y)(Eyx) \supset Exx]$ 11 UG — MISTAKE!

The **UG** is *not* permitted. For 'u' occurs in a line—namely (4)—that was obtained by **EI**; and proviso (ii) on **UG** accordingly *blocks* any later attempt to **UG** on 'u'. There does not seem to be any easy way out. Without applying **EI** to (3) it is difficult to see how (10) can be obtained; and if we do apply **EI** to (3) we cannot get (12).

Now in fact (10) can be obtained by indirect proof without using **EI** on (3), but there is a simpler way to carry out the deduction. Look again at the premises, and at the conclusion. The conclusion is universally quantified, and we would like to establish it by first deducing '$(\exists y)(Eyu) \supset Euu$' and then using **UG**. Our *first* thought was to establish the latter by assuming '$(\exists y)(Eyu)$' and deducing 'Euu' from that assumption; unfortunately, in the process of carrying out that conditional proof we made a move that blocked the **UG** we hoped to employ at the end. But there is another way to establish '$(\exists y)(Eyu) \supset Euu$': we can assume the *negation* of its consequent and try to deduce the *negation* of its antecedent; **CP** will then give us '$\sim Euu \supset \sim(\exists y)(Eyu)$' and **Trans** will then turn that into the conditional we want:

1. $(x)(y)(Exy \supset Eyx)$	Premise	
2. $(x)(y)(Exy \supset Exx)$	Premise	DEDUCE: $(x)[(\exists y)(Eyx) \supset Exx]$
$\rightarrow$ 3. $\sim Euu$	Assumption	
4. $(y)(Euy \supset Euu)$	2 UI	
5. $Euu' \supset Euu$	4 UI	
6. $\sim Euu'$	3, 5 MT	
7. $(y)(Eu'y \supset Eyu')$	1 UI	
8. $Eu'u \supset Euu'$	7 UI	
9. $\sim Eu'u$	6, 8 MT	
10. $(y)(\sim Eyu)$	9 UG	
11. $\sim(\exists y)(Eyu)$	10 QN	
12. $\sim Euu \supset \sim(\exists y)(Eyu)$	3–11 CP	
13. $(\exists y)(Eyu) \supset Euu$	12 Trans	
14. $(x)[(\exists y)(Eyx) \supset Exx]$	13 UG	

There is nothing suspect about the **UG** at line (14) now — **EI** was never employed in the deduction. Notice, too, that the **UG** at line (10) is perfectly correct — the letter 'u''' does not occur in the assumption (3).

We might now amend our earlier rule of thumb for dealing with arguments whose conclusions are universally quantified conditionals:

When the conclusion of an argument is of the form $(\mu)(\ldots\mu\ldots \supset ---\mu---)$, try assuming $\ldots\beta\ldots$ and deducing $---\beta---$, intending to use **CP** *followed by* **UG**. *ALTERNATELY, TRY ASSUMING $\sim(---\beta---)$ AND DEDUCING $\sim(\ldots\beta\ldots)$, INTENDING TO USE* **CP** *FOLLOWED BY* **TRANS** *AND THEN* **UG**.

EXERCISES

I.   Deduce the conclusions of the following arguments from their premises. You may find it helpful to formulate a general strategy for each before attempting it.

1. $(x)[(Ax \ \& \ Bx) \supset Cx]$
   $(x)(Dx \supset Bx)$
   $(\exists x)(Ax \ \& \ Dx)$
   $\therefore (\exists x)(Cx \ \& \ Dx)$

2. $(x)[Hx \supset (Bx \equiv \sim Tx)]$
   $(\exists x)(Hx \ \& \ \sim Bx)$
   $(\exists x)(Hx \ \& \ \sim Tx)$
   $\therefore (\exists x)(Hx \ \& \ Tx)$

3. $(\exists x)(Ix \ \& \ Jx)$
   $(\exists x)[Kx \ \& \ (y)(Iy \supset Lxy)]$
   $\therefore (\exists x)(\exists y) [Iy \ \& \ (Jy \ \& \ Lxy)]$

4. $(x)[(\exists y)(Rxy) \supset (\exists z)(\sim Wz)]$
   $(\exists y)(\exists z)(Ryz \ \& \ Hz)$
   $(x)(\sim Hx \supset Wx)$
   $\therefore (\exists z)(\sim Wz \ \& \ Hz)$

5. $(\exists x)(Ux) \supset (y)(z)(Syz \supset Uy)$
   $(\exists x)(\exists y)[(Px \ \& \ Qx) \ \& \ (Ry \ \& \ Syx)]$
   $(x)(Qx \supset Ux)$
   $\therefore (\exists x)(Rx \ \& \ Ux)$

II. Formulate a strategy, either mentally or in writing, for deducing the conclusion of each of the following arguments from its premises. Then construct the deduction.

1. $(x)[Ax \supset (Bx \ v \ Cx)]$
   $(x)[(Bx \ v \ Cx) \supset Dx]$
   $\therefore (x)[(Ax \ \& \ \sim Bx) \supset (Dx \ \& \ Cx)]$

2. $(x)(y)[(Hx \& Lxy) \supset Lya]$
   $(x)(y)(Lxa \supset Lxy)$
   $\therefore (x)(y)[(Hx \& Lxy) \supset Lyx]$

3. $(x)(Rx \supset {\sim}Gx)$
   $(x)[({\sim}Ex \& {\sim}Gx) \supset Hx]$
   $\therefore (x)(Ex \supset Gx) \supset (y)(Ry \supset Hy)$

4. $(x)(y)(Py \supset Mxy)$
   $(x)[(\exists y)(Mxy) \supset {\sim}Nx]$
   $\therefore (x)(Ox \supset Px) \supset (y)(Oy \supset {\sim}Ny)$

5. $(x)(y)(z)(Sxy \supset Syz)$
   $\therefore (x)(y)[Sxy \supset (x)(Szx)]$

III. Deduce the conclusion of each of the following arguments from its premises; be prepared to combine and modify strategies already used, and to invent some new ones of your own.

1. $(\exists x)(Px \& Lxa)$
   $(y)(Py \supset Lay)$
   $(x)(y)[(Lxa \& Lay) \supset Lxy]$
   $\therefore (\exists x)[Px \& (y)(Py \supset Lxy)]$

2. $(x)(y)(Sxy \supset Syx)$
   $(x)(Sxd \supset {\sim}Tbx)$
   $(x)(y)(Rax \supset Txy)$
   $\therefore Rab \supset {\sim}Sdc$

3. $(\exists x)(Fx) \supset (y)(Ay \supset By)$
   $(\exists x)(Gx) \supset (y)(Cy \supset {\sim}By)$
   $\therefore (\exists x)(Gx \& Fx) \supset (y)(Cy \supset {\sim}Ay)$

4. $(x)(Cx \supset Fx)$
   $\therefore (x)[(\exists y)(Cy \& Dxy) \supset (\exists z)(Fz \& Dxz)]$

5. $(x)(\exists y)(Lxy)$
   $\therefore (y)(\exists z)(Lyz)$

IV. Some of the following arguments are valid and some are invalid. Symbolize each. Then deduce the conclusions of the valid ones from their premises and after studying Chapter 6 show that the rest are invalid.

1. Chicago is a city. Consequently, every town near Chicago is near at least one city. (*c*-Chicago; *Cx-x* is a city; *Tx-x* is a town; *Nxy-x* is near *y*.)

2. Some sentences have truth-values. No pattern of marks traced in the sand by the wind has a truth-value. Therefore, some patterns of marks traced in the sand by the wind are not sentences. (*Sx-x* is a sentence; *Tx-x* has a truth-value; *Px-x* is a pattern of marks traced in the sand by the wind.)

3. Everyone doubts something or other. Anyone who doubts something can be certain of at least one thing. Therefore each person can be certain of at least one thing. (Px-x is a person; Dxy-x doubts y; Cxy-x can be certain of y.)

4. Every applicant will be interviewed by telephone or flown to the factory. No applicants living abroad will be flown to the factory. Thus all applicants living abroad will be interviewed by telephone. (Ax-x is an applicant; Ix-x will be interviewed by telephone; Fx-x will be flown to the factory; Lx-x is living abroad.)

5. If there is a bull market underway, then only glamor stocks rise faster than International Toothpick. A competitor of International Toothpick is rising faster than it. Therefore, if none of International Toothpick's competitors are glamor stocks, no bull market is underway. (Bx-x is a bull market; Ux-x is underway; Gx-x is a glamor stock; Rxy-x is rising faster than y; i-International Toothpick; Cxy-x is a competitor of y.)

6. If everything is in motion, then everything has velocity. Hence if something is in motion, then everything has velocity. (Mx-x is in motion; Vx-x has velocity.)

7. Every particle that has mass exerts a force on everything. Everything that something exerts a force on has mass. So every particle exerts a force on everything that exerts a force on it. (Px-x is a particle; Mx-x has mass; Exy-x exerts a force on y.)

8. Every valid argument Jones presented today was unsound. An argument is sound if and only if it is valid and all its premises are true. Therefore if the Ontological Argument is one of the valid arguments Jones presented today, at least one of its premises is untrue. (Vx-x is valid; Ax-x is an argument; j-Jones; Pxy-x presented y today; Sx-x is sound; Rxy-x is a premise of y; Tx-x is true; a-the Ontological Argument.)

9. Every spy who was undetected eluded at least one guard. Some spies were armed and so was each guard they eluded. No armed spy was detected. Therefore, there exists a guard who is such that he is a spy only if he was undetected. (Sx-x is a spy; Dx-x was detected; Exy-x eluded y; Gx-x is a guard; Ax-x was armed.)

10. If one thing resembles a second thing, then the second resembles the first. If one thing resembles a second thing and the second resembles a third, then the first thing resembles the third. Therefore everything resembles itself. (Rxy-x resembles y.)

11. There is a politician who is despised by all citizens who despise any politician at all. Every citizen despises at least

one politician. Therefore there is a politician who is despised by every citizen. (*Px-x* is a politician; *Dxy-x* despises *y*; *Cx-x* is a citizen.)

12. Those who have both a pen and a sword are mightier than those who have just one of the two. Not all of those with both a pen and a sword are mightier than all tyrants who lack at least one of them. It follows that there is a tyrant who has neither a pen nor a sword. (*Px-x* has a pen; *Sx-x* has a sword; *Mxy-x* is mightier than *y*; *Tx-x* is a tyrant.)

13. For each positive integer *x*, there exists a sentence containing more than *x* words. Any sentence that contains, for each positive integer *x*, more than *x* words is of infinite length. Therefore there exists at least one sentence of infinite length. (*Ix-x* is a positive integer; *Sx-x* is a sentence; *Cxy-x* contains more than *y* words; *Lx-x* is of infinite length.)

*14. If each even integer is such that it satisfies Goldbach's hypothesis if all smaller even integers satisfy that hypothesis, then every even integer satisfies Goldbach's hypothesis. Therefore if for each even integer that does not satisfy Goldbach's hypothesis there exists a smaller even integer that does not satisfy it either, then all even integers satisfy Goldbach's hypothesis. (*Ex-x* is an even integer; *Sx-x* satisfies Goldbach's hypothesis; *Lxy-x* is less than *y*.)

*15. If I am morally permitted to perform any action, then everyone is morally permitted to perform that action. If something awful occurs if everyone performs a certain action then no one is morally permitted to perform that action. So if something awful occurs if I perform a certain action then I am not morally permitted to perform that action. (*Mxy-x* is morally permitted to perform *y*; *Ax-x* is an action; *Hx-x* is a person; *Wx-x* is awful; *Ox-x* occurs; *Pxy-x* performs *y*; *i*-me.)

*16. Any argument of interest to all philosophers is worthy of everyone's attention. Anything worthy of anyone's attention ought to be considered by that person. Therefore if each philosopher is interested in all arguments propounded by those who disagree with him, then any argument propounded by anyone who disagrees with all philosophers ought to be considered by everyone. (*Ax-x* is an argument; *Ixy-x* is interested in *y*; *Px-x* is a philosopher; *Wxy-x* is worthy of *y*'s attention; *Hx-x* is a person; *Oxy-x* ought to consider *y*; *Rxy-x* propounds *y*; *Dxy-x* disagrees with *y*.)

*7.3 Theorems of Quantificational Logic

In Chapter 3 we noticed that our rules of inference and replacement for the sentential calculus could be used not only to deduce conclusions from premises but also, on occasion, to deduce certain sentences from no premises at all. Such sentences were the *theorems* of sentential logic and they corresponded exactly, as we saw in Chapter 4, to the truth-table tautologies. We have since then extended our symbolic apparatus and our rules to cover quantification as well, and have concentrated thus far on establishing validity; but of course quantificational logic has theorems of its own. They correspond to the logical truths discussed in Section 1 of the preceding chapter: the sentences provable in our system of quantificational logic are precisely those sentences that are true (to use the terminology of that chapter) on all interpretations.[8]

The sentence '$(x)(Fx \lor \sim Fx)$', for example, is clearly true no matter what predicate expression 'F' is taken to abbreviate and no matter what universe of discourse the variable 'x' is taken to range over. To show that it is a theorem we have only to construct the following very simple proof:

1. $Fu \lor \sim Fu$ E-M I

2. $(x)(Fx \lor \sim Fx)$ 1 UG

Most interesting theorems require more involved proofs, of course. Since no premises are available to us in such situations, the first lines of our deductions can only be justified by **E-M I** (as above) or as assumptions for **CP**. CP is especially useful in establishing conditionals like '$(\exists y)(x)(Cyx) \supset (x)(\exists y)(Cyx)$':

1. $(\exists y)(x)(Cyx)$	Assumption
2. $(x)(Cvx)$	1 EI
3. Cvu	2 UI
4. $(\exists y)(Cyu)$	3 EG
5. $(x)(\exists y)(Cyx)$	4 UG
6. $(\exists y)(x)(Cyx) \supset (x)(\exists y)(Cyx)$	1–5 CP

(Notice that the **UG** at line (5) is correct: the pseudo-name to which it is being applied, 'u', occurs neither in the assumption nor in the line obtained by **EI**.)

---

[8]The rather complicated demonstration of this claim for a system of quantificational logic essentially equivalent in deductive power to the present one is carried out, for example, in E. Mendelson, *Introduction to Mathematical Logic*, 3rd ed., Monterey: Wadsworth and Brooks, 1987.

Often more complicated strategies are advantageous. If we try proving the theorem '$[(x)(Fx) \lor (y)(Gy)] \supset (z)(Fx \lor Gz)$' by assuming its antecedent and proceeding to deduce its consequent, we are immediately faced with a problem:

$\rightarrow$ 1. $(x)(Fx) \lor (y)(Gy)$ Assumption

(1), as it stands, cannot be instantiated. Before using it we have to do some additional work that does not suggest itself immediately:

$\rightarrow$ 2. $(x)(Fx)$		Assumption
3. Fu		2 UI
4. $Fu \lor Gu$		3 Add
5. $(z)(Fz \lor Gz)$		4 UG
6. $(x)(Fx) \supset (z)(Fz \lor Gz)$		2–5 CP
$\rightarrow$ 7. $(y)(Gy)$		Assumption
8. Gu		7 UI
9. $Gu \lor Fu$		8 Add
10. $Fu \lor Gu$		9 Com
11. $(z)(Fz \lor Gz)$		10 UG
12. $(y)(Gy) \supset (z)(Fz \lor Gz)$		7–11 CP
13. $(z)(Fz \lor Gz) \lor (z)(Fz \lor Gz)$		1, 6, 12 CD
14. $(z)(Fz \lor Gz) \lor \sim(z)(Fz \lor Gz)$		E-M I
15. $[(z)(Fz \lor Gz) \lor (z)(Fz \lor Gz)] \, \& \, [(z)(Fz \lor Gz) \lor \sim(z)(Fz \lor Gz)]$		13, 14 Conj
16. $(z)(Fz \lor Gz) \lor [(z)(Fz \lor Gz) \, \& \, \sim(z)(Fz \lor Gz)]$		15 Dist
17. $[(z)(Fz \lor Gz) \, \& \, \sim(z)(Fz \lor Gz)] \lor (z)(Fz \lor Gz)$		16 Com
18. $\sim(z)(Fz \lor Gz) \lor \sim\sim(z)(Fz \lor Gz)$		E-M I
19. $\sim[(z)(Fz \lor Gz) \, \& \, \sim(z)(Fz \lor Gz)]$		18 De M
20. $(z)(Fz \lor Gz)$		19, 17 DS
21. $[(x)(Fx) \lor (y)(Gy)] \supset (z)(Fz \lor Gz)$		1–20 CP

The proof is considerably less messy if we employ a more familiar strategy and instead begin by assuming the negation of the consequent of the theorem we're trying to establish and aim for the negation of its antecedent, finishing off with a **Trans** step:

→	1. $\sim(z)(Fz \vee Gz)$	Assumption
	2. $(\exists z)[\sim(Fz \vee Gz)]$	1 QN
	3. $\sim(Fv \vee Gv)$	2 EI
	4. $\sim Fv \,\&\, \sim Gv$	3 De M
	5. $\sim Fv$	4 Simp
	6. $(\exists x)(\sim Fx)$	5 EG
	7. $\sim(x)(Fx)$	6 QN
	8. $\sim Gv$	4 Com, Simp
	9. $(\exists y)(\sim Gy)$	8 EG
	10. $\sim(y)(Gy)$	9 QN
	11. $\sim(x)(Fx) \,\&\, \sim(y)(Gy)$	7, 10 Conj
	12. $\sim[(x)(Fx) \vee (y)(Gy)]$	11 De M
	13. $\sim(z)(Fz \vee Gz) \supset \sim[(x)(Fx) \vee (y)(Gy)]$	1–12 CP
	14. $[(x)(Fx) \vee (y)(Gy)] \supset (z)(Fz \vee Gz)$	13 Trans

When the theorem to be established is a biconditional, it is often best to divide the proof into two parts, establishing the two halves of the biconditional separately. To prove '$(x)(Fx \,\&\, Gx) \equiv [(y)(Fy) \,\&\, (z)(Gz)]$', for example, we might begin with:

→	1. $(x)(Fx \,\&\, Gx)$	Assumption
	2. $Fu \,\&\, Gu$	1 UI
	3. Fu	2 Simp
	4. $(y)(Fy)$	3 UG
	5. Gu	2 Com, Simp
	6. $(z)(Gz)$	5 UG
	7. $(y)(Fy) \,\&\, (z)(Gz)$	4, 6 Conj
	8. $(x)(Fx \,\&\, Gx) \supset [(y)(Fy) \,\&\, (z)(Gz)]$	1–7 CP

With that half out of the way, we can proceed in the other direction:

9. (y)(Fy) & (z)(Gz)		Assumption
10. (y)(Fy)		9 Simp
11. Fu		10 UI
12. (z)(Gz)		9 Com, Simp
13. Gu		12 UI
14. Fu & Gu		11, 13 Conj
15. (x)(Fx & Gx)		14 UG

16. [(y)(Fy) & (z)(Gz)] ⊃ (x)(Fx & Gx)   9–15 CP

Then it is simply a matter of putting the two sub-conclusions together:

17. {(x)(Fx & Gx) ⊃ [(y)(Fy) & (z)(Gz)]} &
    {[(y)(Fy) & (z)(Gz)] ⊃ (x)(Fx & Gx)}   8, 16 Conj

18. (x)(Fx & Gx) ≡ [(y)(Fy) & (z)(Gz)]   17 Equiv.

In showing (18) to be a theorem — and hence a logical truth — we have shown the sentences '(x)(Fx & Gx)' and '(y)(Fy) & (z)(Gz)' to be *equivalent*. For clearly two sentences whose corresponding biconditional is a theorem must be true under precisely the same conditions. Thus, the deductive system of the present chapter can be used to establish equivalence as well as validity and theoremhood.

    Of course the proofs of certain theorems call for a combination of strategies. To establish '(x){~ [Gx ≡ (Gx & Fx)] ⊃ ~Fx}' we might argue:

1. Fu	Assumption
2. Gu	Assumption
3. Gu & Fu	2, 1 Conj
4. Gu ⊃ (Gu & Fu)	2–3 CP
5. Gu & Fu	Assumption
6. Gu	5 Simp
7. (Gu & Fu) ⊃ Gu	5–6 CP
8. [Gu ⊃ (Gu & Fu)] & [(Gu & Fu) ⊃ Gu]	4, 7 Conj
9. Gu ≡ (Gu & Fu)	8 Equiv

10. $Fu \supset [Gu \equiv (Gu \& Fu)]$ 1–9 CP

11. $\sim[Gu \equiv (Gu \& Fu)] \supset \sim Fu$ 10 Trans

12. $(x)\{\sim[Gx \equiv (Gx \& Fx)] \supset \sim Fx\}$ 11 UG

And as a sort of last resort we can always employ an *indirect* proof, assuming the contradictory of the theorem we're trying to prove and deducing a contradiction from that assumption:

1. $(\exists y)(x)(Rxx \& \sim Rxy)$	Assumption
2. $(x)(Rxx \& \sim Rxv)$	1 EI
3. $Rvv \& \sim Rvv$	2 UI
4. $(\exists y)(x)(Rxx \& \sim Rxy) \supset (Rvv \& \sim Rvv)$	1–3 CP
5. $\sim Rvv \vee \sim\sim Rvv$	E-M I
6. $\sim(Rvv \& \sim Rvv)$	5 De M
7. $\sim(\exists y)(x)(Rxx \& \sim Rxy)$	6, 4 MT

EXERCISES

I. Show that each of the following is a theorem of our system of quantificational logic:
1. $(\exists x)(Fx \& Gx) \supset [(\exists y)(Fy) \& (\exists z)(Gz)]$
2. $(x)[(\exists y)(\sim Gxy) \supset \sim(z)(Gxz)]$
3. $(\exists x)(Fx \vee Gx) \equiv [(\exists y)(Fy) \vee (\exists z)(Gz)]$
4. $(x)[Mx \supset (\exists y)(Vy)] \equiv [(\exists x)(Mx) \supset (\exists y)(Vy)]$
5. $(x)[Fx \supset (y)(Gxy)] \equiv (x)(y)(Fx \supset Gxy)$
6. $(x)(Fx \vee Ga) \equiv [(y)(Fy) \vee Ga]$

II. Prove the following theorems, and then discuss their relationship to our rules **EI, EG, UI**, and **UG**:
1. $(\exists y)[(\exists x)(Fx) \supset Fy]$
2. $(y)[Fy \supset (\exists x)(Fx)]$
3. $(y)[(x)(Fx) \supset Fy]$
4. $(\exists y)[Fy \supset (x)(Fx)]$

III. *Without* using the rule **QN**, prove the following biconditionals:
1. $\sim(\exists x)(Fx) \equiv (y)(\sim Fy)$
2. $\sim(x)(Fx) \equiv (\exists y)(\sim Fy)$

*7.4 Rules for '='

Certain arguments whose premises or conclusions contain the identity sign, '=', suggest the addition of two final rules. The first allows us, in effect, to introduce the truism 'Each thing is identical with itself' into any deduction, so we call it

Identity Introduction (II)

At any point in any deduction, one may introduce the sentence

$(x)(x = x)$.

Using this rule we can handle the argument 'The numbers 3, 7 and 21 are factors of the number 231; therefore one of the factors of the number 231 is identical with the number 3' along the following lines:

1. *(Fad & Fbd) & Fcd*	Premise	DEDUCE: $(\exists x)(Fxd \ \& \ x = a)$
2. *Fad & Fbd*	1 Simp	
3. *Fad*	2 Simp	
4. $(x)(x = x)$	II	
5. $a = a$	4 UI	
6. *Fad* & $a = a$	3, 5 Conj	
7. $(\exists x)(Fxd \ \& \ x = a)$	6 EG	

But another rule is needed for the argument

No island is a peninsula
Formosa is an island off the coast of China
Formosa and Taiwan are identical
Therefore Taiwan is not a peninsula.

We can begin easily enough:

1. $(x)(Ix \supset \sim Px)$	Premise	
2. *If & Ofc*	Premise	
3. $f = t$	Premise	DEDUCE: $\sim Pt$
4. *If*	2 Simp	
5. *If* $\supset \sim Pf$	1 UI	
6. $\sim Pf$	4, 5 MP	

But now what? It seems entirely clear (on the intuitive level) that our conclusion follows from (6) and (3) so that we are entitled to move to

7. $\sim Pt$ 6, 3 ?

After all, if Formosa is not a peninsula—as (6) tells us—and Formosa and Taiwan are *identical*—as (3) says—then Taiwan must not be a peninsula either. With slogans about substituting "equals" for "equals" in mind, we are naturally led to formulate the following rule:

Identity Elimination (IE)

Where α is any pseudo-name or constant, and γ is any pseudo-name or constant, one may in a deduction infer from two preceding sentences

   $...\alpha...$

and

   $\alpha = \gamma$ (or alternately, $\gamma = \alpha$)

any corresponding sentence

   $...\gamma...$

provided that $...\gamma...$ is just like $...\alpha...$ except for containing γ at one or more places where $...\alpha...$ contains α.

Pseudo-names as well as constants must be allowed if we are to carry out the deduction of 'Any place Bob is visiting this summer is in the Southwest' from 'The Grand Canyon is in the Southwest' and 'If Bob is visiting any place this summer, it is the Grand Canyon'.

1. Sg	Premise	
2. $(x)[(Px \ \& \ Vbx) \supset g = x]$	Premise	DEDUCE:
→ 3. $Pu \ \& \ Vbu$	Assumption	$(x)[(Px \ \& \ Vbx) \supset Sx]$
4. $(Pu \ \& \ Vbu) \supset g = u$	2 UI	
5. $g = u$	3, 4 MP	
6. Su	1, 5 IE	
7. $(Pu \ \& \ Vbu) \supset Su$	3–6 CP	
8. $(x)[(Px \ \& \ Vbx) \supset Sx]$	7 UG	

And **IE** permits replacement of only *some* of the occurrences of the one pseudo-name or constant by the other so that we may argue:

1. $f = t$	Premise	DEDUCE: $t = f$
2. $(x)(x = x)$	II	
3. $f = f$	2 UI	
4. $t = f$	3, 1 IE	

The trick here is to use (1) as the second premise of the **IE** rule and replace just *one* 'f' in the otherwise useless-looking (3). Obviously the same deduction can be generalized to provide a proof for '$(x)(y)(x = y \supset y = x)$', i.e. a proof that the identity relation is symmetric (hereafter, uses of **UI** following those of **II** will be combined with the **II** move):

1. $u = u'$	Assumption
2. $u = u$	II, UI
3. $u' = u$	2, 1 IE
4. $u = u' \supset u' = u$	1-3 CP
5. $(y)(u = y \supset y = u)$	4 UG
6. $(x)(y)(x = y \supset y = x)$	5 UG

A modification of this strategy permits one to prove that identity is transitive as well (see Exercise I (1) below), and reflexivity is taken care of directly by **II**.

Since Formosa is an island and China is not, China is of course not identical with Formosa. To draw the latter conclusion from the premise, **IE** can be combined with an indirect approach:

1. $If \,\&\sim Ic$	Premise	DEDUCE: $f \neq c$
2. $f = c$	Assumption	
3. If	1 Simp	
4. Ic	3, 2 IE	
5. $f = c \supset Ic$	2–4 CP	
6. $\sim Ic$	1 Com, Simp	
7. $f \neq c$	5, 6 MT	

(writing '$f \neq c$' for '$\sim f = c$', as usual).

Sentences involving definite descriptions pick up identity signs as they go into symbols and so provide additional examples of **IE** in use. Notice its role, for example, in the following deduction of 'Only a leader of the reform faction will be nominated' from 'The leader of the party is a leader of the reform faction' and 'The nominees will all be leaders of the party':

1. $(\exists x)\{[Lxp \ \& \ (y)(Lyp \supset x = y.] \ \& \ Lxf\}$ Premise

2. $(x)(Nx \supset Lxp)$ Premise

DEDUCE: $(x)(Nx \supset Lxf)$

→	3. Nu	Assumption
	4. $Nu \supset Lup$	2 UI
	5. Lup	3, 4 MP
	6. $[Lvp \ \& \ (y)(Lyp \supset v = y)] \ \& \ Lvf$	1 EI
	7. $Lvp \ \& \ (y)(Lyp \supset v = y)$	6 Simp
	8. $(y)(Lyp \supset v = y)$	7 Com, Simp
	9. $Lup \supset v = u$	8 UI
	10. $v = u$	9, 5 MP
	11. Lvf	6 Com, Simp
	12. Luf	11, 10 IE
13. $Nu \supset Luf$		3–12 CP
14. $(x)(Nx \supset Lxf)$		13 UG

The careful reader will have noticed that we have been especially timorous in discussing **IE**. We have shown how it can be used to justify quite naturally the steps required for a number of deductions, but we have not treated this rule with the same measure of confidence and finality as earlier ones — and there is a good reason for this.

IE licenses, in effect, the substitution of one term for another whenever we can establish the corresponding identity sentence. One wonders, then, how **IE** may be justified or formulated so as to block replacement of 'Venus' with 'Hesperus' in

Poets sometimes play on the fact that Venus and Hesperus are one and the same,

on the basis of the true identity sentence

Venus is identical with Hesperus.

Such a replacement yields a disconcerting result:

Poets sometimes play on the fact that Hesperus and Hesperus are one and the same.

The problem of accounting for the invalidity of these arguments while licensing those given earlier has long been a subject of philosophical research.[9] We mention only that the contexts involved in the exercises below are ones in which there is general agreement that application of **IE** is perfectly acceptable. The reader should be on his guard, however, when he is considering application of **IE** in contexts encountered outside textbook examples.

EXERCISES

I. Prove each of the following theorems:
    1. $(x)(y)(z)[(x = y \ \& \ y = z) \supset x = z]$
    2. $(x)(y)(z)[(x = y \ \& \ x = z) \supset z = y]$
    3. $(x)(y)[(Fx \ \& \ \sim Fy) \supset x \neq y]$
    4. $(x)(y)[x = y \supset (Fx \equiv Fy)]$

II. Deduce the conclusions of each of the following arguments from their premises:
    1. Taiwan is either an island or a peninsula. Formosa is not a peninsula. Formosa and Taiwan are identical. Therefore Taiwan is an island.
    2. Samuel Clemens was a humorist but Samuel Morse was not. Samuel Clemens is identical with Mark Twain. Therefore Samuel Morse is not identical with Mark Twain.
    3. Lewis Carroll is the author of *Alice in Wonderland.* Anyone who authored *Alice in Wonderland* is the sole author of *The Game of Logic.* Lewis Carroll is Charles Dodgson. Therefore Charles Dodgson is the sole author of *The Game of Logic.*
    4. At least one person is Jones' sister. All of Jones's sisters visited him in the hospital. At most one person visited Jones in the hospital. Therefore exactly one person visited Jones in the hospital.
    5. Jones has at least one sister. All of Jones's sisters visited him in the hospital. At most one person visited Jones in the hospital. Anyone who visited Jones in the hospital brought flowers. Of course all of Jones' sisters are persons. Therefore Jones's only sister brought flowers.

---

[9]See, for example, W.V. Quine, *From a Logical Point of View,* 2nd ed., Cambridge: Harvard Univ. Press, 1961, Chapter VIII("Reference and Modality"); L. Linsky, *Referring,* London: Routledge and Kegan Paul, 1967, Chapters 2-4; and the references cited in footnote 12 of Section 5.3.

6. At least one well-qualified person will be hired. At least one experienced person will be hired. At most one person will be hired. Therefore the person who will be hired is both well-qualified and experienced.

7. $(x)(y)[(Rxy \ \& \ Ryx) \supset x = y]$
   $(x)(y)(Rxy \supset Ryx)$
   $\therefore (x)[(\exists y)(Rxy \lor Ryx) \supset Rxx]$

8. $(x)(y)[(Rxy \ \& \ Ryx) \supset x = y]$
   $(x)(y)(Rxy \supset Ryx)$
   $\therefore (x)\{[(\exists y)(Rxy) \lor (\exists z)(Rxz)] \supset Rxx\}$

9. The author of *Waverley* is bald. Scott is the author of *Waverley*. Therefore Scott is bald.

III. From the single premise 'If every member of one set is a member of a second, and every member of the second is a member of the first, then the sets in question are identical' (sometimes called the *Axiom of Extensionality*), deduce each of the following.

   1. For any two sets there exists at most one set whose members are precisely those things that are members of both of the others.

   2. If there exists a set with no members, then there is exactly one set with no members.

   3. If x and y are sets, then x and y are identical if and only if each member of either is a member of the other.

*IV. Show that the sentence

   I want a girl just like the girl that married dear old Dad

   when translated with the description having *primary* occurrence implies the translation in which it has *secondary* occurrence.

APPENDIX A

NORMAL FORMS AND CIRCUITS

By a **literal** we shall mean any sentence letter or its negation; thus 'A' and '$\sim B$' are literals but '$\sim\sim A$' and '$A \& B$' are not. By a **disjunction of literals** we mean any disjunction all of whose disjuncts are literals but in which no literal occurs more than once. In this appendix, internal parentheses for extended disjunctions will be omitted, and hence not only '$A \vee \sim B$' but '$\sim A \vee B \vee C$' are disjunctions of literals. But '$A \vee A$' is not, and neither is '$\sim\sim A \vee B \vee (A \& C)$', since the first and third disjuncts are not literals. We shall likewise omit internal parentheses for extended conjunctions, and shall mean by a **conjunction of literals** any conjunction all of whose conjuncts are literals but in which no literal occurs more than once. Both '$A \& \sim B$' and '$\sim A \& B \& C$' are thus conjunctions of literals, but '$A \& A$' and '$A \& \sim\sim B \& (A \vee C)$' are not. Moreover, '$A \& \sim A$' is a conjunction of literals, though the same *letter* occurs more than once.

We are now in a position to define two concepts of sentential logic that will be of importance to us here.

A truth-functional sentence is in **conjunctive normal form (CNF)** if and only if it is either (i) a literal, (ii) a disjunction of literals, or (iii) a conjunction all of whose conjuncts are either literals or disjunctions of literals but in which no conjunct occurs more than once.[1]

Thus 'A', '$A \vee \sim B$', '$A \& (C \vee B \vee \sim D)$', '$(A \vee B) \& (B \vee \sim C)$', and '$A \& B$' are in CNF, but '$(C \vee D) \& (C \vee D)$', '$(A \& B) \vee D$', and '$\sim\sim B \& (A \vee D)$' are not.

---

[1] The notion of conjunctive normal form employed here differs from that of Chapter 4 because of our present insistence on lack of duplication.

A truth-functional sentence is in **disjunctive normal form (DNF)** if and only if it is either (i) a literal, (ii) a conjunction of literals, or (iii) a disjunction all of whose disjuncts are either literals or conjunctions of literals but in which no disjunct occurs more than once.

Thus, '$\sim A$', 'A & $\sim B$', 'A v (C & B & $\sim D$)', '(A & B) v (B & $\sim C$)', and 'A v $\sim B$' are in DNF, but '(C & D) v (C & D)', '(A v B) & D', and '$\sim \sim B$ v (A & D)' are not.

Notice then that any literal is in *both* CNF and DNF. Moreover, disjunctions of literals are in DNF as well as CNF, and conjunctions of literals are in CNF as well as DNF. However, '(A & $\sim B$) v D' is in DNF only, and (A v $\sim B$) & D' is in CNF only. On the other hand, 'A v (B & (C v D))' and '$\sim \sim A$ & (B v C)' are in neither form. It should be further noted that any sentence in either of the two forms will have the connective '$\sim$' attaching only to single sentence letters. Thus '$\sim$(A & B)' is neither in CNF nor in DNF, and the same goes for '$\sim$(A v B)'.

Any truth-functional sentence whatever is either *in* CNF or is *equivalent* to (or, by Section 4.2, *interdeducible* with) a sentence in CNF. In the latter case, it may be *reduced* to the CNF sentence by repeated application of the rules of replacement in Chapter 3, together with some additional rules that will simplify our work. That is, we shall employ **DN**, **Com**, **Assoc**, as well as expanded versions of **DeM** and **Dist**, to accommodate extended conjunctions and disjunctions:

De Morgan: $\sim (p_1 \, \& \, \ldots \, \& \, p_n) :: \sim p_1 \, v \, \ldots \, v \sim p_n$
$\sim (p_1 \, v \, \ldots \, v \, p_n) :: \sim p_1 \, \& \, \ldots \, \& \sim p_n$

Distribution: $p \, \& \, (q_1 \, v \, \ldots \, v \, q_n) :: (p \, \& \, q_1) \, v \, \ldots \, v \, (p \, \& \, q_n)$
$p \, v \, (q_1 \, \& \, \ldots \, \& \, q_n) :: (p \, v \, q_1) \, \& \, \ldots \, \& \, (p \, v \, q_n)$

In addition, we shall employ the **Law of Idempotence** (abbreviated **Idem**), whose legitimacy is easily established by a two-row truth-table.

Idempotence: $p :: p \, v \, p$
$p :: p \, \& \, p$

Clearly a conjunction with a sentence of the form 'p v $\sim p$' as a conjunct is equivalent to the other conjunct alone since the whole will be true if and only if that latter conjunct is true, and false otherwise. Similarly, a disjunction with a sentence of the form 'p & $\sim p$' as a disjunct is equivalent to the other disjunct by itself. Thus, a conjunct of the form 'p v $\sim p$' and a disjunct of the form 'p & $\sim p$' are both superfluous and may therefore be eliminated. We may add, then, two more rules, **Tautology Elimination (TE)** and **Contradiction Elimination (CE)**:

$TE: p :: p \, \& \, (q \lor \sim q)$
$CE: p :: p \lor (q \, \& \, \sim q)$

We shall also employ the familiar **Impl** and **Equiv** principles for eliminating '⊃' and '≡' respectively.

To reduce a sentence not in CNF to one that is, we proceed in five basic steps (though in simple cases one or more of these may not be necessary). First, use **Impl** and **Equiv** to eliminate any occurrences of '⊃' and '≡'. Second, use **De M** repeatedly to drive all curls that do not attach to single letters as far inside as possible. Third, use **DN** to eliminate excess curls. Fourth, use **Com, Assoc,** and **Dist** to obtain (if necessary) conjunctions of disjunctions, and fifth, use **Idem** and **CE** to eliminate any redundancies. As an example, let us consider:

(1) $\sim \{[B \, \& \sim ((C \, \& \sim B) \lor \sim D)] \lor (A \supset D)\}.$

At the first step, we eliminate '⊃':

$\sim \{[B \, \& \sim ((C \, \& \sim B) \lor \sim D)] \lor (\sim A \lor D)\}.$ Impl

Next, we use **De M** to drive curls inside:

$\sim [B \, \& \sim ((C \, \& \sim B) \lor \sim D)] \, \& \sim (\sim A \lor D)$ De M
$[\sim B \lor \sim \sim ((C \, \& \sim B) \lor \sim D)] \, \& \, (\sim \sim A \, \& \sim D)$ De M (twice)

We now apply **DN** to obtain:

$[\sim B \lor ((C \, \& \sim B) \lor \sim D)] \, \& \, (A \, \& \sim D)$ DN (twice)

Now at our fourth step we apply **Com** and **Dist** to the right disjunct of the lefthand side (for brevity, we shall combine **Com** with applications of other rules into one step):

$[\sim B \lor ((\sim D \lor C) \, \& \, (\sim D \lor \sim B))] \, \& \, (A \, \& \sim D)$ Com, Dist

Clearly another **Dist** move is required on the lefthand side:

$[(\sim B \vee (\sim D \vee C)) \& (\sim B \vee (\sim D \vee \sim B))] \& (A \& \sim D)$ Dist

Note that this entire expression is of the form '$(p \& q) \& (r \& s)$', and we know by **Assoc** that placement of these parentheses is immaterial. So in accordance with our earlier decision to omit them, we now drop them out, citing **Assoc** as our justification.[2]

$(\sim B \vee (\sim D \vee C)) \& (\sim B \vee (\sim D \vee \sim B)) \& A \& \sim D$ Assoc

The inner parentheses of the two disjunctions are also dispensable, and dropping them now gives:

$(\sim B \vee \sim D \vee C) \& (\sim B \vee \sim D \vee \sim B) \& A \& \sim D$ Assoc

Finally, the literal '$\sim B$' occurs twice in the disjunction '$\sim B \vee \sim D \vee \sim B$'. So **Com** and **Idem** will eliminate one of them, and the result is a sentence in CNF:

(1′) $(\sim B \vee \sim D \vee C) \& (\sim B \vee \sim D) \& A \& \sim D$ Com, Idem

(1′) is a conjunction each of whose conjuncts is either a literal or a disjunction of literals. Of course a reduction to CNF is rarely as involved as the above; it was used here so that all five steps could be fully illustrated.

Often, a sentence reduced to CNF can be further reduced to a *simpler* CNF. An elementary example would be:

(2) $(A \vee B) \& (A \vee \sim B) \& (\sim A \vee B)$.

Though (2) is in CNF, it is equivalent to another, less complex sentence in CNF. Let us use **Dist** on the two left conjuncts; we thus obtain

$(A \vee (B \& \sim B)) \& (\sim A \vee B)$ Dist

---

[2]We are obviously using a different version of **Assoc** here than in Chapter 3. There, it was employed to *move about* parentheses, whereas here it is employed to get rid of them. But since they *can* be moved about without affecting truth-conditions, they are dispensable, and **Assoc** is the rule that makes this idea explicit.

By **CE**, the left side reduces to '*A*', giving

A & (~A v B)   CE

Dist, **Com**, and **CE** again now yield the simpler CNF.

(A & ~A) v (A & B)   Dist
(2') A & B   Com, CE

The foregoing considerations apply equally to DNF. Any truth-functional sentence is either *in* DNF or else is *equivalent* to a sentence in DNF, and moreover the same replacement rules and reduction process may be used to find, for any sentence not in DNF, one that is. Consider for example:

(3) (A & ~B) ⊃ [C ≡ (C v D)].

Proceeding as we did before, we derive successively:

~(A & ~B) v [C ≡ (C v D)]   Impl
~(A & ~B) v [(C & (C v D)) v (~C & ~(C v D))]   Equiv
(~A v ~~B) v [(C & (C v D)) v (~C & (~C & ~D))]   De M (twice)
(~A v B) v [(C & (C v D)) v (~C & (~C & ~D))]   DN
(~A v B) v [((C & C) v (C & D)) v (~C & (~C & ~D))]   Dist
~A v B v (C & C) v (C & D) v (~C & ~C & ~D)   Assoc
(3') ~A v B v C v (C & D) v (~C & ~D)   Idem (twice)

Clearly (3') is in DNF, since each disjunct is either a literal or else a conjunction of literals. However, a simpler DNF is obtainable. We commute the third and fourth disjuncts and then distribute.

~A v B v (C & D) v ((C v ~C) & (C v ~D))   Com, Dist
~A v B v (C & D) v (C v ~D)   Com, TE
~A v B v C v ~D v (C & D)   Com, Assoc
~A v B v C v ((~D v C) & (~D v D))   Dist
~A v B v C v (~D v C)   Com, TE
~A v B v C v C v ~D   Com, Assoc
(3") ~A v B v C v ~D   Idem

Normal forms have many uses. In Chapter 4 a similar notion of CNF is employed to establish some theoretical results concerning the concepts and methods of the preceding two chapters. In addition, certain logical features of a complex truth-functional sentence are often more easily discerned if it is reduced to an equivalent CNF or DNF – for example, whether it is a tautology, contradiction, or neither. In general, the truth-conditions of a sentence are displayed more perspicuously if it is in normal form than otherwise.

EXERCISES

I. Put each of the following into *either* CNF *or* DNF – whichever seems most readily obtainable. Then see if there is a *simpler* normal form.
    1. $(A \mathrel{\&} \sim B) \supset (C \lor \sim A)$
    2. $\sim A \mathrel{\&} (\sim A \lor B)$
    3. $\sim [D \lor (B \mathrel{\&} \sim C)]$
    4. $D \equiv \sim (E \mathrel{\&} F)$
    5. $[(A \supset B) \supset A] \supset A$
    6. $\sim [A \supset (B \mathrel{\&} A)]$
    7. $(A \lor B) \supset [(A \supset (B \mathrel{\&} C)) \supset (C \mathrel{\&} \sim A)]$

We now examine some special cases of normal forms.

A truth-functional sentence is a **conjunctive Boolean expansion** if and only if it is in conjunctive normal form and, moreover, every sentence letter occurring (with or without an accompanying negation sign) in *any* of its conjuncts occurs in *all* its conjuncts (perhaps occurring negated in some of these conjuncts and unnegated in the others).[3]

Thus, '$A \lor \sim C$', '$(\sim A \lor \sim B) \mathrel{\&} (\sim A \lor B) \mathrel{\&} (A \lor B)$', and '$(E \lor \sim F \lor D) \mathrel{\&} (\sim E \lor F \lor D)$' are all conjunctive Boolean expansions (hereafter CBE). Every CBE is in CNF, though of course the reverse does not hold. Rather than attempt a definition as tortuous as the above for *disjunctive* expansions, we shall try another approach – one that brings out the complementary nature of the two kinds of expansion.

A truth-functional sentence is a **disjunctive Boolean expansion** if and only if it can be obtained from a CBE by replacing each '&' with 'v' and vice versa.

---

[3]Sentences with only one component letter are a limiting case: 'A', '$\sim B$' and '$\sim C \mathrel{\&} C$' count as CBE's.

Correspondingly, then, '$A \& \sim C$', '$(\sim A \& \sim B) \vee (\sim A \& B) \vee (A \& B)$', and '$(E \& \sim F \& D) \vee (\sim E \& F \& D)$' are all disjunctive Boolean expansions (DBE). Clearly a DBE is in DNF.

Every truth-functional sentence either is a CBE (a DBE) or else is equivalent to a CBE (and a DBE) containing the same number of letters. Reduction of a sentence to a Boolean expansion can be achieved using the same replacement rules used earlier for normal forms. However, there is another way of obtaining a Boolean expansion. One may go directly to the truth-table for the sentence, and produce its expansion from the assignment of T's and F's in its column. Consider for example

A	B	$A \equiv B$
T	T	T
T	F	F
F	T	F
F	F	T

A	B	C	$(A \vee B) \supset (A \& \sim C)$
T	T	T	F
T	T	F	T
T	F	T	F
T	F	F	T
F	T	T	F
F	T	F	F
F	F	T	T
F	F	F	T

Now a glance at the table tells us that '$A \equiv B$' is true in just two cases: when 'A' and 'B' are both true (that is, '$A \& B$' is true) or when 'A' and 'B' are both false (that is, when '$\sim A \& \sim B$' is true). Thus '$A \equiv B$' is true under just the same conditions as the disjunction '$(A \& B) \vee (\sim A \& \sim B)$', which is a DBE. To find a DBE, therefore, we need only to look at those rows in which a T is assigned; for each such row we conjoin the component letters, negating those to which an F is assigned in that row, and disjoin all the conjunctions thus formed. If just one row has a T, we of course will not have a disjunction but instead a conjunction of literals — any sentence having a T in just the third row of the lefthand table may be reexpressed as the DBE '$\sim A \& B$'. But where two or more rows have a T, the DBE will have exactly as many disjuncts as there are rows assigned T.

Turning now to '$(A \vee B) \supset (A \& \sim C)$', it will be true just in case either '$A \& B \& \sim C$' is true (second row) or '$A \& \sim B \& \sim C$' is true (fourth row) or '$\sim A \& \sim B \& C$' is true (seventh row) or '$\sim A \& \sim B \& \sim C$' is true (last row). Hence the desired DBE is:

$(A \& B \& \sim C) \vee (A \& \sim B \& \sim C) \vee (\sim A \& \sim B \& C) \vee (\sim A \& \sim B \& \sim C)$.

To obtain a CBE, consider first the truth-table column for '$A \vee B$'.

It has an **F** in *only* the last row, and since '*A* ∨ *B*' is a disjunction of literals in which no letter occurs more than once, it is a CBE. We can similarly produce CBE's for sentences whose columns have more than one **F**. For each **F** row, we disjoin the component letters, but this time negating those to which a **T** is assigned, and then conjoin these disjunctions. Thus there will be exactly as many conjuncts of a CBE as there are rows assigned **F**. For '*A* ≡ *B*', then, we look at the second and third rows, disjoining the component letters in each case and negating those to which **T** is assigned. Conjoining the two disjunctions yields the CBE '(~*A* ∨ *B*) & (*A* ∨ ~*B*)'. By the same token, the CBE for '(*A* ∨ *B*) ⊃ (*A* & ~*C*)' is:

(~*A* ∨ ~*B* ∨ ~*C*) & (~*A* ∨ *B* ∨ ~*C*) & (*A* ∨ ~*B* ∨ ~*C*) & (*A* ∨ ~*B* ∨ *C*).

　　The symbolic notation developed in Chapter 1 for sentential logic has other applications as well. Thus far, letters like '*A*' and '*B*' have been taken to represent particular sentences, and variables '*p*', '*q*', and so on to range over those sentences. The connectives '&', '∨', and '~' have served as abbreviations for certain key uses of 'and', 'or', and 'not' respectively, and the letters **T** and **F** have been employed for the two truth-values. However, all of these expressions may be interpreted in other, highly useful ways. We shall now examine briefly such an interpretation — one which pertains to the behavior of electrical circuits.

　　A **two-terminal switching circuit** is an array of wires and switches with a wire leading into the circuit and a wire leading out. The letters '*A*', '*B*', and so on will be used to represent switches, and each switch must either be *open* or *closed*. Current passes through the switch *only* when it is closed. A simple, one-switch circuit may be illustrated as follows:

(1)　　

Switch *A* is indicated by the closely spaced dots, and is in the closed position. In open position, the circuit appears like this.

Some circuits require a switch that is closed just in case another switch is open, and vice versa. In such a case, if one of the switches is labeled '*A*', the other is represented as '~*A*'. Moreover, circuits sometimes require two or more switches to operate in tandem — opening and closing together — and thus controlled by a master switch. Those switches of a circuit linked together in such a way are all represented by the same letter. Consider then the circuit:

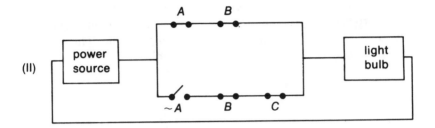

(II)

Whenever *A* is closed, ~*A* is open, and vice versa. The two switches labeled '*B*', however, are always *both* open or (as above) *both* closed. Since ~*A* is open, current clearly cannot get to the light bulb via the lower route. But since *all* switches are closed on the other route, current can nonetheless reach the bulb.

Now a **parallel circuit** is one in which switches are placed as follows:

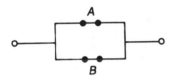

Clearly current will flow from one terminal to the other just in case *at least one* of the switches is *closed*. Since a disjunction is true just in case one disjunct is true, parallel circuits are analogues of disjunctions. The above circuit may be represented in our notation by '*A* v *B*'.

A **series circuit** is just what the name implies — two or more switches are connected serially, as in

Here, current will flow if and only if *both* switches are closed — just as a conjunction is true if and only if both conjuncts are. This circuit may therefore be represented by '*A* & *B*'.

A **series-parallel circuit** is any circuit that is either series or parallel or a combination of the two. Circuit (II) above is such a combination; it consists of two series circuits in parallel. Current thus reaches the bulb if and only if *either A* and *B* are both closed *or* if ~*A*, *B*, and *C* are all closed. The circuit is therefore expressible by '(*A* & *B*) v (~*A* & *B* & *C*)', which incidentally is in DNF.

Given any sentence of our notation whose only connectives are '&', 'v', and '~' and in which '~' attaches *only* to single letters, we can construct a corresponding circuit that will pass current under conditions exactly analogous to those on which the sentence is true. The sentence '(*A* & ~*B*) v [*B* & (*C* v *A*)]' corresponds to the circuit:

Here, current can flow from terminal to terminal, since all switches but *B* are closed. While the current cannot get through the lower series circuit, it can get through the upper. By the same token, the sentence '(*A* & ~*B*) v [*B* & (*C* v *A*)]' is true when '*A*' and '*C*' are true but '*B*' is false, for on that assignment the left disjunct is true.

But now let us suppose the switches are positioned differently — and analogously, we suppose a different truth-value assignment to the component letters.

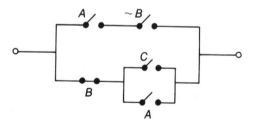

Current cannot get through the upper series, since both switches there are open; similarly, it cannot pass through the lower series, because both switches in the component parallel circuit are open. So current will not flow when switches *A* and *C* are open and *B* is closed. Likewise, '(*A* & ~*B*) v [*B* & (*C* v *A*)]' is false when '*A*' and '*C*' are false and '*B*' is true, for each component conjunction has at least one false conjunct.

It should be obvious by now that since circuits are expressible by the compound sentences of our notation, the complete behavior of a circuit for all possible positions of its switches is displayed by the truth-table for its corresponding sentence. And this means that we can *design* circuits to behave in a desired fashion by first constructing the truth-table and then finding a sentence having that column of **T**'s and **F**'s; from there, the required circuit is easily produced. As an elementary example, suppose the light in Sam's basement is connected to a switch in the kitchen and one in the living room. And further suppose that (for whatever reason) Sam wishes it to be on just in case exactly one of the switches is thrown. Where 'K' stands for the kitchen switch and 'L' for the living-room switch, the behavior of the required circuit is specified by the following table:

K	L	
T	T	F
T	F	T
F	T	T
F	F	F

(In the columns for 'K' and 'L', a **T** in effect means *closed*, an **F** means *open*. But in the third column, a **T** represents *current flows* while an **F** represents *current does not flow*.) To find a sentence having that arrangement of **T**'s and **F**'s, we need only produce a Boolean expansion in the way described earlier. The DBE in this case is '$(K \& \sim L) \lor (\sim K \& L)$', and hence a circuit that will do the job is

(III)

Here we have two series circuits in parallel. But suppose the layout of the house is such that two parallel circuits in series would be more convenient to install. In such case, we need only produce the CBE from our table, '$(\sim K \lor \sim L) \& (K \lor L)$. Hence another circuit that will do the *same* job is

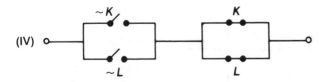

(IV)

Just as '$(K \& \sim L) \lor (\sim K \& L)$' and '$(\sim K \lor \sim L) \& (K \lor L)$' are equivalent sentences, so (III) and (IV) may be described as "equivalent" circuits. The two sentences are *true* under the same conditions, and *current will flow* through the two circuits under the same conditions.

It is not always easy to find the truth-table that will lay out the behavior we desire of a circuit. But once found, it is a simple task to find the circuit, since Boolean expansions provide a mechanical technique for producing the required sentences. Suppose the Supreme Court of Transylvania consists of three judges: Allslop, Bickley, and Chief Justice Coffman. Each has a switch that when closed allows current to flow in a circuit wired to a light atop the courthouse. Each closes his switch when his verdict is "guilty" and leaves it open otherwise. The rules say that a defendant is found guilty only if *at least two* judges so decree, but that Coffman and only he retains veto power. That is, even if *both* of the others close their switches, current will not flow to the light unless Coffman closes his as well. However, in any other circumstance where at least two switches are closed, current reaches the light. To find the appropriate circuit, we first construct a truth-table using obvious letters for the three switches:

A	B	C	
T	T	T	T
T	T	F	F
T	F	T	T
T	F	F	F
F	T	T	T
F	T	F	F
F	F	T	F
F	F	F	F

The second row illustrates Coffman's veto power; current does not flow even though other justices have closed their switches. Only in the first, third, and fifth circumstances does the light go on. The corresponding DBE, then, is:

$(A \& B \& C) \lor (A \& \sim B \& C) \lor (\sim A \& B \& C).$

Thus a circuit that will satisfy our conditions would be:

Were an electrician to install such a circuit, however, he would in all likelihood be fired. For there are many *simpler* circuits – that is, with fewer switches and wires – that are equivalent. To find such a circuit, we can try to reduce the above DBE to a simpler normal form sentence. We first commute (three times) to bring 'C' to the left of each disjunct and then distribute, obtaining

C & [(A & B) v (A & ~B) v (~A & B)]   Com, Dist

We now proceed to distribute the part '(A & B) v (A & ~B)':

C & [(A & (B v ~B)) v (~A & B)]   Dist

We now obtain:

C & [A v (~A & B)]   TE
C & [(A v ~A) & (A v B)]   Dist
C & (A v B)   Com, TE

Thus we have our equivalent circuit.

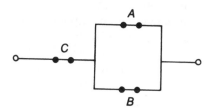

So while the DBE (or of course CBE) can always be found with no mental exertion whatever, it rarely yields the simplest circuit for the task at hand.

EXERCISES

I. Construct circuits for each of the following.
    1. A v [B & (~B v (C & ~A))]
    2. (A v ~A) v (B & ~B)
    3. {{[(A v B) & (A v ~B) & B] v C} & D} v (C & ~B)

II. Design a circuit for a voting indicator that lights if and only if either voter A votes *yes* or else *exactly one* of voters B and C vote *yes*. Can you find a circuit with just five switches that will do the job?

Two-terminal switching circuits are "two-way" circuits in that the electrical current can move in either direction, but the circuits employed in computers and similar machines allow current to move in just one direction and are often called **sequential circuits.** Such circuits therefore have both "inputs" and "outputs," and the current passing from the former to the latter is to be thought of as consisting of discrete "pulses" of two kinds, one of which shall be labelled '1' ("high voltage," "high pulse," etc.) and the other '0' ("low voltage," "low pulse"). Sequential circuits operate synchronously, that is, with each input being activated by a 1-pulse or a 0-pulse simultaneously with the others.

A sequential circuit is a network of wires and "gates" (or "nodes"), each of which is either in state 1 or state 0 at any given time of operation. The gates are basically devices that convert one or more input pulses to an output pulse in a manner logically analogous to the truth conditions that define our truth-functional connectives. More specifically, a NOT-gate (or "inverter") converts a 0-pulse to a 1-pulse and vice versa:

input ————————[~]—————————output

and thus may be expressed by the standard truth-table for negation, using '1' instead of 'T' and '0' in place of 'F'. Similarly, an AND-gate converts two (or more) input pulses to an output pulse according to the truth-table for conjunction; that is, if both (or all) inputs are 1-pulses, the output pulse is also 1, otherwise the output pulse is 0. An OR-gate yields an output pulse of 1 just in case at least one of the input pulses is a 1. An IFF-gate converts a pair of 1-pulses to a 1-pulse and does the same for a pair of 0-pulses; otherwise its output is a 0-pulse. These gates may be represented as follows (the letters 'A' and 'B' here represent not English sentences but input sources):

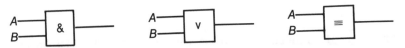

and the familiar tables below describe their behavior completely:

A	B	$A \& B$	$A \lor B$	$A \equiv B$
1	1	1	1	1
1	0	0	1	0
0	1	0	1	0
0	0	0	0	1

We are now in a position to examine a simple example of a sequential circuit having three inputs and two outputs (of course it could have any number of each):

inputs   outputs

The outputs are represented here as *functions of* the inputs — meaning that whether a given output pulse is 1 or 0 is determined solely by what the input pulses are at sources *A, B* and *C*. This of course is exactly analogous to how compound sentences of our notation stand to their components; the former's truth value is uniquely determined by those for its component ·sentence letters. By tracing the wiring paths back from the outputs to the inputs, then, we can find compound sentences which express the behavior of the circuit's outputs. The main connective of the compound will correspond to the first gate encountered in the tracing back. Thus, the output labelled '*f(A, B, C)*' may be expressed by ' ~*A* v (*B* & *C*)' and the other output by '*B* ≡ (*A* & *C*)'. We could of course get by without an IFF-gate since we know that '*p* ≡ *q*' is equivalent to '(*p* & *q*) v (~*p* & ~*q*)', and hence it could be replaced by a sub-circuit like:

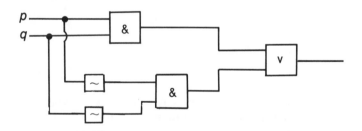

but it will simplify matters if we use the IFF-gate instead. Summarizing, then, a compound expression like '*B* ≡ (*A* & *C*)' on its usual (Chapter Two) interpretation expresses a sentence whose truth-value is uniquely determined by the truth-values of its component sentences; here, however, it has the distinct but parallel interpretation wherein it expresses an output whose electrical pulse at any given moment is uniquely determined by the pulses of the circuit's inputs.

To perform some significant work with sequential circuits, we will need another device usually called a *delay*. It is basically a very simple "memory" unit which briefly stores information for subsequent use. The initial output of a delay is always a 0-pulse; thereafter, its input at any given moment becomes its output at the next moment (if the input at time t is a 1-pulse, so is its output at time $t + 1$). We will represent a delay by the following simple diagram:

When connected to a sequential circuit, the pulse of one of the circuit's outputs is fed into the delay's input, and the delay's output pulse is then supplied to one of the circuit's inputs at the next moment of operation. When circuits and delays are so arranged, the result is called a *finite automaton,* and digital computers may be looked at as finite automata composed of millions of circuits and delays operating at millions of pulses per second.[4] If we connect the above sequential circuit with a delay, the result is a very simple finite automaton:

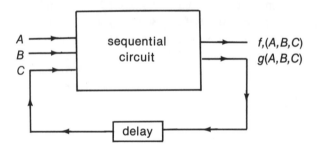

The output at $g(A,B,C)$ at a given time t is thus channelled back through the delay and becomes an "internal" input at source C for time $t + 1$. The delay must be synchronized with the other two input sources so that all three pulses will arrive at the circuit precisely at $t + 1$.

A finite automaton is hence a deterministic system with finite sets of inputs, internal states (i.e., delay outputs) and output states, and its operation over a given duration is determined by its input "history" as follows: the initial internal state is 0 on each wire, thereafter the input and

---

[4]It should be noted however that many of the computers in use nowadays employ sequential circuits whose design differs from the type presented here, but they are still of course based on the same logical principles.

internal states at any time t determine the internal state at $t + 1$. The output at any such t is uniquely determined by the input and internal states at t in a manner precisely analogous to the determination of a compound sentence's truth-value given those of its component sentences.

It is not difficult to design a finite automaton that will perform a relatively simple task like computing the sum of any pair of natural numbers. The numbers to be added will each be expressed in terms of a unique set of input pulses, their sum as a set of output pulses, and the internal states will correspond roughly to the intermediate calculations (or "partial sums"). It is to this task that we now turn; we will briefly sketch the design of a simple "adder." But to do this, we must first take note of a special notation for representing the natural numbers.

In the *binary-number system*, each natural number is expressed as a unique string of 1's and 0's. The principle underlying this representation is that each such number is the sum of a unique set of powers of 2, i.e. of 2^0, 2^1, 2^2, 2^3, etc. (or more briefly, 1, 2, 4, 8, ...). In binary notation, the digits in the binary expression are ordered from left to right according to the descending powers of 2, and '1' indicates a power of 2 that occurs in the sum of the integer in question while each '0' marks one that does not. For example, the numbers 42 and 47 respectively are expressed by the strings '101010' and '101111' (since $42 = 32 + 8 + 2$ and $47 = 32 + 8 + 4 + 2 + 1$), as the following chart should make clear:

Descending powers of 2:	32	16	8	4	2	1
42 =	1	0	1	0	1	0
47 =	1	0	1	1	1	1

Some further examples (note that the expression for each even integer ends in '0' while those for the odd integers all end in '1'):

Decimal	Binary	Decimal	Binary
1	1	13	1101
2	10	18	10010
5	101	24	11000
8	1000	50	110010
10	1010	103	1100111

The point of using this notation is simple: the digits of a binary expression may be correlated straightforwardly with the 1-pulses and 0-pulses that move through a sequential circuit, and hence every natural number will correspond to a unique *sequence* of pulses that are fed one by one into the circuit at one of its inputs. The first input pulse corresponds to the rightmost digit of the number, with the last pulse being the first digit. Hence the binary expression '1010' would correspond to the sequence of inputs beginning first with a 0-pulse, followed by a 1-pulse, followed by a 0-pulse and ending with a 1-pulse. Our adder will thus require two external inputs—one for each of the numbers being added—and one external output for the sum of the two numbers. The sum of course is emitted as a sequence of pulses the first of which will be its rightmost digit.

However, in addition we will need an output feeding into a delay whose output in turn will feed into a third input source. The reason for this last bit of apparatus has to do with *adding* two numbers expressed in binary notation. Grade school arithmetic tells us that the sum of two even numbers is always even, and similarly for the sum of two odd ones, but the sum of an even and an odd is itself odd. Since a binary number whose last digit is '0' expresses an even number and one whose last digit is '1' expresses an odd number, it follows that if the two binary expressions to be added both end in '0' or both end in '1', their sum will end in '0'. Otherwise, its last digit will be '1'. This same reasoning holds when comparing the next to last digits of the two numbers, and the digits to their left, and so on. Hence, adding 8 and 5 to obtain 13 looks like this:

$$
\begin{array}{r}
101 \\
+\,1000 \\
\hline
1101
\end{array}
$$

However, there is a complication with regard to these subsequent digits that must be addressed, viz. that when we are adding two 1's there is a "carry" of 1 to the next digits. To see this, let us look at how the additions '1 + 1 = 2' and '7 + 5 = 12' appear in binary notation:

$$
\begin{array}{r}
1 \\
+\,1 \\
\hline
10
\end{array}
\qquad
\begin{array}{r}
111 \\
+\,101 \\
\hline
1100
\end{array}
$$

When '1' is "added" to itself the result is a '0', but a '1' must be carried to the next column on the left or else where there is no such column (as in the 1 + 1 case) included in the sum written below. In the 7 + 5 addition, a '1' is carried to the second column from the right, the result being that the carried '1' and the middle '1' in '111' yield a '0' which when combined with the '0' in '101' yields a second '0' in the sum. But since two '1's were "added" in that column, there is another carry of 1 this time to the leftmost column. The result is a '1' in the sum for that column with a final carry giving us the leftmost '1' in '1100'.

The reason for including the delay in our adder should now be apparent; it will handle pulses that correspond to carries. A carry of 1 means that the delay's output will be a 1-pulse to the internal input, where it will enter the circuit simultaneously with the pulses at the other two inputs representing the next digit of the integers being summed. When there is no carry, the delay's output will be 0.

Let the two binary numbers to be added be represented as the strings of digits '$a_n \ldots a_1$' and '$b_n \ldots b_1$'. We will place a '0' on the left of both since their sum might have $n + 1$ digits. Of course the two might not have the same number of digits to begin with, and in such a case a '0' will be added as often as needed to the shorter one to bring it to the length of the other. So if we are to add, say, '101' and '1110', we imagine them being expressed for our present purpose as '00101' and '01110' respectively.

We will suppose that input A of our machine will handle '$a_n \ldots a_1$' (the pulse corresponding to 'a_1' of course entering the circuit first) with input B taking care of '$b_n \ldots b_1$'. Input C, then, will receive the output from the delay. One of the circuit's outputs, $f(A,B,C)$, will emit the sequence of pulses that constitute the sum of the two numbers and we will express that sum as '$f_{n+1} f_n \ldots f_1$'. The other output, $g(A, B, C)$, performs the task of "carrying" and will be fed into the delay. Now consider how our machine will process the first or rightmost digit of the two numbers being summed. A "truth-table" tells us what we need to know:

(V)	A	B	$f(A,B,C)$	$g(A, B, C)$
	1	1	0	1
	1	0	1	0
	0	1	1	0
	0	0	0	0

Since there is no carry at the beginning of our calculation, the delay input, C, is not included in this table (its nominal output pulse of 0 at this point will not affect the assignments for the two righthand columns). But for the remaining digits of the two numbers, a carry is always a possibility, and hence we must specify our output pulses in terms of three input sources and eight possible input combinations. Our discussion to this point should explain why the '1's and '0's are assigned as they are in the two output columns (in the column headings, the subscript 'i' ranges from 1 to n inclusive):

(VI)	A (a_i pulses)	B (b_i pulses)	C (from delay)	$f(A,B,C)$ (f_i pulses)	$g(A,B,C)$ (to delay)
	1	1	1	1	1
	1	1	0	0	1
	1	0	1	0	1
	1	0	0	1	0
	0	1	1	0	1
	0	1	0	1	0
	0	0	1	1	0
	0	0	0	0	0

We need a third table to specify the output pulse corresponding to digit 'f_{n+1}' of the sum; here only the delay input is relevant as the other input pulses are of course 0. Obtaining this table is easy:

(VII)	C (from delay)	$f(A,B,C)$ (f_{n+1})
	1	1
	0	0

From the output columns in Table (VI) we can design a sequential circuit that will carry out the task of determining the sum of any two natural numbers. The procedure is much the same as the one used for the two-terminal circuits discussed earlier. We do not need further circuitry corresponding to Tables (V) and (VII) for they are just fragments of Table (VI). That is, if we mentally delete the rows in which "from delay" is assigned 1, the remaining assignments for $f(A,B,C)$ and $g(A,B,C)$ are exactly the same as those in Table (V)—and of course the initial input from the delay is always 0. Similarly, Table (VII) in effect is a compressed version of rows seven and eight of Table (VI), and the assignments to $f(A,B,C)$ are the same in both cases.

We will now find compound sentences which express the output columns of Table (VI), and from there it will be an easy matter to design the required sequential circuit. For each column, we obtain the corresponding DBE and then find a simpler but equivalent form. For $f(A,B,C)$ we have:

$(A \mathbin{\&} B \mathbin{\&} C) \vee (A \mathbin{\&} \sim B \mathbin{\&} \sim C) \vee (\sim A \mathbin{\&} B \mathbin{\&} \sim C) \vee (\sim A \mathbin{\&} \sim B \mathbin{\&} C)$	
$[C \mathbin{\&} ((A \mathbin{\&} B) \vee (\sim A \mathbin{\&} \sim B))] \vee [\sim C \mathbin{\&} ((A \mathbin{\&} \sim B) \vee (\sim A \mathbin{\&} B))]$	Com, Dist
$[C \mathbin{\&} (A \equiv B)] \vee [\sim C \mathbin{\&} \sim [(A \mathbin{\&} \sim B) \mathbin{\&} \sim (\sim A \mathbin{\&} B)]]$	Equiv, DN, DeM
$[C \mathbin{\&} (A \equiv B)] \vee [\sim C \mathbin{\&} \sim [(\sim A \vee B) \mathbin{\&} (A \vee \sim B)]]$	DeM, DN
$[C \mathbin{\&} (A \equiv B)] \vee [\sim C \mathbin{\&} \sim [(A \supset B) \mathbin{\&} (B \supset A)]]$	Com, Impl
$[C \mathbin{\&} (A \equiv B)] \vee [\sim C \mathbin{\&} \sim (A \equiv B)]$	Equiv
$C \equiv (A \equiv B)$	Equiv

Turning now to $g(A,B,C)$, we have:

$(A \mathbin{\&} B \mathbin{\&} C) \vee (A \mathbin{\&} B \mathbin{\&} \sim C) \vee (A \mathbin{\&} \sim B \mathbin{\&} C) \vee (\sim A \mathbin{\&} B \mathbin{\&} C)$	
$[(A \mathbin{\&} B) \vee (C \mathbin{\&} \sim C)] \vee [C \mathbin{\&} ((A \mathbin{\&} \sim B) \vee (\sim A \mathbin{\&} B))]$	Com, Dist
$(A \mathbin{\&} B) \vee [C \mathbin{\&} \sim [\sim (A \mathbin{\&} \sim B) \mathbin{\&} \sim (\sim A \mathbin{\&} B)]]$	CE, DN, DeM

$(A \& B) \vee [C \& \sim[(\sim A \vee B) \& (A \vee \sim B)]]$ DeM, DN
$(A \& B) \vee [C \& \sim[(A \supset B) \& (B \supset A)]$ Com, Impl
$(A \& B) \vee [C \& \sim(A \equiv B)]$ Equiv

From our two sentences, '$C \equiv (A \equiv B)$' and '$(A \& B) \vee [C \& \sim(A \equiv B)]$', we can obtain their corresponding circuits. Combining these circuits into one and attaching the delay mechanism gives us our adder:

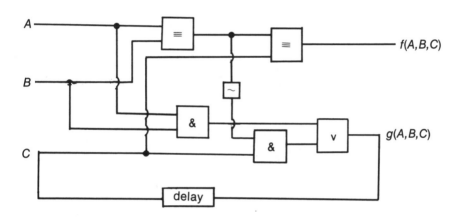

EXERCISES

1. Add each pair of binary numbers:

   110   11001
   + 100   + 1111101

2. A NAND-gate corresponds to a sentence of the form '$\sim(p \& q)$' and hence has an output pulse of 1 just in case one or both of the input pulses are 0. A NOR-gate corresponds to a sentence of the form '$\sim(p \vee q)$', and thus has a 1-pulse as its output just in case both input pulses are 0. They may be represented as follows:

Find a circuit corresponding to '$(A \supset \sim B) \& (C \& D)$' containing three gates, one of which is a NAND-gate and another a NOR-gate.

3. An EXCL-OR-gate has a 1-pulse as output just in case exactly one of its input pulses is 1, and so corresponds to the sentence, 'Either p or q but not both'. We may represent it by placing a 'o' to the right of the expression for an IFF-gate (as in the diagrams in exercise 2). Using EXCL-OR-gates, design an adding circuit that contains one less gate than the circuit illustrated above.

APPENDIX B
TRUTH TREES[1]

If '*A*' is true, the disjunction '*A* v *B*' is also true regardless of the truth-value of '*B*'. This may be put by saying that the truth of '*A*' is a *sufficient* condition for the truth of '*A* v *B*' — the latter's truth follows directly from that of the former. However, '*A* v *B*' may be true even though '*A*' is false, and hence the truth of '*A*' is *not* a *necessary* condition for the truth of '*A* v *B*'. That is, '*A*' does not *have* to be true in order for '*A* v *B*' to be true. Obviously, the same remarks apply to '*B*'; its truth is a sufficient — but not a necessary — condition for the truth of '*A* v *B*'. With conjunctions, however, the situation is reversed. The truth of '*A*' is not a sufficient condition for the truth of '*A* & *B*' because a conjunction is true only when *both* conjuncts are. But the truth of '*A*' clearly is a necessary condition for the truth of '*A* & *B*' since the latter cannot be true unless '*A*' is. Likewise, the truth of '*B*' is necessary but not sufficient for the truth of '*A* & *B*'. So, while the truth of a single disjunct is a sufficient (but not a necessary) condition for the truth of the disjunction itself, the truth of a single conjunct is a necessary (but not a sufficient) condition for the truth of the conjunction.

It will be useful here to have a means of displaying such relationships between compound sentences and their components. Where each component is such that its truth is alone sufficient for the truth of the compound, we shall "branch out" those components. The disjunction '*A* v *B*' thus gives rise to this diagram:

---

[1]The truth-tree method presented here is a version first used by Richard C. Jeffrey in his *Formal Logic: Its Scope and Limits*, New York: McGraw-Hill Company, 1967. A similar discussion may be found in Hugues Leblanc and William A. Wisdom, *Deductive Logic*, Boston: Allyn and Bacon, Inc., 1972.

The fork here indicates that the compound will be true if and only if *at least one* of the components displayed at the tips is true.

Where each component is such that its truth is a necessary condition for the truth of the compound, we shall write the components directly below the compound, one underneath the other. So for '*A* & *B*' we write:

A & B
  A
  B

Writing '*A*' and '*B*' in a column indicates that each *must* be true for the compound to be true (and though neither the truth of '*A*' nor that of '*B*' is alone sufficient for the truth of '*A* & *B*', their *joint* truth is of course sufficient).

We will call such diagrams — for reasons which will soon be apparent — **truth trees**, and will say that in our two examples '*A* v *B*' and '*A* & *B*' have been **decomposed**, that is, broken down into sentence letters or their negations. Once the conventions concerning branching and columns are understood, it is clear that such diagrams spell out all possible ways in which the compounds may be true. The decomposition of more complex sentences requires of course larger trees.

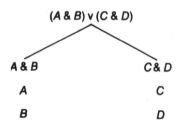

Here, the main connective is 'v', so we begin with a fork — one for each disjunct — and since both disjuncts are conjunctions we decompose each by writing its conjuncts beneath. Having done this, the truth-conditions of the compound are displayed clearly: it will be true just in case both '*A*' and '*B*' are true (left branch) or both '*C*' and '*D*' are true (right branch). That is, in the corresponding 16-row truth-table, those rows in which '*A*' and '*B*' are both assigned **T** and those in which '*C*' and '*D*' are both assigned **T** will also have a **T** assigned to '(*A* & *B*) v (*C* & *D*)'. Any other assignment of truth-values to the four component letters results in a falsehood.

Let us now tackle a more complicated example, the conjunction '$(A \lor B) \& (C \lor D)$'. Since it is a conjunction, we begin by listing the conjuncts underneath:

$(A \lor B) \& (C \lor D)$
$\quad A \lor B$
$\quad C \lor D$

But since both conjuncts are themselves compound, each must be decomposed. Working on '$A \lor B$' first, we obtain:

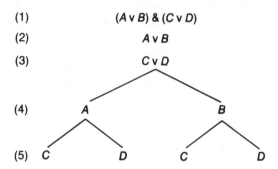

(1) $(A \lor B) \& (C \lor D)$

(2) $A \lor B$

(3) $C \lor D$

(4) A B

(5) C D C D

Here, '$A \lor B$' is decomposed at (4), and '$C \lor D$' at (5). Notice that '$C \lor D$' must be decomposed at *two* points, once for each disjunct of '$A \lor B$'. For the entire compound to be true, '$A \lor B$' must be true (as (2) tells us), but the fork between (3) and (4) shows us that for it to be true either 'A' or 'B' must be true. So both of these possibilities must be accounted for when decomposing the other disjunction if we are to delineate *all* conditions under which the compound will be true. Decomposing '$C \lor D$' at (5), then, completes our work. The resulting tree thus has four *branches*, two passing upwards from (5) through 'A' on the left, the other two through 'B' on the right. Each lays out a condition under which the original compound will be true, and together are exhaustive of such conditions. So '$(A \lor B) \& (C \lor D)$' will be true just in case:

'C' and 'A' are both true (leftmost branch)
'D' and 'A' are both true (left-middle branch)
'C' and 'B' are both true (right-middle branch)
'D' and 'B' are both true (rightmost branch).

Plainly the above tree accomplishes the same end as a 16-row truth-table but with considerably less space and effort; all conditions under which

'$(A \lor B)$ & $(C \lor D)$' is true have been laid out and may be discerned by tracing through each branch of the tree.

Truth trees typically afford us a quicker, more facile method of establishing both validity and invalidity than do truth-tables. They thus constitute an alternative short-cut technique to that presented in Section 2.4. We know that an argument is (truth-table) valid just in case there is no assignment of truth-values to its component letters according to which all premises come out true but the conclusion false — in other words, just in case there is no assignment on which the premises and the *negation* of the conclusion all come out true. Now truth trees provide us with an efficient means of testing a set of sentences for *consistency* — for whether there is an assignment making *all* sentences of the set true (cf. Section 2.3). So we can determine whether an argument is valid or not by constructing a corresponding truth tree for its premises and negated conclusion, and then noting whether one of its branches lays out a truth-value assignment on which all are true (in other words, on which the premises are true and the conclusion false). As an example, let us consider the argument

$(A$ & $B) \supset {\sim}C$
$D \lor (A$ & $B)$
$\therefore C \supset (D \lor E).$

To construct a tree, we first list the premises along with the *negated* conclusion.

(1) $(A$ & $B) \supset {\sim}C$
(2) $D \lor (A$ & $B)$
(3) ${\sim}[C \supset (D \lor E)]$

Next, we pick one of these (it does not matter which we begin with) and decompose it. Beginning with (3), we note that it is a negated conditional, and since a conditional is false only when its antecedent is true and consequent false, its negation will be *true* in just such a circumstance. Thus a sentence of the form '${\sim}(p \supset q)$' is true just when the sentence represented by 'p' is true and that by 'q' false (hence '${\sim}q$' true). That is, the truth of 'p' and of '${\sim}q$' are both necessary conditions for the whole to be true. The negation of a conditional is thus decomposed columnwise:

${\sim}(p \supset q)$
p
${\sim}q$

Decomposing (3), then, we obtain:

(1) $(A \& B) \supset \sim C$
(2) $D \vee (A \& B)$
(3) $\checkmark \sim [C \supset (D \vee E)]$
(4) C
(5) $\sim (D \vee E)$

The checkmark to the left of (3) means simply that the third sentence has been decomposed. The use of checks will enable us to keep track of which compounds have been decomposed at a given point and which have not. Now since (5) itself is a compound, it is decomposed before moving on to (1) and (2), for ultimately we want each of the three sentences with which we began to be broken down into sentence letters or their negations. (5) is the negation of a disjunction and so will be true just in case 'D' and 'E' are both false, that is, '$\sim D$' and '$\sim E$' are both true. Sentences of the form '$\sim (p \vee q)$' are hence decomposed by writing '$\sim p$' and '$\sim q$' directly underneath. Our tree now becomes:

(1) $(A \& B) \supset \sim C$
(2) $D \vee (A \& B)$
(3) $\checkmark \sim [C \supset (D \vee E)]$
(4) C
(5) $\checkmark \sim (D \vee E)$
(6) $\sim D$
(7) $\sim E$

We may now turn to (2). Since it is a disjunction, a fork is called for after step (7):

(1) $(A \& B) \supset \sim C$
(2) $\checkmark D \vee (A \& B)$
(3) $\checkmark \sim [C \supset (D \vee E)]$
(4) C
(5) $\checkmark \sim (D \vee E)$
(6) $\sim D$
(7) $\sim E$

(8) D $A \& B$

Let us take stock of the situation here. If the checked compounds (2) and (3) are *both* true, then *either* the seven sentences from (2) down through '*D*' on the left branch are all true *or* the seven sentences from (2) down through '*A & B*' on the right branch are all true. But the first of these alternatives is impossible, since for '*D*' to be true '~*D*' at (6) must be false. That is, there is no assignment of truth-values on which *all* sentences on the left branch will be true inasmuch as '*D*' and '~*D*' would *per impossible* have to be true. So the left branch does not provide us with a condition under which (2) and (3) will both be true; in tracing the branch upwards from '*D*' at (8) we stumble into '~*D*' at (6). This being the case, we can "close off" the left branch from future consideration since we now know that *no* extension of the branch beyond '*D*' could ever correspond to a genuine truth-value assignment. We shall call any branch in which a sentence letter and its negation both occur a **closed branch**, and shall indicate such a branch by placing an '×' beneath its last member. Thus we mark the closure of the branch from (2) through '*D*' at (8) as follows:

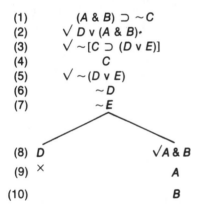

At the same time, the conjunction on the right branch has been decomposed. Note however that this branch is not closed. It is still *open* at this point since '*A & B*' has been decomposed into '*A*' and '*B*', neither of whose negations occur earlier in the branch.

Finally, we decompose (1). Since a conditional is false *only* when its antecedent is true and consequent false, it will be true if *either* the antecedent is false (hence its negation true) *or* the consequent is true. Conditionals thus require a fork and our tree has now acquired some new growth:

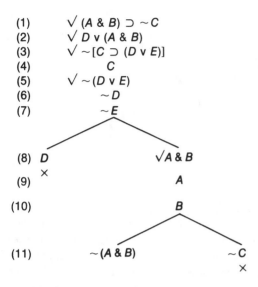

(1)   √ (A & B) ⊃ ~C
(2)   √ D v (A & B)
(3)   √ ~[C ⊃ (D v E)]
(4)   C
(5)   √ ~(D v E)
(6)   ~D
(7)   ~E

(8) D   √A & B
     ×
(9)   A

(10)   B

(11)   ~(A & B)   ~C
                                ×

Another closed branch has made its appearance. The right branch at (11) cannot yield a truth-value assignment making all of (1), (2), and (3) true since 'C' at (4) and '~C' at (11) would *per impossible* both have to be true. We must still however contend with the left branch. A sentence of the form '~(p & q)' will be true just in case the sentence represented by 'p' is false ('~p' true) or that by 'q' is false ('~q' true). We thus have

~(p & q)

~p   ~q

Our completed tree, then, looks like this:

(1)   √ (A & B) ⊃ ~C
(2)   √ D v (A & B)
(3)   √ ~[C ⊃ (D v E)]
(4)   C
(5)   √ ~(D v E)
(6)   ~D

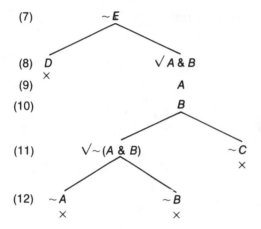

(7) ~E

(8) D   √ A & B
    ×
(9)   A

(10)   B

(11) √ ~(A & B)   ~C
                          ×

(12) ~A   ~B
    ×   ×

The new branches at (12) are both closed since '*A*' and '*B*' occur earlier in their paths. *Every* branch of the tree, then, is closed, and this means that no truth-value assignment will render all of (1), (2), and (3) true. Hence there is no assignment making the premises of our argument true and its conclusion false. The argument is therefore *valid*, and moreover has been so established with far less effort than a 32-row truth-table would have required.

Let us call a tree which tests the premises and negated conclusion of an argument for joint truth a **corresponding tree** of that argument. For most arguments there are many such trees since the component sentences may be decomposed in a variety of orders, but of course any one of them will be sufficient for our purpose. In such a tree, then, the premises and negated conclusion are (with a minor exception to be noted later) fully decomposed—no compound sentences containing '&', 'v', '⊃' or '≡' remain unchecked. We may now define the concept of validity as it pertains to tree methods:

DEFINITION A truth-functional argument is **tree valid** if and only if, for any of its corresponding trees, all branches are closed.

However, if after all decomposition has occurred, *at least one* branch does not close, then the argument is invalid. Tree validity is nothing more than the truth-table validity of Chapter Two in different garb. If every branch closes, then no assignment of truth-values to the component letters yields true premises and false conclusion—hence there is no truth-table *row* in which all premises are assigned **T**'s and the conclusion assigned **F**.

Rules for producing trees were introduced as they were required in the above example. We shall now collect them together along with some new ones. In all, we shall have nine rules: four governing sentences with '&', 'v', '⊃' and '≡' as main connectives, four for the negations of such sentences, and one for eliminating double curls. The justification of the rules of course resides ultimately in the truth-tables for each of these compounds.

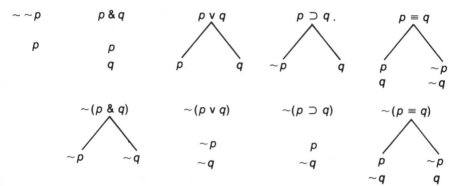

From the above remarks on validity it follows that a corresponding tree for an invalid argument will have *at least one open branch* – a branch that does not contain *both* a sentence letter and its negation. Such a branch will trace out an assignment of truth-values to the component letters of the argument which makes the premises and negated conclusion all true. As an example let us consider

$(L \, \& \, M) \supset (H \, \& \sim W)$
$L \vee W$
$\therefore H \vee \sim M$

Decomposing the conclusion first once again, we obtain:[2]

$$\sqrt{}\ (L \, \& \, M) \supset (H \, \& \sim W)$$
$$\sqrt{}\ L \vee W$$
$$\sqrt{}\sim (H \vee \sim M)$$
$$\sim H$$
$$\sqrt{}\sim\sim M$$

---

[2]The line-numbering used in our earlier example was for expository purposes only; it is not essential to tree construction and hereafter will be omitted.

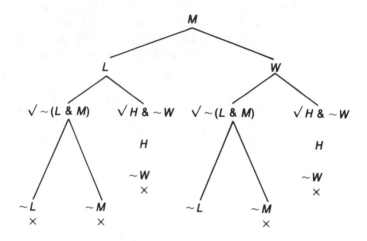

The third branch from the right fails to close since 'L' does not occur earlier in that branch (it does of course occur on the *left* side, but for that reason is not in the upward path leading from the righthand '$\sim L$'). This branch therefore delineates an assignment of truth-values to the letters 'L', 'W', 'M', and 'H' on which the two premises and negated conclusion come out true, and consequently establishes the argument as invalid. The assignment in question may be found by tracing through the open branch, noting whether the sentence letters in its path occur by themselves or negated. Thus, the first three sentences will all be true (hence the premises true and conclusion false) when 'L' is false, 'W' is true, 'M' is true, and 'H' is false.

Some further points about this tree are worth mentioning. First, unlike the preceding one, this tree requires a sentence to be decomposed twice, viz. the first premise. This of course is because the branches issuing from 'L' and 'W' are both open at that point, and thus an exhaustive search for a truth-value assignment making the original three sentences true requires that the truth-conditions of the first of them be set down for each alternative. Second, the double negative '$\sim \sim M$'—like any other compound—must be checked after its decomposition. Lastly, note that the third branch from the left at the bottom is closed even though 'W'—the contradictory of its very last member—does not occur earlier on the branch. It suffices for just one of the conjuncts of '$H \,\&\, \sim W$' to have a contradictory elsewhere on the branch, and '$\sim H$' occurs at the fourth line.

A further example illustrates some less common features a truth tree may exhibit. The validity of the argument

$A \supset \sim E$
A
$B \,\&\, \sim (C \,\&\, D)$
$\therefore E \lor (B \,\&\, A)$

is shown by a corresponding tree in which both branches are closed:

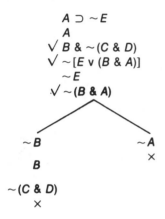

The right branch is closed since '*A*' occurs as a premise. So when a single letter or its negation occurs as a premise, it contributes to the closing of a branch just as if it had been generated through decomposition. However, since it requires no decomposition, it is *not* checked off. Notice that the first premise also has not been checked; this is because once the third and fourth lines were decomposed, all branches were closed. Thus, it is a superfluous premise—the conclusion following from the second and third premises *alone*. Further, since the right branch closes at '~*A*', the third premise is decomposed only for the left branch. Since both '*B*' and '~*B*' have now been produced, the branch is now closed and the tree is completed. Decomposition of '~(*C* & *D*)' would at this stage be pointless, and so it is left unchecked. Hence, *not all* trees will have every compound checked, for decomposition is unnecessary in certain cases. And where a premise or conclusion consists of a single letter or its negation, it will not be checked either.

The argument forms associated with such elementary rules of inference as **Simplification**, **Conjunction**, *Modus Ponens*, and **Hypothetical Syllogism** may be shown valid via truth trees (see Sections 2.3, 3.1, and 3.2):

Simplification	Conjunction	*Modus Ponens*	Hypothetical Syllogism

$\checkmark p \,\&\, q$ p $\checkmark p \supset q$ $\checkmark p \supset q$

$\sim p$ q p $\checkmark q \supset r$

p $\checkmark \sim (p \,\&\, q)$ $\sim q$ $\checkmark \sim (p \supset r)$

q p

$\times$

For Conjunction:

$\sim p \quad\quad \sim q$
$\times \quad\quad\quad \times$

For Modus Ponens:

$\sim p \quad\quad q$
$\times \quad\quad\quad \times$

For Hypothetical Syllogism:

$\sim r$

$\sim q \quad\quad r$
$\quad\quad\quad \times$

$\sim p \quad\quad q$
$\times \quad\quad\quad \times$

Trees for the remaining elementary rules are left as exercises for the reader. The fallaciousness of "affirming the consequent" (see Sections 2.3 and 3.2) is shown by the tree below ("denying the antecedent" is left to the reader):

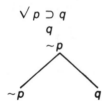

$\checkmark p \supset q$

q

$\sim p$

$\sim p \quad\quad q$

In each of our worked examples, the negated conclusion of the argument being tested was decomposed first. As stated earlier, however, it does not matter which sentence we begin with, and it is sometimes more convenient to decompose one of the premises first. The general rule to follow is: *decompose first those sentences requiring no forks* (sentences of the forms '$\sim\sim p$', '$p \,\&\, q$', '$\sim (p \lor q)$', and '$\sim (p \supset q)$', for the tree will be far less bulky and cumbersome if branching is kept to a minimum. Moreover, when faced with a choice of decomposing compounds both of which require forks, it is obviously wise to see if one of them will lead immediately to a closed branch. Decomposing it first will simplify later work.

The tree method is essentially a means of setting out the truth-conditions for each member of a set of sentences by systematically eliminating their connectives until only sentence letters and their negations remain. If at any point in the decomposition process *all* branches are closed, we may stop our work since the set of sentences with which we began has been shown not to admit of a truth-value assignment making all of them true. On the other hand, if at least one branch remains open after all compounds have been decomposed (whether they be in the original set or produced in the decomposition process), then there is at least one truth-value assignment on which all sentences in the set are true. To retrieve such an assignment, one need only trace upwards through the branch, assigning an **F** to each negated letter and a **T** to the rest.

Since trees provide us with a test for an argument's validity, we should also expect a test of some kind for determining whether a solitary sentence is a tautology. But what sort of test? Suppose we construct a tree for a compound sentence with the result that all branches are closed. What does that tell us about the sentence? It tells us that there is no assignment of truth-values to the component letters on which the sentence is true; in other words, we know that its truth-table column has no **T**'s and is thus a *contradiction*. Since the negation of a tautology is always a contradiction (and vice versa), we have an easy test: first, negate the sentence to be tested, then construct a tree for that negation. If all branches close, then the (unnegated) sentence is a tautology. If not all of them close, it is not. More formally, a truth-functionally compound sentence is a *tautology* if and only if *a tree constructed for its negation has nothing but closed branches*. The tree below shows that the compound sentence '$(A \equiv \sim B) \supset [(B \supset \sim A) \& (A \lor B)]$' is a tautology:

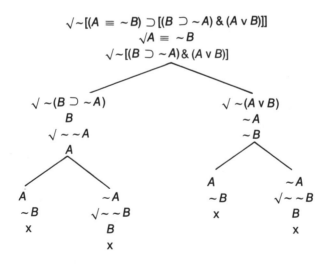

EXERCISES

I. Using the tree method, determine whether the arguments in Exercise I of Section 2.4 are valid.

II. Using the tree method, show the validity of the argument forms corresponding to the rules: **Addition, Disjunctive Syllogism,** *Modus Tollens*, and **Constructive Dilemma**.

III. Determine whether any of the following are tautologies:
   1. $A \vee \sim(B \& A)$
   2. $(A \& \sim B) \vee \sim[\sim(B \& C) \& (A \& C)]$
   3. $(A \& \sim B) \& [\sim A \vee (B \& A)]$

IV. Explain how the tree method may be used to determine whether or not a pair of sentences are *equivalent*. Then determine whether any of the following pairs are equivalent:
   1. $\sim(E \vee \sim F);$ $\quad F \supset \sim E$
   2. $\sim P \vee \sim(R \& Q);$ $\quad P \supset (R \supset Q)$
   3. $A \vee (B \& \sim C);$ $\quad (A \vee \sim C) \& (A \vee B)$

APPENDIX C

MATHEMATICAL INDUCTION

Once upon a time, Sir John Cecil, first holder of the Stopes Chair of Archeology, entered the sacred tomb of an ancient pharoh and a curse was placed on him. And our story goes that when such a curse falls on an academic intruder, it falls on his successor as well. If these be facts—that a curse befell the initial occupant of the Stopes Chair, and that each occupant is such that if a curse befalls him then it befalls his successor as well—we are bound to conclude that all who have occupied the Stopes Chair are accursed.

Somewhat similar inferences are common in mathematics, where the rule of inference in question can be formulated precisely. For example:

The number 1, plus its square, is an even number. And each positive integer is such that if the sum of it and its square is even, then the same is true of its successor. For if $m + m^2$ is even then clearly $(m + m^2) + 2(m + 1)$ is also, and $(m + m^2) + 2(m + 1) = (m + m^2) + (2m + 2) = (m + 1) + (m^2 + 2m + 1) = (m + 1) + (m + 1)^2$. We are clearly entitled to conclude that the sum of any positive integer and its square is even—for we have shown that this is true of the first positive integer, and that it carries over from each to the next.

In general, if we show that the number 1 satisfies a certain condition and show also that whenever *any* positive integer satisfies that condition then the number after it satisfies that condition as well, we will be entitled to conclude that *all* positive integers satisfy the condition in question. For the second of our claims in effect ensures that each will have inherited it from its predecessor, beginning with 2's having inherited it from 1.

To formulate the general rule of inference at work here, let '...x...' represent any sequence of English words, mathematical symbols, and so on, that becomes a sentence if each occurrence of '*x*' in it is replaced by a numeral.

The Principle of Weak Mathematical Induction.

From

    ...1...

and

    For each positive integer *m*, if ...*m*... then ...*m* + 1...

one may infer

    For each positive integer *n*, ...*n*...

When we use this principle to show that all positive integers satisfy a certain condition, our work will always divide into two parts, as did our proof that the sum of a number and its square is always even. First we will show that the number 1 satisfies the condition (this part of our work is called the **base case**); and second, we will show that if any number *m* ($m \geqslant 1$) satisfies the condition then the number after it, $m + 1$, satisfies the condition also (this part is called the **induction step**). The principle of mathematical induction then justifies our inferring from these two sub-results that all positive integers satisfy the condition in question. (It's as if with the first we assigned to Edward I, and with the second to his heirs in perpetuity, the title to certain lands.)

To see how this works in practice, let's follow another example through in detail. We have seen how to show that for each positive integer n, $n + n^2$ is even. This means, of course, that for each such n, $\frac{1}{2}(n + n^2)$ will be a positive integer. Suppose we compute the first few numbers of this sort.

n	n^2	$n + n^2$	$\frac{1}{2}(n + n^2)$
1	1	2	1
2	4	6	3
3	9	12	6
4	16	20	10
5	25	30	15

The series so generated (1, 3, 6, 10, 15, ...) will be familiar to anyone who has ever added numbers together in sequence:

$1 = 1$

$3 = 1 + 2$

$6 = 1 + 2 + 3$

$10 = 1 + 2 + 3 + 4$

$15 = 1 + 2 + 3 + 4 + 5$

So it is hard to resist *conjecturing*, at least, that

For each positive integer n, $\frac{1}{2}(n + n^2) = 1 + \cdots + n$.

And once the conjecture has been formulated, a proof by weak mathematical induction is not hard to construct. Remember that in using induction, we must establish two things: that $\frac{1}{2}(1 + 1^2) = 1$; and that for each positive integer m, if $\frac{1}{2}(m + m^2) = 1 + \cdots + m$ then $\frac{1}{2}((m + 1) + (m + 1)^2) = 1 + \cdots + (m + 1)$. The first of these is quite trivial:

BASE CASE $\frac{1}{2}(1 + 1^2) = 1$, for

$\frac{1}{2}(1 + 1^2) = \frac{1}{2}(1 + 1)$

$\qquad\qquad = 2/2$

$\qquad\qquad = 1$

The second part requires a bit more work:

INDUCTION STEP   Each positive integer m is such that if $\frac{1}{2}(m + m^2) = 1 + \cdots + m$ then $\frac{1}{2}((m + 1) + (m + 1)^2) = 1 + \cdots + (m + 1)$. For let m be any positive integer and assume, as the hypothesis for the induction step, that $\frac{1}{2}(m + m^2) = 1 + \cdots + m$. Now

$\frac{1}{2}((m + 1) + (m + 1)^2) = \frac{1}{2}((m + 1) + (m^2 + 2m + 1))$

$\qquad\qquad\qquad\qquad = \frac{1}{2}(m + m^2 + 2m + 2)$

$\qquad\qquad\qquad\qquad = \frac{1}{2}(m + m^2) + \frac{1}{2}(2m + 2)$

$\qquad\qquad\qquad\qquad = \frac{1}{2}(m + m^2) + (m + 1)$

So $\frac{1}{2}((m + 1) + (m + 1)^2) = \frac{1}{2}(m + m^2) + (m + 1)$. But we know, from our assumption, that $\frac{1}{2}(m + m^2) = 1 + \cdots + m$. Consequently $\frac{1}{2}((m + 1) + (m + 1)^2) = 1 + \cdots + m + (m + 1)$, completing our proof.

Beyond its obvious usefulness in establishing generalizations concerning numbers, weak mathematical induction can often be employed in establishing general results in logic. In Section 4.2, for example, we use it (tacitly) to show that for each positive integer n, any line of any deduction whose line-number is no greater than n must have a **T** in each row of its truth-table in which all the premises, $\mathscr{P}_1, \ldots, \mathscr{P}_k$, of that deduction have **T**'s. For the base case, we show that this is true of all lines of deductions from $\mathscr{P}_1, \ldots, \mathscr{P}_k$ whose line-numbers are no greater than 1; and for the induction step we show that if all lines of all deductions from $\mathscr{P}_1, \ldots, \mathscr{P}_k$ whose line-numbers are no greater than m (m any positive integer) have **T**'s when $\mathscr{P}_1$ through $\mathscr{P}_k$ do, then the same holds of all lines of such deductions whose line-numbers are no greater than $m + 1$.

As another example, let us establish Chapter 2's claim that

For each positive integer n, there are exactly 2^n ways of assigning truth-values **T** and **F** to n distinct sentence letters.

For the base case we must show that there are exactly 2^1 (that is, 2) ways of assigning truth-values to 1 sentence letter, and this is clear enough: we can assign a **T** to it or an **F**, but there is no third possibility.

For the induction step, let m be any positive integer and assume, as the hypothesis for our induction, that there are exactly 2^m ways of assigning truth-values **T** and **F** to m distinct sentence letters. Suppose, now, we have $m + 1$ sentence letters, say 'A_1', ..., 'A_m', 'A_{m+1}'. By the induction hypothesis there are exactly 2^m ways of assigning **T** and **F** to 'A_1', ..., 'A_m'. And for *each* of these 2^m assignments we have two choices of assignment to 'A_{m+1}': we may assign it a **T** or we may assign it an **F**. In all, then, there are 2×2^m ways of assigning truth-values to the $m + 1$ sentence letters in question; and since $2 \times 2^m = 2^{m+1}$, our induction is complete.

EXERCISES

1. Where n is any positive integer, $n + n^2 = n(n + 1)$. Show, by weak induction, that

   For each positive integer n, either n or $n + 1$ is even,

   thus providing an alternate proof that $n + n^2$ is always even.
2. Prove by weak induction that for each positive integer n, $1^2 + \cdots + n^2 = \frac{1}{6}(n(n + 1)(2n + 1))$.
3. Prove that for each positive integer n, $1^3 + \cdots + n^3 = (1 + \cdots + n)^2$.
4. Prove that for each positive integer n, there are exactly 2^{2^n} distinct truth-functions (see Section 2.5) of n arguments.
5. Without using any results from Chapter 4, prove that every theorem of the deductive system developed in Section 3.1 is a tautology (in

the sense of Section 2.1) by using induction to establish

For each positive integer n, each line of each n-line proof is a tautology.

*6. The following principle seems self-evident:

The Well-Ordering Principle.

Every nonempty collection of positive integers contains a *smallest* member.

Show that the conclusion of the principle of weak mathematical induction follows from the two premises of that principle by using the well-ordering principle.

Occasionally in mathematics, and often in logic, it is useful to employ the following principle (where '---x---' represents any sequence of English words, symbols, etc., that becomes a sentence when each occurrence of 'x' in it is replaced with a numeral):

The Principle of Strong Mathematical Induction

From

--- 1 ---

and

For each positive integer m, if --- k --- where k is any positive integer such that $k \leqslant m$, then --- $m + 1$ ---

one may infer

For each positive integer n, --- n ---.

Unlike the weak form of induction, here the induction hypothesis will be assumed to hold not just for m but for all positive integers less than m as well. To illustrate the use of strong induction, let us say that a positive integer is *factor-free* just in case the only positive integers that divide it without remainder are itself and 1. Then we can prove

For each positive integer n, n either is factor-free or else is the product of factor-free positive integers.

For the base case, it is clear that 1 is factor-free since no other positive integers divide it without remainder. So 1 is either factor-free or the product of factor-free positive integers.

For the (strong) induction step, let m be any positive integer and assume as our induction hypothesis that *every* positive integer k less than or equal to m is either factor-free or the product of factor-free integers. If $m + 1$ is factor-free there is nothing to show, so say instead that $m + 1$ is divisible by some positive integer i greater than 1 and less than $m + 1$. Let j be the number of times i goes into $m + 1$, so that $m + 1 = i \times j$. Since i and j are both greater than 1 and less than $m + 1$, the induction hypothesis tells us that each is either itself factor-free or else is the product of factor-free positive integers. So $i \times j$ is the product of factor-free positive integers. But $i \times j = m + 1$, so our proof is complete.

As a second example, let us call a sentence formulated in the symbolic language of Section 1.3 *curl-free* if and only if the connective ' $\sim$ ' does not occur in it. Then we can show:

For each positive integer n, every curl-free sentence containing n occurrences of sentence letters takes the value **T** in the first row of its truth-table (i.e., the row in which every letter occurring in the sentence is assigned the value **T**).

In the base case, note that every curl-free sentence containing just one occurrence of a letter is itself simply that letter. So of course it takes the value **T** in the first row of its table where that letter is assigned the value **T**.

For the induction step, let m be any positive integer and assume that for every positive integer k less than or equal to m, every curl-free sentence containing k letters takes the value **T** in row one of its truth-table. Now consider any sentence $\mathscr{S}$ containing $m + 1$ occurrences of letters. Since no curls occur in $\mathscr{S}$, its main connective must be either '&' or 'v'. So for some sentences $\mathscr{Q}$ and $\mathscr{R}$, $\mathscr{S}$ is either the sentence '$\mathscr{Q} \& \mathscr{R}$' or the sentence '$\mathscr{Q}$ v $\mathscr{R}$' Because $\mathscr{S}$ contains $m + 1$ letters in all, $\mathscr{Q}$ and $\mathscr{R}$ must each contain m or fewer letters. By the induction hypothesis, then, both $\mathscr{Q}$ and $\mathscr{R}$ take the value **T** in row one of their truth-tables. So since conjunctions and disjunctions take the value **T** when their components do, $\mathscr{S}$ must take the value **T** in row one of its truth-table as well, completing our proof.

Of course the result we have just established is stated as a fact about all positive integers. But since every curl-free sentence contains for *some* positive integer n exactly n occurrences of letters, we may reformulate the result without mentioning numbers explicitly; we have in fact shown:

Every curl-free sentence takes the value **T** in row one of its truth-table.

EXERCISES

1. A positive integer is said to be *prime* if and only if it is greater than one and is factor-free. Show how to modify our proof above concerning factor-free positive integers to establish that for each positive integer n, either $n = 1$, or n is prime, or n is the product of prime numbers.

2. Prove that for each positive integer n, every curl-free sentence of the language of Section 1.3 takes the value **F** in the last row of its truth-table, that is, the row in which all letters occurring in the sentence are assigned the value **F**. (Note that it follows from this result, and the one established in the text, that no curl-free sentence can be either a tautology or a contradiction, that is, every tautology—and every contradiction—contains at least one occurrence of ' $\sim$ '.)

3. By a '*B*'-*sentence*, let us mean a sentence formulated in the language of Section 1.3 (or 1.4—it makes no difference) in which the only sentence letter occurring is '*B*'. Show: for each positive integer n, every '*B*'-sentence containing exactly n occurrences of the letter '*B*' is equivalent (according to truth-tables) to one of the sentences '*B*', ' $\sim B$', '$B \lor \sim B$' and '$B \& \sim B$'.

4. Redo Exercise 5 from the preceding Exercise Set, this time using *strong* induction.

5. Use the well-ordering principle (Exercise 6, preceding Set) to show that the conclusion of the principle of strong induction follows from the premises of that principle.

*6. Suppose we have established ---1--- and also that for each positive integer m, if ---k--- for each positive integer k such that $k \leq m$ then ---$m + 1$---. Do we need the principle of *strong* induction to conclude that ---n--- for each positive integer n, or does weak induction suffice if we choose our substitution for '...n...' carefully? [Hint: take '...n...' to be '---1--- and .... and ---n---'.]

APPENDIX D

AXIOMATIC SYSTEMS
A Preface to Their Study

Any collection of signs, together with a collection of sequences of those signs, constitutes what we shall here regard as a **formal language**. We call the former the **primitive symbols** of the language in question, and the latter its **well-formed formulas** ('wffs', for short).

EXAMPLE   The Formal Language *N*.

Primitive Symbols: '|' (the 'stroke') and '–' (the 'dash').

Wffs: those finite sequences of such symbols that consist of one or more strokes, followed by a single dash, followed by one or more strokes.

Note that in the specification of a formal language, no "meanings" are assigned to the primitive symbols.

Of course, full-blooded natural languages such as English involve more than just "alphabets" (primitive symbols) and "sentences" (wffs), but the more anemic *formal* languages serve as adequate bases for the systems we are interested in constructing here.

Given such a formal language, we may select certain of its wffs as **axioms**, and lay down certain **rules of inference** (or **transformation rules**) permitting passage from some of its wffs to others. And the resulting package, consisting of the formal language, together with the axioms and the rules, is an **axiomatic system**.

EXAMPLE   The Axiomatic System **N**.

Formal Language: *N*

Axiom: |–|

Rule: From any wff

$\mathscr{S}-\mathscr{T}$

one may pass to the corresponding wff

$|\mathscr{S}-\mathscr{T}|$

In discussing such a system we make use not of a formal language, but of a natural language, ordinary English. Since the latter is used for talking *about* the former, we call it our **metalanguage**. In addition to ordinary English expressions, our metalanguage contains **metalinguistic variables**, '$\mathscr{S}$', '$\mathscr{T}$', and so on, that is, variables which range over sequences of primitive symbols of the system being discussed. Thus, in our statement of N's rule, '$\mathscr{S}-\mathscr{T}$' is not itself a wff of N, but rather an expression (in the metalanguage) used for talking about wffs of N — in the present case, those sequences of primitive symbols consisting of any number of strokes followed by a dash followed by any number of strokes.

A first application of the rule of our sample system N to its axiom gives us '$||-||$'; another application of the rule, this time to the result '$||-||$' of the first, provides '$|||-|||$'; and further applications will deliver still more wffs. These wffs, generated from the axiom by successive applications of the rule, are the *theorems* of our system, and the sequence of wffs involved in the generation of each is a *proof* of it, within that system. Thus such sequences as[1]

```
|-|      |-|       |-|        |-|
        ||-||     ||-||      ||-||
                 |||-|||    |||-|||
                          ||||-||||
```

are proofs in the system N, and the last line of each is one of N's theorems. These notions may be defined for axiomatic systems generally: where $\mathscr{U}$ is any axiomatic system, a **proof** in $\mathscr{U}$ is a sequence of wffs of $\mathscr{U}$ in which each wff is either an axiom of $\mathscr{U}$ or the result of applying one of $\mathscr{U}$'s rules to one or more earlier wffs in the sequence; and a **theorem** of $\mathscr{U}$ is any wff which is the last line of some proof in $\mathscr{U}$.

To show that a wff *is* a theorem of a given system, then, it suffices to construct a proof of it, that is, a proof of which it is the last line. But how are we to show that certain wffs are *not* provable in particular systems? How, for example, could we show that '$|-||||$' and '$||||-||$' are *not* theorems of N?

---

[1] We here dispense with the requirement earlier in the text that the line numbers and justifications also appear in the deduction.

A method that is often useful is provided by the following general result:

(R) If (i) every axiom of an axiomatic system $\mathscr{U}$ has some feature, and (ii) the rules of $\mathscr{U}$ "preserve" that feature in the sense that they can only lead from wffs with the feature to others that also have it, then every theorem of $\mathscr{U}$ has the feature in question.

To establish this result, notice that every theorem of an axiomatic system $\mathscr{U}$ is the last line of some proof in $\mathscr{U}$. It will do, then, to show that every line of every proof in $\mathscr{U}$ must have the feature in question, and this is easily established by mathematical induction.[2] By (i) the first lines of all proofs in $\mathscr{U}$ have the feature. And by (ii) whenever all lines of all m-line proofs have the feature so do all lines of all $(m + 1)$-line proofs—for the $(m + 1)$st lines of such proofs must either be axioms themselves, and so have the feature by (i), or else follow from earlier lines by one of $\mathscr{U}$'s rules and so have the feature by (ii).

Now we can return to one of our earlier questions and, with the aid of (R), show that '$|-||||$' is not a theorem of N.

One of the features possessed by N's axiom is that of *being composed of an even number of strokes*. And N's rule clearly preserves this feature, for it lets us add to a wff precisely two strokes (one at each end). The general result (R) assures us, then, that all theorems of N have this feature—all theorems of N are composed of an even number of strokes. So those wffs of N composed of an *odd* number of strokes ('$||-|||||$', '$||||-|||||$', and so on) are not provable in N. In particular, '$|-||||$', being composed of *five* strokes, cannot be a theorem of N.

But what of '$||||-||$'? Since it contains an even number of strokes, the preceding considerations do not suffice to establish its unprovability in N. To do that, we shall have to find a *different* feature, one which (R) can be used to show is possessed by all theorems of N but which '$||||-||$' lacks.

To describe one such feature it is convenient to provide an **interpretation** for the wffs of N, that is, to assign "meanings" to them by the assignment of meanings to certain sequences of N's primitive symbols. Temporarily, let us agree to regard a sequence of m strokes ($m = 1, 2, 3, \ldots$) as standing for the mth positive integer, and read '$-$' as 'is identical with'. Then N's axiom has the feature of *being true on this interpretation*, since it is true that the number one is identical with the number one. This

---

[2]See Appendix C.

feature is also preserved by **N**'s rule, for if it is true that the number m is identical with the number n, then it is true that $1 + m$ is identical with $n + 1$. Of course this feature, since it involves a specific interpretation of **N**'s wffs, is considerably more complicated than the one we looked at before. But the applicability of (R) is not blocked by complexity: all theorems of **N** must have this feature, that is, all theorems of **N** must be true on the interpretation suggested. However '||||–||' is *not* true on that interpretation, since four is not identical with two, so '||||–||' cannot be a theorem of **N**.

Indeed, *no* wff of **N** in which the dash is flanked by a different number of strokes on the left than on the right is provable in **N**, for each such wff is false on the interpretation described. On the other hand, any wff *true* on this interpretation — any wff composed, for some positive integer m, of m strokes, followed by a dash, followed by m strokes again — *is* a theorem of **N**, since each such wff is obtainable from **N**'s axiom by $m - 1$ applications of **N**'s rule. Our interpretation thus *characterizes* the theorems of **N**, for the latter are precisely the wffs of **N** true on that interpretation.

Consequently, **N** is **decidable** in the sense that there exists a mechanical procedure for determining of an arbitrary wff of **N** whether or not it is provable in **N**. One has only to count strokes, and see if the same number occur on the left as on the right.

EXERCISES

1. Consider another interpretation of the wffs of **N**, one according to which sequences of strokes are understood as before but '–' is read 'is less than or equal to'. Are all the theorems of **N** true on this interpretation? Explain.
2. Are all the wffs of **N** that are true on the interpretation described in the preceding exercise also theorems of **N**? Explain.
3. Would the answers to these questions change if we altered the interpretation so that '–' is read 'is less than'? If '–' is read 'is the sole prime factor of'? If '–' is read 'is the square of'? If '–' is read 'is an even number identical with'? (In each case, say why or why not.)
4. Construct a new axiomatic system, one just like **N** except for having additional axioms and/or rules, whose theorems can be characterized as being precisely those wffs true on the interpretation suggested in Exercise 1.

Of course the system **N** is too simple to be of much value except as an example. But more significant results are obtainable, and more interesting notions emerge, when we turn instead to

The Axiomatic System **D**.

Primitive Symbols: '~', '&', 'v', '(', ')', 'A', 'B', 'C',..., 'Y', 'Z', 'A₁', 'A₂',...

Wffs: (i) Each letter ('A', 'B', etc.) is a wff;
(ii) the result of prefixing '~' to a wff is a wff;
(iii) the result of placing '&' or 'v' between any two wffs, enclosing the whole in parentheses, is a wff;
nothing is a wff unless it can be shown to be so on the basis of (i)–(iii) above.

Axioms: Provided $\mathscr{P}$ is a wff,

$(\mathscr{P} \vee \sim\mathscr{P})$

is an axiom.[3]

Rules: **Conj, Simp, Add, DS, Com, Assoc, Dist, De M**, and **DN**.

D is, for all intents and purposes, the same system as the one developed in Section 1 of Chapter 3. Its presentation here differs from our earlier presentation in just one minor detail: the so-called "rule" **E-M I** has been dropped, the wffs it supplied being taken instead as axioms.

D's formal language was introduced in Chapter 1, and the *standard* interpretation (by means of truth-tables) of that language was treated in detail in Chapter 2. In Section 1 of Chapter 3 we established a number of **D**'s theorems, and in Chapter 4 obtained some metatheorems relating **D** to its interpretation. When, for example, the letters 'A', 'B', 'C', and so on, are taken to stand for particular English sentences, the theorems of **D** are not just true but *tautologies,* a fact which is sometimes put by saying that **D** is **weakly sound** with respect to the interpretation in question.[4] The converse, we saw, is true as well: **D** is **weakly complete** with respect to the interpretation of Chapter 2 in the sense that each wff of **D** which is a tautology is also a theorem of **D**. And these two results concerning **D**'s soundness and completeness together establish the *decidability* of **D**, truth-tables providing a mechanical procedure for determining which wffs of **D** are among its theorems and which are not.

In the case of **D**, and many other axiomatic systems as well, one is

---

[3] In Chapters 2 and 3 we adopted the convention of omitting outer parentheses in an expression standing alone. Here, however, we adhere strictly to the above definition of a wff and retain all parentheses.

[4] The general result (R) obtained above makes it especially evident that **D** is weakly sound: **D**'s axioms are all tautologies and its rules preserve tautologousness, so by (R) all theorems of **D** are tautologies. Compare Exercise 5, pp. 294-295.

interested not just in theorems and proofs, but in *deductions*. A wff $\mathscr{C}$ is **deducible from** wffs $\mathscr{P}_1, \ldots, \mathscr{P}_k$ in an axiomatic system $\mathscr{U}$ if and only if there is a finite sequence of wffs of $\mathscr{U}$ beginning with $\mathscr{P}_1, \ldots, \mathscr{P}_k$ and ending with $\mathscr{C}$ in which each intermediate wff is either one of $\mathscr{U}$'s axioms or the result of applying one of $\mathscr{U}$'s rules to earlier wffs in the sequence. (The sequence of wffs in question, of course, is called a **deduction in** $\mathscr{U}$ **of** $\mathscr{C}$ **from** $\mathscr{P}_1, \ldots, \mathscr{P}_k$.) In Chapter 3 we were primarily interested in constructing deductions in **D**, and in Chapter 4 were able to establish *strong* completeness and soundness results complementing the weaker ones: where $\mathscr{P}_1, \ldots, \mathscr{P}_k$ and $\mathscr{C}$ are wffs of **D**, if there is no assignment of truth-values to letters according to which $\mathscr{P}_1, \ldots, \mathscr{P}_k$ are assigned T's but $\mathscr{C}$ is assigned an **F**, then $\mathscr{C}$ is deducible in **D** from $\mathscr{P}_1, \ldots, \mathscr{P}_k$; and conversely, if $\mathscr{C}$ is deducible in **D** from $\mathscr{P}_1, \ldots, \mathscr{P}_k$, then there is no assignment of truth-values according to which $\mathscr{P}_1, \ldots, \mathscr{P}_k$ are assigned T's but $\mathscr{C}$ an **F**.

It is also of interest to ask whether there exists a theorem of **D** whose negation is provable as well, saying that **D** is **inconsistent with respect to** '$\sim$' if there is such a wff and **consistent with respect to** '$\sim$' if there is not. In light of our earlier work, the question is easily answered: if there were such a wff then both it and its negation would have to be tautologies, for all theorems of **D** are; and it is clearly impossible for the negation of any tautology to be a tautology itself.

More generally, a system is said to be **absolutely inconsistent** if *all* its wffs are theorems, and **absolutely consistent** otherwise, that is, so long as at least one of its wffs is unprovable. The appropriateness of these definitions is best brought out by noticing that they coincide with the earlier ones (tied to '$\sim$') in the case of a number of systems like **D**.[5]

(C) Let $\mathscr{U}$ be any axiomatic system whose wffs include those of **D** and whose rules include **Add** and **DS**. Then $\mathscr{U}$ is absolutely inconsistent if and only if it is inconsistent with respect to '$\sim$' (and so, is absolutely consistent if and only if consistent with respect to '$\sim$').

Half of this result is obvious: if $\mathscr{U}$ is absolutely inconsistent then *all* of its wffs are theorems and so in particular both 'A' and '$\sim A$' are, whence $\mathscr{U}$ is inconsistent with respect to '$\sim$'.

For the other half, assume that $\mathscr{U}$ is inconsistent with respect to '$\sim$'. Then there is at least one wff $\mathscr{P}$ such that both $\mathscr{P}$ and its negation are theorems of $\mathscr{U}$. Clearly, then, a proof of any wff $\mathscr{Q}$ can be constructed in $\mathscr{U}$; one need merely append the proof of $\mathscr{P}$ to the proof of its negation, and

---

[5] The *usefulness* of the notions of absolute consistency and inconsistency lies, of course, in the fact that they are applicable even to systems such as **N** which contain no primitive symbol '$\sim$'.

follow with appropriate applications of **Add** and **DS**:

$$\begin{array}{c} \vdots \\ \sim\mathscr{P} \\ \vdots \\ \mathscr{P} \\ \mathscr{P} \lor \mathscr{Q} \\ \mathscr{Q} \end{array}$$

[Add]
[DS]

If any such system were inconsistent with respect to '$\sim$', then, all of its wffs would be provable and so it would be complete, in a totally trivial way, with respect to any interpretation.

Of course it is possible to construct axiomatic systems equivalent in deductive power to **D** which make do with considerably fewer axioms. Let us say that a wff $\mathscr{P}^*$ of **D** is a **substitution instance** of a wff $\mathscr{P}$ of **D** just in case $\mathscr{P}^*$ can be derived from $\mathscr{P}$ by substituting for each letter in $\mathscr{P}$ some wff of **D** (the same wff replacing the same letter throughout). Recall from Chapter 2 that if $\mathscr{P}$ is a tautology ('$(A \lor \sim A)$', say) then the associated sentence form ('$(p \lor \sim p)$', in the case of this example) must be tautologous as well, so that any other wff of the same form as $\mathscr{P}$ must also be a tautology. Now each substitution instance $\mathscr{P}^*$ of any wff $\mathscr{P}$ is of the same form as $\mathscr{P}$. Consequently, *every substitution instance of a tautology is itself a tautology.* And since the theorems of **D** are (according to the soundness and completeness results obtained earlier) precisely those wffs of **D** that are tautologies, it follows that *every substitution instance of a theorem of* **D** *is itself a theorem of* **D**. The following rule of inference, then, can be added to **D** without altering its stock of theorems:

Substitution (Subs). From any theorem $\mathscr{P}$, one may infer any substitution instance $\mathscr{P}^*$ of $\mathscr{P}$.

And when that addition is made, most of **D**'s axioms become obtainable from others via the new rule, and so all but one of them can be dropped:

The Axiomatic System **D***.

Formal Language: That of **D**

Axiom: $(A \lor \sim A)$

Rules: Those of **D**, plus **Subs**

Since the missing axioms of **D** can all be recovered from **D***'s axiom by **Subs**, all theorems of **D** are theorems of **D***; and since **Subs** brings no new theorems with it, all theorems of **D*** are theorems of **D**. For similar reasons, a wff $\mathscr{C}$ is deducible in **D** from wffs $\mathscr{P}_1,\ldots,\mathscr{P}_k$ if and only if it is deducible in **D*** from those same premises. As a result, **D*** has all the features we've shown **D** to have — it is sound and complete, in both weak and strong senses, decidable, consistent with respect to '~', and absolutely consistent.

D* has an additional feature as well: it is **absolutely complete** in the sense that were any nontheorem of **D*** taken as additional axiom, the resulting system would be absolutely inconsistent. For let $\mathscr{P}$ be any wff not provable in **D***; then $\mathscr{P}$ cannot be a tautology, since all tautologies are theorems. There exists, then, an assignment of **T**'s and **F**'s to the letters occurring in $\mathscr{P}$ according to which $\mathscr{P}$ itself takes the value **F**. Let $\mathscr{P}*$ be the wff that results from $\mathscr{P}$ by substituting the wff '$(A \vee \sim A)$' for each letter in $\mathscr{P}$ assigned a **T** and the wff '$(A \,\&\, \sim A)$' for each letter in $\mathscr{P}$ assigned an **F**. Then $\mathscr{P}*$ receives the value **F** for *all* assignments of values to 'A', so $\mathscr{P}*$ is a contradiction. The negation of $\mathscr{P}*$, then, is a tautology, and is consequently a theorem of **D***. But $\mathscr{P}*$ is a substitution instance of $\mathscr{P}$, and so is obtainable (by **Subs**) in any system obtained from **D*** by adding $\mathscr{P}$ as new axiom. Any such system, then, must be inconsistent with respect to '~', since both $\mathscr{P}*$ and its negation are provable in it. And by our earlier result (C) that system must then be absolutely inconsistent as well.

By introducing the rule **Subs**, then, we can pare down **D**'s infinitely many axioms to the single axiom of **D***. And by complicating the latter, we can make do with fewer rules. If **D***'s axiom is replaced with

$$(\sim(\sim(\sim A \vee B) \vee (C \vee (D \vee E))) \vee (\sim(\sim D \vee A) \vee (C \vee (E \vee A)))).$$

for example, we need only **De M**, **Subs**, and the rule permitting passage from a wff $\mathscr{P}$ and a disjunction

$$(\sim\mathscr{P} \vee \mathscr{Q})$$

to $\mathscr{Q}$. With that single axiom and those three rules, we can prove all tautologies and deduce from any wffs $\mathscr{P}_1,\ldots,\mathscr{P}_k$ all wffs deducible from them in **D***.[6]

---

[6]Compare C.A. Meredith, "Single axioms for the systems (C,N), (C,O) and (A,N) of the two-valued propositional calculus", *Journal of Computing Systems,* vol. 1(1953), pp. 155-164.

Of course, our treatment of '&', 'v' and '~' can be extended to '⊃'. If the latter is added to our list of primitive symbols, and the characterization of our wffs is altered in the obvious way, addition of the rule **Impl** suffices to allow us to obtain extended soundness and completeness results (relative to the usual truth-table interpretation of '⊃'). Alternative axiomatizations have also been found, of course. It is possible to make do with just **Subs** and **MP** as rules, for example, so long as our axioms include:[7]

$(((A \supset B) \supset C) \supset ((C \supset A) \supset (D \supset A)))$

$((A \& B) \supset A)$

$((A \& B) \supset B)$

$(A \supset (B \supset (A \& B)))$

$((\sim A \supset \sim B) \supset (B \supset A))$

$(A \supset (A \vee B))$

$(B \supset (A \vee B))$

$((A \supset C) \supset ((B \supset C) \supset ((A \vee B) \supset C)))$

We leave it to the reader to pursue these and other matters in more advanced texts devoted to the study of axiomatic systems. Ultimately, of course, the apparatus of predicate letters, constants, variables, and quantifiers is added to the formal language of our system **D**, versions of **UI**, **UG** and the like are incorporated into the axioms and rules, and all of quantificational logic is developed axiomatically. Though we stop short of this development here, it ought to be noted that some of the most interesting results in modern logic lie in this area. Soundness and completeness results, relative to the kind of interpretation introduced briefly in Chapter 6 above, are obtainable. And quantificational logic, unlike the sentential calculi discussed in this Appendix, turns out to be *undecidable*— no mechanical methods for determining which wffs are provable, nor which wffs are deducible from which others, can be devised.[8]

---

[7]Compare A.N. Prior, *Formal Logic,* Oxford: Clarendon Press, 1962, pp. 302-303. If '≡' is added to our formal language, it suffices to add $((A \equiv B) \supset ((A \supset B) \& (B \supset A)))$ and $(((A \supset B) \& (B \supset A)) \supset (A \equiv B))$ to our axioms.

[8]For the axiomatic development of quantificational logic, and proofs of these and other results, see for example E. Mendelson, *Introduction to Mathematical Logic,* 3rd ed., Monterey: Wadsworth and Brooks, 1987, or G. Hunter, *Metalogic,* Berkeley: University of California, 1971.

ANSWERS TO SELECTED EXERCISES

CHAPTER ONE

Pages 8-9

I-2. If Lassie is a human being, then she's mortal
Lassie is mortal
∴ Lassie is a human being

I-5. All human beings are reptiles
All reptiles are mammals
∴ All human beings are mammals

Page 15

I-2. True. If 'either-or' is construed exclusively, it is undetermined.

I-3. Undetermined; though the left disjunct is false, the right disjunct is undetermined, making the disjunction itself undetermined.

II-2. False, since the left conjunct is false.

II-3. True, since the left disjunct is true.

Pages 21–22

I-3. ' ~(O & I)', where 'O' abbreviates 'Four is prime' and 'I' abbreviates 'Five is prime'. Also correct: ' ~ O ∨ ~I'.

I-6. 'I & ~(O ∨ X)', also 'I & (~O & ~X)', where 'O' and 'I' are as above and 'X' abbreviates 'Six is prime'.

III-3. Derivable: 'A' for 'p', '~(B & C)' for 'q' and 'D' for 'r'.

III-6. Not derivable, different substitutions for the two occurrences of 'p'.

IV-2. Not derivable, different substitutions for the two occurrences of 'p'. '~~A' is substituted for the lefthand 'p' but 'A' for the righthand one. (Yes, '~~A' and 'A' say the same thing, but they are *different* sentences.)

Pages 27–28

I-3. $V \vee \sim(H \vee S)$ [also of course $V \vee (\sim H \& \sim S)$]

I-6. $(T \vee V) \& \sim(A \& R)$

I-9. $V \& (\sim D \vee P)$

I-11. $(W \& J) \& (I \vee O)$

Pages 34–35

I-4. $(\sim F \& \sim C) \supset \sim(B \vee E)$

I-7. $(T \& C) \supset G$

I-10. $J \supset (I \equiv \sim F)$

I-14. $(\sim P \& \sim C) \supset [F \equiv (B \vee D)]$

IV. No, sentences of these forms are true under the same conditions, viz. when and only when p and q differ in truth-value. Likewise, these forms are true under the same conditions as the corresponding exclusive disjunction, and hence they provide us with two more ways of translating exclusive disjunctions into our symbolic notation.

Page 38

3. Truth-functional; using obvious notation: $C \& \sim H$.

11. Not a compound sentence; translation: T.

14. Non-truth-functional; substituting a true sentence like 'John Kennedy was the 35th U.S. president' for '$7 + 5 = 12$' changes sentence (14) from a truth to a falsehood.

17. If the quotation is regarded as a sentence *name* then the sentence is not compound. Otherwise, it is truth-functional.

CHAPTER TWO

Page 45

I-2.

B	$\sim B$	$\sim B \& B$	$B \vee (\sim B \& B)$
T	F	F	T
F	T	F	F

Neither a tautology nor a contradiction.

I-7. To conserve space, columns for the component compound sentences are displayed directly beneath such components as they occur in the whole. The column for the latter is displayed beneath its main connective.

C	D	E	(C v D)	&	(C v ~D)	~D	&	~C	&	v	~C	&	~E
T	T	T	T	T	T	F	F	F	F	F	F	F	F
T	T	F	T	T	T	F	F	F	F	F	F	F	T
T	F	T	T	T	T	T	F	F	F	F	F	F	F
T	F	F	T	T	T	T	F	F	F	F	F	F	T
F	T	T	T	F	F	F	F	T	T	T	T	F	F
F	T	F	T	F	F	F	F	T	F	T	T	T	T
F	F	T	F	F	T	T	F	T	T	T	T	F	F
F	F	F	F	F	T	T	F	T	F	T	T	T	T

Contradiction.

III-2.

p	q	p v q	q v p
T	T	T	T
T	F	T	T
F	T	T	T
F	F	F	F

III-6.

p	q	r	p v (q & r)		(p v q) & (p v r)		
T	T	T	T	T	T	T	T
T	T	F	T	F	T	T	T
T	F	T	T	F	T	T	T
T	F	F	T	F	T	T	T
F	T	T	T	T	T	T	T
F	T	F	F	F	T	F	F
F	F	T	F	F	F	F	T
F	F	F	F	F	F	F	F

IV-1. True. A conjunction is true *only* when both its conjuncts are, so if a conjunction takes a **T** in every row of its truth-table, so must those conjuncts.

Page 50

III-4.

F	*E*	*F* v *E*	*E* & *F*	(*F* v *E*) ⊃ (*E* & *F*)
T	T	T	T	T
T	F	T	F	F
F	T	T	F	F
F	F	F	F	T

Neither a tautology nor a contradiction.

IV-3. $(\sim B \vee \sim R) \supset (P \supset I)$

Pages 57–58

			Prem. 1			Prem. 2		Concl.		
I-4. *E*	*A*	*R*	~(*E* & *A*) v *R*			~(*R* & *E*)		~*A* v ~*R*		
T	T	T	F	T	T	F	T	F	F	F
T	T	F	F	T	F	T	F	F	T	T
T	F	T	T	F	T	F	T	T	T	F
T	F	F	T	F	T	T	F	T	T	T
F	T	T	T	F	T	T	F	F	F	F
F	T	F	T	F	T	T	F	F	T	T
F	F	T	T	F	T	T	F	T	T	F
F	F	F	T	F	T	T	F	T	T	T

Truth-table invalid (row5).

III. It is a tautology. Given that the argument is valid, it is impossible for the premise to be true and the conclusion false. But the disjunction consists of the conclusion and the *negation* of the premise as disjuncts, and therefore it is impossible for those disjuncts to be jointly false and hence for the disjunction itself to be false.

Pages 62–63

II-4.

	Concl.		Prem. 1		Prem. 2	Prem. 3	
L	M	C	L ⊃ (M & C)		M v L	~M v C	
T	T	T	T	T	T	F	T
T	T	F	F	F	T	F	F
T	F	T	F	F	T	T	T
T	F	F	F	F	T	T	T
F	T	T	T	T	T	F	T
F	T	F	T	F	T	F	F
F	F	T	T	F	F	T	T
F	F	F	T	F	F	T	T

Truth-table valid.

II-5.

			Prem. 1				Prem. 2	Concl.	
P	F	V	(~P v ~F) ⊃ V				P ⊃ F	P ⊃ ~V	
T	T	T	F	F	F	T	T	F	F
T	T	F	F	F	F	T	T	T	T
T	F	T	F	T	T	T	F	F	F
T	F	F	F	T	T	F	F	T	T
F	T	T	T	T	F	T	T	T	F
F	T	F	T	T	F	F	T	T	T
F	F	T	T	T	T	T	T	T	F
F	F	F	T	T	T	F	T	T	T

Truth-table invalid (row 1).

III. First question: yes, it is truth-table valid. Since the original argument cannot have true premises and false conclusion, neither can this one since its conclusion is a conditional consisting of the second premise of the original argument as antecedent and its conclusion as consequent. Hence that conditional can never be false when its sole premise (the first premise of the original one) is true.

Second question: yes, it is a tautology. It is equivalent to the corresponding conditional of the original argument (see exercise I-1 at the end of Section 2.2).

Pages 68–69

I-3. Valid. The only condition under which the conclusion is false is when both A and B are true. But then the only way for the first two premises to be true is for E and F to be true. However, under these conditions the third premise has two false disjuncts and so must be false. Hence there is no assignment on which all premises are true when the conclusion is assumed to be false.

I-6. Invalid.

A	B	C	D	$A \supset (B \supset C)$	$D \supset (B \& A)$	$C \supset D$	$C \equiv A$
T	F	F	F	T	T	T	F

Pages 77–78

II.

p	q	p/q	$\sim p$	p/p	$p \& q$	$(p/q)/(p/q)$
T	T	F	F	F	T	T
T	F	T	F	F	F	F
F	T	T	T	T	F	F
F	F	T	T	T	F	F

A sentence "stroked" with itself is equivalent to the negation of that sentence. Moreover, the column for conjunction is the exact opposite of that for the stroke. So, given the result about negation, we need only "stroke" 'p/q' with itself to obtain a sentence equivalent to the conjunction of 'p' and 'q'.

CHAPTER THREE

Pages 85–86

I-3.
1. $N \vee (S \vee \sim A)$	Premise	
2. $\sim N \& \sim P$	Premise	
3. $\sim S \& \sim Q$	Premise	DEDUCE: $\sim N \& (\sim S \& \sim A)$
4. $\sim N$	2 Simp	
5. $S \vee \sim A$	1, 4 DS	
6. $\sim S$	3 Simp	
7. $\sim A$	5, 6 DS	
8. $\sim S \& \sim A$	6, 7 Conj	
9. $\sim N \& (\sim S \& \sim A)$	4, 8 Conj	

I-6.
1. $R \& S$ Premise   DEDUCE:
2. R 1 Simp $(R \vee {\sim}S) \& (T \vee {\sim}T)$
3. $R \vee {\sim}S$ 2 Add
4. $T \vee {\sim}T$ E-M I
5. $(R \vee {\sim}S) \& (T \vee {\sim}T)$ 3, 4 Conj

IV. According to the previous Exercise, where '*' abbreviates the English expression 'and possibly' all arguments of either of the following forms are valid:

$$p * q \qquad p$$
$$\therefore p \qquad\qquad q$$
$$\qquad\qquad\qquad \therefore p * q.$$

So if, for example, the conclusion of the argument

$$R \& P$$
$$\therefore P \& R$$

were deducible from its premise by means of our five rules, then '$P * R$' would be obtainable from '$R * P$' in equally unassailable fashion: one would have only to replace each occurrence of '&' in the former deduction with '*', replacing appeals to **Simp** [respectively, **Conj**] with appeals to the validity of all arguments of the first [respectively, second] form above. But if '$P * R$' could be obtained from '$R * P$' in such a way, '$P * R$' would have to be true whenever '$R * P$;' is. Since the latter is not the case (let 'R' be 'Life exists on earth'; let 'P' be 'Life exists on the moon'), then, '$P \& R$' cannot be deduced from '$R \& P$' using just our five rules.

Pages 93–95

1.
1. ${\sim}(P \& C)$ Premise
2. P Premise   DEDUCE: ${\sim}C$
3. ${\sim}P \vee {\sim}C$ 1 De M
4. ${\sim}{\sim}P$ 2 DN
5. ${\sim}C$ 3, 4 DS

4.
1. $(F \vee S) \& (F \vee T)$ Premise
2. ${\sim}F$ Premise   DEDUCE: $S \& T$

Since the second premise is the negation of the left disjunct of both disjunctions occurring in the first premise, it is easy enough to do this problem by using **DS** twice:

3. $F \vee S$ 1 Simp
4. S 3, 2 DS
5. $(F \vee T) \& (F \vee S)$ 1 Com
6. $F \vee T$ 5 Simp
7. T 6, 2 DS
8. $S \& T$ 4, 7 Conj

But this is one of those problems where **Dist** can cut down the length considerably if we spot the chance to use it:

1. $(F \vee S) \& (F \vee T)$ Premise
2. $\sim F$ Premise   DEDUCE: $S \& T$
3. $F \vee (S \& T)$ 1 Dist
4. $S \& T$ 3, 2 DS

12. 1. $M \vee \sim\sim M$ Premise   DEDUCE: M
     2. $M \vee \sim M$ E-M I
     3. $(M \vee \sim M) \& (M \vee \sim\sim M)$ 2,1 Conj
     4. $M \vee (\sim M \& \sim\sim M)$ 3 Dist
     5. $(\sim M \& \sim\sim M) \vee M$ 4 Com
     6. $\sim\sim M \vee \sim\sim\sim M$ E-M I
     7. $\sim(\sim M \& \sim\sim M)$ 6 De M
     8. M 5, 7 DS

*15. 1. $\sim E \vee (\sim W \vee \sim A)$ Premise
      2. $(W \& A) \vee (\sim W \& \sim A)$ Premise
                                    DEDUCE: $\sim E \vee (\sim W \& \sim A)$
      3. $\sim E \vee \sim(W \& A)$ 1 De M
      4. $[\sim E \vee \sim(W \& A)] \vee (\sim W \& \sim A)$ 3 Add
      5. $\sim E \vee [\sim(W \& A) \vee (\sim W \& \sim A)]$ 4 Assoc
      6. $\sim E \vee [(\sim W \& \sim A) \vee \sim(W \& A)]$ 5 Com
      7. $[\sim E \vee (\sim W \& \sim A)] \vee \sim(W \& A)$ 6 Assoc

We have now shown that either our conclusion is true or '$\sim(W \& A)$' is. In lines (8)–(11), we'll show that either our conclusion is true or '$W \& A$' is. Conjoining these and distributing, we'll find at step (13) that either the conclusion is true or else the contradiction '$(W \& A) \& \sim(W \& A)$' is. With lines (14)–(17) it is then easy to eliminate (as in Exercise 12 above) the contradictory disjunct.

8. $[(W \& A) \lor (\sim W \& \sim A)] \lor \sim E$	2 Add
9. $(W \& A) \lor [(\sim W \& \sim A) \lor \sim E]$	8 Assoc
10. $(W \& A) \lor [\sim E \lor (\sim W \& \sim A)]$	9 Com
11. $[\sim E \lor (\sim W \& \sim A)] \lor (W \& A)$	10 Com
12. $\{[\sim E \lor (\sim W \& \sim A)] \lor (W \& A)\} \&$	
$\qquad \{[\sim E \lor (\sim W \& \sim A)] \lor \sim (W \& A)\}$	11, 7 Conj
13. $[\sim E \lor (\sim W \& \sim A)] \lor [(W \& A) \& \sim (W \& A)]$	12 Dist
14. $[(W \& A) \& \sim (W \& A)] \lor [\sim E \lor (\sim W \& \sim A)]$	13 Com
15. $\sim (W \& A) \lor \sim \sim (W \& A)$	E-M I
16. $\sim [(W \& A) \& \sim (W \& A)]$	15 De M
17. $\sim E \lor (\sim W \& \sim A)$	14, 16 DS

Page 98

I-2.
1. $F \lor \sim F$	E-M I
2. $\sim F \lor F$	1 Com
3. $(\sim F \lor F) \lor \sim E$	2 Add
4. $\sim F \lor (F \lor \sim E)$	3 Assoc
5. $\sim F \lor (\sim E \lor F)$	4 Com

Pages 106–109

II-2.
1. $A \supset R$	Premise	
2. $\sim (R \& S)$	Premise	DEDUCE: $S \supset \sim A$
3. $\sim R \lor \sim S$	2 De M	
4. $R \supset \sim S$	3 Impl	
5. $A \supset \sim S$	1, 4 HS	
6. $\sim \sim S \supset \sim A$	5 Trans	
7. $S \supset \sim A$	6 DN	

II-7.
1. $L \supset N$	Premise	
2. $(\sim N \supset \sim L) \supset (M \supset P)$	Premise	
3. $\sim P$	Premise	DEDUCE: $\sim M$
4. $\sim N \supset \sim L$	1 Trans	
5. $M \supset P$	2, 4 MP	
6. $\sim M$	5, 3 MT	

II-12. 1. $K \equiv Q$ Premise
    2. $(K \& V) \equiv \sim(Q \vee V)$ Premise

                                DEDUCE: $K \supset \sim V$

    3. $[(K \& V) \supset \sim(Q \vee V)] \& [\sim(Q \vee V) \supset (K \& V)]$ 2 Equiv
    4. $(K \& V) \supset \sim(Q \vee V)$ 3 Simp
    5. $(K \& V) \supset (\sim Q \& \sim V)$ 4 De M
    6. $(K \& V) \supset (\sim V \& \sim Q)$ 5 Com
    7. $\sim(K \& V) \vee (\sim V \& \sim Q)$ 6 Impl
    8. $[\sim(K \& V) \vee \sim V] \& [\sim(K \& V) \vee \sim Q]$ 7 Dist
    9. $\sim(K \& V) \vee \sim V$ 8 Simp
  10. $(K \& V) \supset \sim V$ 9 Impl
  11. $K \supset (V \supset \sim V)$ 10 Exp
  12. $K \supset (\sim V \vee \sim V)$ 11 Impl

Clearly '$\sim V \vee \sim V$' and '$\sim V$' say the same thing. By establishing '$(\sim V \vee \sim V) \supset \sim V$' in lines (13)-(19) below, we'll set up an **HS** with (12) that will give us our conclusion.

  13. $\sim V \vee \sim\sim V$ E-M I
  14. $\sim V \vee \sim\sim V$ E-M I
  15. $(\sim V \vee \sim\sim V) \& (\sim V \vee \sim\sim V)$ 13, 14 Conj
  16. $\sim V \vee (\sim\sim V \& \sim\sim V)$ 15 Dist
  17. $(\sim\sim V \& \sim\sim V) \vee \sim V$ 16 Com
  18. $\sim(\sim V \vee \sim V) \vee \sim V$ 17 De M
  19. $(\sim V \vee \sim V) \supset \sim V$ 18 Impl
  20. $K \supset \sim V$ 12, 19 HS

III-6. 1. $F \& (T \supset S)$ Premise
    2. $F \supset \sim S$ Premise   DEDUCE: $\sim T$
    3. F 1 Simp
    4. $\sim S$ 2, 3 MP
    5. $(T \supset S) \& F$ 1 Com
    6. $T \supset S$ 5 Simp
    7. $\sim T$ 6, 4 MT

III-9. 1. $\sim F \supset (T \vee B)$ Premise
        2. $T \supset S$ Premise

DEDUCE: $(\sim S \& \sim F) \supset B$

        3. $\sim F \supset (B \vee T)$ 1 Com
        4. $\sim F \supset (\sim \sim B \vee T)$ 3 DN
        5. $\sim F \supset (\sim B \supset T)$ 4 Impl
        6. $(\sim F \& \sim B) \supset T$ 6 Exp
        7. $(\sim F \& \sim B) \supset S$ 6, 2 HS
        8. $\sim S \supset \sim (\sim F \& \sim B)$ 7 Trans
        9. $\sim S \supset (\sim \sim F \vee \sim \sim B)$ 8 De M
       10. $\sim S \supset (\sim \sim F \vee B)$ 9 DN
       11. $\sim S \supset (\sim F \supset B)$ 10 Impl
       12. $(\sim S \& \sim F) \supset B$ 11 Exp

CHALLENGE: See how much *shorter* a deduction you can find.

III-13. 1. $B \vee K$ Premise
        2. $Q \vee P$ Premise
        3. $(B \vee Q) \supset (K \& P)$ Premise

DEDUCE: $K \& P$

        4. $\sim (B \vee Q) \vee (K \& P)$ 3 Impl
        5. $(K \& P) \vee \sim (B \vee Q)$ 4 Com
        6. $(B \vee K) \vee Q$ 1 Add
        7. $B \vee (K \vee Q)$ 6 Assoc
        8. $B \vee (Q \vee K)$ 7 Com
        9. $(B \vee Q) \vee K$ 8 Assoc
       10. $(Q \vee P) \vee B$ 2 Add
       11. $B \vee (Q \vee P)$ 10 Com
       12. $(B \vee Q) \vee P$ 11 Assoc
       13. $[(B \vee Q) \vee K] \& [(B \vee Q) \vee P]$ 9, 12 Conj
       14. $(B \vee Q) \vee (K \& P)$ 13 Dist
       15. $(K \& P) \vee (B \vee Q)$ 14 Com
       16. $[(K \& P) \vee (B \vee Q)] \& [(K \& P) \vee \sim (B \vee Q)]$ 15, 5 Conj
       17. $(K \& P) \vee [(B \vee Q) \& \sim (B \vee Q)]$ 16 Dist
       18. $[(B \vee Q) \& \sim (B \vee Q)] \vee (K \& P)$ 17 Com
       19. $\sim (B \vee Q) \vee \sim \sim (B \vee Q)$ E-M I
       20. $\sim [(B \vee Q) \& \sim (B \vee Q)]$ 19 De M
       21. $K \& P$ 18, 20 DS

Page 113

I-5. 1. $O \supset U$ Premise   DEDUCE: $(B \& O) \supset U$
→2. $B \& O$ Assumption
  3. $O \& B$ 2 Com
  4. O 3 Simp
  5. U 1, 4 MP

  6. $(B \& O) \supset U$ 2-5 CP

Pages 117–118

I-4. 1. $S \supset (V \& P)$ Premise
                DEDUCE: $(S \supset V) \& (S \supset P)$
→2. S Assumption
  3. $V \& P$ 1, 2 MP
  4. V 3 Simp

  5. $S \supset V$ 2-4 CP
→6. S Assumption
  7. $V \& P$ 1, 6 MP
  8. $P \& V$ 7 Com
  9. P 8 Simp

  10. $S \supset P$ 6-9 CP
  11. $(S \supset V) \& (S \supset P)$ 5, 10 Conj

II-4. 1. $F \supset {\sim}S$ Premise
    2. ${\sim}V \supset {\sim}S$ Premise   DEDUCE:
                    $(F \vee {\sim}V) \supset {\sim}S$

Usually when doing a problem by **Conditional Proof,** the strategy that works best is to assume the *antecedent* of the conditional we are trying to establish and then deduce its consequent. On this particular problem, that approach leads to a fairly messy deduction:

$$
\begin{array}{lll}
\rightarrow \text{3.} & F \vee \sim V & \text{Assumption} \\
\text{4.} & \sim S \vee \sim S & \text{3, 1, 2 CD} \\
\text{5.} & \sim S \vee \sim\sim S & \text{E-M I} \\
\text{6.} & (\sim S \vee \sim S) \,\&\, (\sim S \vee \sim\sim S) & \text{4, 5 Conj} \\
\text{7.} & \sim S \vee (\sim S \,\&\, \sim\sim S) & \text{6 Dist} \\
\text{8.} & (\sim S \,\&\, \sim\sim S) \vee \sim S & \text{7 Com} \\
\text{9.} & \sim\sim S \vee \sim\sim\sim S & \text{E-M I} \\
\text{10.} & \sim(\sim S \,\&\, \sim\sim S) & \text{9 De M} \\
\text{11.} & \sim S & \text{8, 10 DS}
\end{array}
$$

12. $(F \vee \sim V) \supset \sim S$ 3-11 CP

Occasionally a better strategy is to assume the *negation* of the consequent of the conditional we're after and try to show the *negation* of its antecedent. Notice how much more smoothly things go here, for example, if we assume '$\sim\sim S$' and deduce '$\sim(F \vee \sim V)$', counting on the rule **Trans** to turn the conditional we get into the one we really want:

$$
\begin{array}{lll}
\rightarrow \text{3.} & \sim\sim S & \text{Assumption} \\
\text{4.} & \sim F & \text{1, 3 MT} \\
\text{5.} & \sim\sim V & \text{2, 3 MT} \\
\text{6.} & \sim F \,\&\, \sim\sim V & \text{4, 5 Conj} \\
\text{7.} & \sim(F \vee \sim V) & \text{6 De M}
\end{array}
$$

8. $\sim\sim S \supset \sim(F \vee \sim V)$ 3-7 CP
9. $(F \vee \sim V) \supset \sim S$ 8 Trans

III-6.

1. $(\sim A \vee \sim B)\ \&\ (C \supset A)$	Assumption
2. $D \supset B$	Assumption
3. D	Assumption
4. B	2, 3 MP
5. $\sim\sim B$	4 DN
6. $\sim A \vee \sim B$	1 Simp
7. $\sim B \vee \sim A$	6 Com
8. $\sim A$	7, 5 DS
9. $(C \supset A)\ \&\ (\sim A \vee \sim B)$	1 Com
10. $C \supset A$	9 Simp
11. $\sim C$	10, 8 MT
12. $D \supset \sim C$	3-11 CP
13. $\sim D \vee \sim C$	12 Impl
14. $\sim C \vee \sim D$	13 Com
15. $(D \supset B) \supset (\sim C \vee \sim D)$	2-14 CP
16. $[(\sim A \vee \sim B)\ \&\ (C \supset A)] \supset [(D \supset B) \supset$ $(\sim C \vee \sim D)]$	1-15 CP

Pages 122–124

I-2.

1. $D \supset M$	Premise
2. $\sim D \supset M$	Premise DEDUCE: M
3. $\sim M$	Assumption
4. $\sim D$	1, 3 MT
5. D	2, 3 MT, DN
6. $D\ \&\ \sim D$	5, 4 Conj
7. $\sim M \supset (D\ \&\ \sim D)$	3-6 CP
8. $\sim D \vee \sim\sim D$	E-M I
9. $\sim(D\ \&\ \sim D)$	8 De M
10. M	7, 9 MT, DN

II-1. 1. $L \vee (M \& N)$ Premise
    2. $M \supset (\sim L \supset \sim O)$ Premise
    3. O Premise   DEDUCE: L
    →4. $\sim L$ Assumption
      5. $M \& N$ 1, 4 DS
      6. M 5 Simp
      7. $\sim L \supset \sim O$ 2, 6 MP
      8. $\sim O$ 7, 4 MP
      9. $O \& \sim O$ 3, 8 Conj
    10. $\sim L \supset (O \& \sim O)$ 4-9 CP
    11. $\sim O \vee \sim \sim O$ E-M I
    12. $\sim (O \& \sim O)$ 11 De M
    13. L 10, 12 MT, DN

Since 'O' appeared *outside* the three-sided box and '$\sim O$' *inside* it, we could have saved a few steps by abandoning our plan to use indirect proof at step (8) and closing off the **CP**-assumption immediately:

    9. $\sim L \supset \sim O$ 4-8 CP
    10. $\sim \sim L$ 9, 3 DN, MT
    11. L 10 DN

IV-2. 1. $[(A \vee B) \& (C \vee D)] \supset E$ Premise
                        DEDUCE: $(A \& D) \supset E$
    →2. $A \& D$ Assumption
      3. A 2 Simp
      4. $A \vee B$ 3 Add
      5. D 2 Com, Simp
      6. $C \vee D$ 5 Add, Com
      7. $(A \vee B) \& (C \vee D)$ 4, 6 Conj
      8. E 1, 7 MP
    9. $(A \& D) \supset E$ 2-8 CP

IV-8.   1. $(A \vee L) \vee O$ Premise
     2. $L \equiv O$ Premise   DEDUCE: $O \vee A$
     →3. $\sim O$ Assumption
      4. $A \vee L$ 1, 3 Com, DS
      5. $(L \supset O) \& (O \supset L)$ 2 Equiv
      6. $L \supset O$ 5 Simp
      7. $\sim L$ 6, 3 MT
      8. A 4, 7 Com, DS
     9. $\sim O \supset A$ 3-8 CP
    10. $O \vee A$ 9 Impl, DN

IV-11.   1. $(E \& W) \equiv (E \& A)$ Premise
                                     DEDUCE: $E \supset (W \equiv A)$
     →2. E Assumption

The conclusion is a conditional, so we assume its antecedent and intend to deduce its consequent, '$W \equiv A$'. Since the latter is a biconditional, we will use two **CP**s to obtain it. Because we can anticipate using both "halves" of (1), we can save a step by applying **Equiv** to (1) before beginning either remaining **CP**:

     3. $[(E \& W) \supset (E \& A)] \&$
         $[(E \& A) \supset (E \& W)]$ 1 Equiv
     → 4. W Assumption
       5. $E \& W$ 2, 4 Conj
       6. $(E \& W) \supset (E \& A)$ 3 Simp
       7. $E \& A$ 6, 5 MP
       8. A 7 Com, Simp
     9. $W \supset A$ 4-8 CP
    →10. A Assumption
     11. $E \& A$ 2, 10 Conj
     12. $(E \& A) \supset (E \& W)$ 3 Com, Simp
     13. $E \& W$ 12, 11 MP
     14. W 13 Com, Simp
     15. $A \supset W$ 10-14 CP
     16. $(W \supset A) \& (A \supset W)$ 9, 15 Conj
     17. $W \equiv A$ 16 Equiv
    18. $E \supset (W \equiv A)$ 2-17 CP

In fact, (1) and (18) are two different ways of saying the same thing, as you can show by next taking (18) as your premise and deducing (1) as your conclusion; the deduction is quite similar to this one.

CHAPTER FOUR

Page 128

2. Anything provable in **D+** will be provable in **D**, for at any point in a proof in **D+** where its extra rule is used to pass from a line of the form

$$n.\ p\ \&\ {\sim}p$$

to a sentence 'q' in one step we could always have obtained 'q' in **D** in five steps:

$n + 1.$	p	n, Simp
$n + 2.$	$p \vee q$	$n + 1$, Add
$n + 3.$	${\sim}p \& p$	n, Com
$n + 4.$	${\sim}p$	$n + 3$, Simp
$n + 5.$	q	$n + 2, n + 4$ DS.

Thus if **D+** were inconsistent, that is, if some theorem of **D+** were a contradiction, then that same contradiction would be provable in **D**. Since **D** is consistent, no contradictions are theorems of **D**. So none can be theorems of **D+**, either.

5. Here is a deduction of '${\sim}{\sim}A$' from 'A' using only the rules of **D−**:

1.	A	Premise	DEDUCE: ${\sim}{\sim}A$
2.	$A \vee {\sim}{\sim}A$	1 Add	
3.	${\sim}{\sim}A \vee A$	2 Com	
4.	${\sim}A \vee {\sim}{\sim}A$	E-M I	
5.	${\sim}{\sim}A \vee {\sim}A$	4 Com	
6.	$({\sim}{\sim}A \vee A) \& ({\sim}{\sim}A \vee {\sim}A)$	3, 5 Conj	
7.	${\sim}{\sim}A \vee (A \& {\sim}A)$	6 Dist	
8.	$(A \& {\sim}A) \vee {\sim}{\sim}A$	7 Com	
9.	${\sim}(A \& {\sim}A)$	4 De M	
10.	${\sim}{\sim}A$	8, 9 DS	

7. **B** is *not* weakly complete, for the tautology '$(R \& P) \vee \sim(P \& R)$' is not provable using just **B**'s rules. One way of showing this is to consider a new set of tables:

p	$\sim p$		p	q	$p \& q$		p	q	$p \vee q$
t	f		t	t	t		t	t	t
f	t		t	f	t		t	f	t
			f	t	f		f	t	t
			f	f	f		f	f	f

Let us say that a sentence is *peculiar* if and only if it always receives the value **t** according to these tables, regardless of the way ts and fs are assigned to the letters occurring in it. Then it is easy to show that each sentence of the form '$p \vee \sim p$' is peculiar, and that the rules **Conj, Simp, Add** and **DS** lead from peculiar sentences only to other sentences that are also peculiar. Since proofs in **B** must be initiated with **E-M I**, then, all proofs in **B** *begin* with peculiar sentences; and since applications of **B**'s other rules to peculiar sentences can only result in other sentences that are also peculiar, *every* line of every proof in **B** must be peculiar. Since the theorems of **B** are the *last* lines of proofs in **B**, then, all theorems of **B** are peculiar. But '$(R \& P) \vee \sim(P \& R)$' is not peculiar:

R	P	$R \& P$	$P \& R$	$\sim(P \& R)$	$(R \& P) \vee \sim(P \& R)$
f	t	f	t	f	f

So '$(R \& P) \vee \sim(P \& R)$' cannot be a theorem of **B**.

Notice that these same tables can be used to show, by a similar argument, that '$P \& R$' is not deducible in **B** from '$R \& P$', thus providing an alternate solution to Exercise IV on p. 86.

Pages 136–137

1. Since the inconsistent deductive system **E** comes from our system **D** only by adding rules, not by dropping any, everything provable in **D** is provable in **E**. So since all tautologies are provable in **D**, they remain provable in **E** which is thus also weakly complete.

   By a similar line of reasoning, if the conclusion of any argument is deducible in **D** from the premises of that argument, it will also be deducible in **E** from those premises, for all of **D**'s rules are available for use in **E**. Since **D** is strongly complete, then, so must be **E**.

Metatheorems III and IV consequently both hold for **E** as well as for **D**. But Metatheorems I and II do *not* hold for **E**.

First, since **E** is inconsistent at least one contradiction, say $\mathscr{C}$, is a theorem of **E**. Being a contradiction, $\mathscr{C}$ is not a tautology. So not all theorems of **E** are tautologies, that is, **E** is not weakly sound.

Nor is **E** strongly sound. Since the contradiction $\mathscr{C}$ is a *theorem* of **E**, it will be deducible in **E** from *any* premise, e.g., 'A v $\sim A$'. But of course the argument with premise 'A v $\sim A$' and conclusion $\mathscr{C}$ is *not* truth-table valid.

4. The deductive system **C** which comes from our system **D** by dropping the rules **Simp** and **DS** is *not* strongly complete. To see this, let us call a sentence *complex* just in case it contains one or more occurrences of 'v' or '&'. (The complex sentences, then, are the conjunctions, the disjunctions, and the sentences obtainable from conjunctions and disjunctions by prefixing to them strings of one or more curls.) Now it is easy, albeit tedious, to verify that all sentences obtainable by **E-M I** are complex, and that applications of any of **C**'s other rules to complex sentences can only produce additional sentences that are also complex.

Of course, 'R & P' is complex. Any line of a deduction in **C** in which 'R & P' is the sole premise must then also be complex. Consequently, no line of any such deduction can be the plain 'R'. So the argument with premise 'R & P' and conclusion 'R' is one which is clearly truth-table valid but whose conclusion cannot be deduced from its premise in **C**.

Additional examples may be easily found. For example, 'R' is not deducible in **C** from the two premises '$(P \& Q)$ v R' and '$\sim(P \& Q)$', either.

CHALLENGE: Let **C+** come from our system **D** by dropping the rule **Simp** but *not* **DS**. Is **C+** strongly complete?

CHAPTER FIVE

Page 151

2. Tc v Fh

9. $(\exists x)\,(Px \& Ox)$

12. $(x)\,(Fx \supset Jx)$

14. $\sim(\exists x)\,(Dx \& \sim Gx)$ [also of course: $(x)\,(Dx \supset Gx)$]

Pages 154–156

I-4. False, object a is either F or K but does not have *both* H and G, i.e. the substitution instance '$(Fa \vee Ka) \supset (Ha \& Ga)$' is false.

II-4. $(Wa \vee Wc) \supset (\exists x) (Px \& Dx)$

II-8. $(x) (Vx \supset Sx)$

II-14. $(x) [(Mx \& \sim Tx) \supset Dx]$

II-19. The locution 'not only . . . but – – –' is translated by '&'. By far the most natural understanding of the plural nouns 'radicals' and 'revolutionaries' in such a context is in terms of 'some' rather than 'every'. Finally, note that two existential quantifiers are needed; the translation '$(\exists x) [(Rx \& Vx) \& Cx]$' asserts that there are individuals who are *both* radicals and revolutionaries who attended the convention. But (19) does not say this. Instead, we have a conjunction of two I-sentences:

$$(\exists x) (Rx \& Cx) \& (\exists x) (Vx \& Cx).$$

Page 162

I-2. Open sentence; 'x' occurs free.

I-4. Open sentence; 'z' occurs free.

II-4. Not a substitution instance; a *bound* occurrence of 'x' has been replaced by a constant (the 'x' in 'Fx').

Pages 168–169

2. $(x) (Lcx \supset Hax)$

6. $(x) (Mx \supset Scx)$

7. $(x) (Sx \supset Ax) \supset (\exists y) (Ly \& Py)$

12. $(x) [(Cx \& Dx) \supset [\sim Rx \vee (\exists y) (Ey \& Ay)]]$

Pages 177–179

II-2. $(x) [Px \supset (\exists y) (Lxy)]$

II-3. $(\exists x) [Px \& (y) (Lxy)]$

II-12. $(\exists x) [Px \& (y) (Py \supset \sim Cxy)]$

II-14. $(x) [(Px \& Fxc) \supset Lcx]$

II-19. Presumably, our man doesn't just hate some dogs but hates dogs in general. So we have:

$(x) [(x$ is a man $\& x$ hates every dog$) \supset x$ has f for a friend)
$(x) [(Mx \& (y) (Dy \supset Hxy)) \supset Fxf]$

III-3. This says that for each valid argument, anything it is derivable from is a valid argument form.

$(x) [Vx \supset (y) (Dxy \supset Fy)]$ also of course:
$(x) (y) [(Vx \& Dxy) \supset Fy]$

III-8. Within the main quantifier's scope we have basically a four-term conjunction, with the fourth conjunct being rather complex:

$(\exists x) \{[(Nx \& Px) \& Bxfe] \& (y) [(y$ is a number which is a square root of some number less than 300$) \supset x > y]\}$.

$(\exists x) \{[(Nx \& Px) \& Bxfe] \& (y) [(Ny \& (\exists z) ((Nz \& z < a) \& y = \sqrt{z})) \supset x > y]\}$

Assuming that only the natural numbers are in our universe of discourse, (8) is false. 17 is the only prime between 14 and 18, and it is also the greatest value for y. But 17 is not greater than itself.

III-10. Does this sentence say: any wind that is ill blows no good to anyone? Or does it say: any wind that blows no good to anyone is an ill one? The latter is the more natural reading, and as such we have:

$(x) [(Wx \& (y) (z) ((Py \& Gz) \supset \sim Bxzy)) \supset Ix]$ also:
$(x) [Wx \supset [(y) (z)((Py \& Gz) \supset \sim Bxzy) \supset Ix]]$.

IV-2. False. This says that for any positive integer we pick, it is greater than at least one positive integer. But this claim does not hold for 1.

IV-4. False. This asserts the existence of a positive integer greater than all *other* positive integers (all integers not identical to it). But there is no greatest positive integer.

IV-7. False. This says that for any positive integer we pick, we can find positive integers it is between. But this condition fails for 1.

Page 185

4. $(\exists x)(\exists y)(Sxm \& Sym \& x \neq y)$

6. $(An \& Lnl) \& (x)[(Ax \& Lxl) \supset x = n]$

10. $(\exists x)[Px \& (y)(z)[(Py \& Pz) \supset (y = x \lor z = x \lor y = z)]]$

Page 192

I-2. $(\exists x)[Axf \& (y)(Ayf \supset y = x) \& x = m]$

I-5. $(\exists x)[Sxn \& (y)((Syn \& y \neq x) \supset Rxy) \& \sim Tx]$

I-8. If the description 'my baby' is regarded as having primary occurrence in this sentence, then it is saying: there is one and only one baby belonging to me and everyone loves her (him, it). But if it is taken to have secondary occurrence, it is saying: everyone is such that they love the one and only one baby belonging to me. The former appears to be the more plausible reading since our inclination is to count (8) as false (or lacking in truth-value) if in fact I have no single baby. On the primary reading (8) would be false in such a case, but on the secondary one it might yet be true. Thus we have:

$(\exists x)[Bx \& Lx \& (y)((By \& Ly) \supset y = x) \& (z)(Pz \supset Lzx)]$

CHAPTER SIX

Pages 205–207

I-2. universe of discourse: animals
interpret 'Ax' as 'x is human'
interpret 'Bx' as 'x is a mammal'
interpret 'Cx' as 'x is a vertebrate'

I-6. universe of discourse: positive integers
      interpret 'Rxy' as '$x < y$'
      interpret 'a' as '2'
      interpret 'b' as '3'

I-8. universe of discourse: human beings
      interpret 'Fx' as 'x is a man'
      interpret 'Gx' as 'x is a father'
      interpret 'Rxy' as 'x is a son of y'

I-10. universe of discourse: human beings
      interpret 'Fx' as 'x is a farmer'
      interpret 'Gx' as 'x quits farming'
      interpret 'Hx' as 'x is a person'
      interpret 'Jx' as 'x starves'

III-2. universe of discourse: persons in a small boat
      interpret 'Fx' as 'x rushes to starboard at 3 P.M.'
      interpret 'Gx' as 'x falls overboard'

Page 212

I-2. $Aa \supset Ba$
      Ca & $\sim Ba$
      $\therefore Aa$ & $\sim Ca$

Aa	Ba	Ca
F	F	T

I-6. Rab
      $\therefore Raa$ v Rbb

Rab	Raa	Rbb
T	F	F

I-8. $[Fa \supset ((Ga$ & $Raa)$ v $(Gb$ & $Rab))]$ & $[Fb \supset ((Ga$ & $Rba)$ v $(Gb$ & $Rbb))]$
      $\therefore [Fa \supset ((Ga \supset Raa)$ & $(Gb \supset Rab))]$ & $[Fb \supset ((Ga \supset Rba)$ & $(Gb \supset Rbb))]$

Fa	Fb	Ga	Gb	Raa	Rab	Rba	Rbb
F	T	T	T	–	–	F	T

I-10. $[(Fa \supset Ga)$ & $(Fb \supset Gb)] \supset [(Ha \supset Ja)$ & $(Hb \supset Jb)]$
      $\therefore [(Fa$ & $Ga)$ v $(Fb$ & $Gb)] \supset [(Ha \supset Ja)$ & $(Hb \supset Jb)]$

Fa	Ga	Fb	Gb	Ha	Ja	Hb	Jb
T	T	T	F	T	F	–	–

II-1. $[Fa \lor (Ga \& Ha)] \& [Fb \lor (Gb \& Hb)] \& [Fc \lor (Gc \& Hc)]$
$[Ea \& (Ga \& Ha)] \lor [Eb \& (Gb \& Hb)] \lor [Ec \& (Gc \& Hc)]$
$[Ea \& (Ga \supset \sim Ha)] \lor [Eb \& (Gb \supset \sim Hb)] \lor [Ec \& (Gc \supset \sim Hc)]$
$\therefore (Ea \lor Fa) \& (Eb \lor Fb) \& (Ec \lor Fc)$

Ea	Eb	Ec	Ga	Gb	Gc	Ha	Hb	Hc	Fa	Fb	Fc
F	T	T	T	T	F	T	T	–	F	–	T

CHAPTER SEVEN

Pages 219–220

2. 1. $(\exists x)(Fx \& Gx)$ Premise   DEDUCE: $(\exists y)(Fy)$
   2. $Fv \& Gv$ 1 EI
   3. Fv 2 Simp
   4. $(\exists y)(Fy)$ 3 EG

7. 1. $(\exists x)[Sx \& \sim(Sx \& Tx)]$ Premise
       DEDUCE: $(\exists x)(Sx) \& (\exists y)(\sim Ty)$
   2. $Sv \& \sim(Sv \& Tv)$ 1 EI
   3. Sv 2 Simp
   4. $(\exists x)(Sx)$ 3 EG
   5. $\sim(Sv \& Tv)$ 2 Com, Simp
   6. $\sim Sv \lor \sim Tv$ 5 De M
   7. $\sim Tv$ 6, 3 DN, DS
   8. $(\exists y)(\sim Ty)$ 7 EG
   9. $(\exists x)(Sx) \& (\exists y)(\sim Ty)$ 4, 8 Conj

10. 1. $\sim(\exists x)(Px \& \sim Wx)$ Premise
    2. Pa Premise   DEDUCE: Wa

Since the quantifier in (1) appears *inside* the '$\sim$', **EI** cannot be applied. So we need to find another approach. When nothing else suggests itself, we can always assume the negation of our conclusion and hope something turns up:

$\rightarrow$3. $\sim Wa$ Assumption
  4. $Pa \& \sim Wa$ 2, 3 Conj
  5. $(\exists x)(Px \& \sim Wx)$ 4 EG

6. $\sim Wa \supset (\exists x)(Px \& \sim Wx)$ 3-5 CP
7. Wa 6, 1 MT, DN

Pages 229–230

6.   1. $(x)[(Bx \lor Cx) \supset Dx]$ Premise
      2. $(\exists y)(\sim Cy \lor \sim By)$ Premise
      3. $(\exists z)[\sim(Ez \lor \sim Cz)]$ Premise   DEDUCE: $(\exists x)(Dx)$
      4. $\sim Cv \lor \sim Bv$ 2 EI

Since (1) is universally quantified and we can **UI** to any pseudoname or constant any time, there's no need to rush: we can let (1) sit while we deal with the existentially quantified premises. With **EI** we do need to be careful, of course. So having used '*v*' in instantiating (2) we must choose a *different* pseudoname when we apply **EI** to (3):

      5. $\sim(Ev' \lor \sim Cv')$ 3 EI
      6. $\sim Ev' \& Cv'$ 5 De M, DN
      7. Cv' 6 Com, Simp

It now appears that it will be more useful to instantiate (1) to '*v'*' than to '*v*':

      8. $(Bv' \lor Cv') \supset Dv'$ 1 UI
      9. $Bv' \lor Cv'$ 7 Add, Com
      10. Dv' 8, 9 MP
      11. $(\exists x)(Dx)$ 10 EG

Notice that line (4) was never used; we could have saved one step by ignoring the second premise of this argument.

10.   1. $(x)(Fx \supset Ix)$ Premise   DEDUCE: $(x)(Fx) \supset (y)(Iy)$

Since the conclusion is a conditional, we should assume its antecedent (which happens to contain a quantifier) and show its consequent:

$$\begin{array}{lll} \rightarrow\!2. & (x)(Fx) & \text{Assumption} \\ 3. & Fu & \text{2 UI} \\ 4. & Fu \supset Iu & \text{1 UI} \\ 5. & Iu & \text{4, 3 MP} \\ 6. & (y)(Iy) & \text{5 UG} \\ \hline 7. & (x)(Fx) \supset (y)(Iy) & \text{2-6 CP} \end{array}$$

It is important to see that the **UG** at step (6) does *not* run afoul of proviso (i) on **UG** because the pseudoname '*u*' involved does *not* occur in the **CP**-assumption (2) still operative.

15.

1.	$(x)[Ex \supset (Mxf \lor Mxb)]$	Premise
2.	$(x)(Mxf \supset Mxb)$	Premise
		DEDUCE: $(x)(Ex \supset Mxb)$
→ 3.	Eu	Assumption
4.	$Eu \supset (Muf \lor Mub)$	1 UI
5.	$Mub \lor Muf$	4, 3 MP, Com
6.	$Muf \supset Mub$	2 UI
7.	$Mub \lor \sim Muf$	6 Impl, Com
8.	$(Mub \lor Muf) \& (Mub \lor \sim Muf)$	5, 7 Conj
9.	$Mub \lor (Muf \& \sim Muf)$	8 Dist
10.	$\sim Muf \lor \sim \sim Muf$	E-M I
11.	$\sim(Muf \& \sim Muf)$	10 De M
12.	Mub	9, 11 Com, DS
13.	$Eu \supset Mub$	3-12 CP
14.	$(x)(Ex \supset Mxb)$	13 UG

A somewhat shorter and considerably *simpler* deduction can be found if you use **CP** to establish '$\sim Mub \supset \sim Eu$' by assuming '$\sim Mub$' at step (3) instead. Try it!

Page 232

1. 1. $\sim(x)(Fx \lor Gx)$ Premise   DEDUCE: $(\exists y)(\sim Fy)$

It would be a mistake to try to instantiate (1) as it stands: the scope of the universal quantifier is *not* the entire line (since the curl occurs outside the quantifier) and **UI** cannot be applied. So we first use **QN** to change (1) into a sentence that *can* be instantiated, though it will be **EI**, not **UI**, that can then be employed:

2. $(\exists x)[\sim(Fx \vee Gx)]$ 1 QN
3. $\sim(Fv \vee Gv)$ 2 EI
4. $\sim Fv \,\&\, \sim Gv$ 3 De M
5. $\sim Fv$ 4 Simp
6. $(\exists y)(\sim Fy)$ 5 EG

4.   1. $(\exists x)[\sim(Fx \,\&\, Gx)]$ Premise
    2. $\sim(x)(Fx \,\&\, Gx) \supset \sim(z)(Hz)$ Premise
    3. $(\exists z)(\sim Hz) \supset (y)(Gy)$ Premise
                    DEDUCE: $\sim(\exists y)(\sim Gy)$

    4. $\sim(x)(Fx \,\&\, Gx)$ 1 QN
    5. $\sim(z)(Hz)$ 2, 4 MP
    6. $(\exists z)(\sim Hz)$ 5 QN
    7. $(y)(Gy)$ 3, 6 MP
    8. $\sim\sim(y)(Gy)$ 7 DN
    9. $\sim(\exists y)(\sim Gy)$ 8 QN

Pages 240–243

I-5.   1. $(\exists x)(Ux) \supset (y)(z)(Syz \supset Uy)$ Premise
    2. $(\exists x)(\exists y)[(Px \,\&\, Qx) \,\&\, (Ry \,\&\, Syx)]$ Premise
    3. $(x)(Qx \supset Ux)$ Premise
                  DEDUCE: $(\exists x)(Rx \,\&\, Ux)$

Our general strategy must be to instantiate everything we can, being careful to do the **EI**s first, and then sort through the resulting lines in the hope of discovering some thing we can show is both R and U. At the same time we should resolve not to fall into the trap set by (1): the quantifiers in (1) are buried in its antecedent and consequent, and are thus beyond the reach of **EI** and **UI.**

    4. $(\exists y)[(Pv \,\&\, Qv) \,\&\, (Ry \,\&\, Syv)]$ 2 EI
    5. $(Pv \,\&\, Qv) \,\&\, (Rv' \,\&\, Sv'v)$ 4 EI
    6. $Pv \,\&\, Qv$ 5 Simp
    7. Qv 6 Com, Simp
    8. $Qv \supset Uv$ 3 UI
    9. Uv 8, 7 MP

Since the scope of the existential quantifier in (1) is not the entire line, **EI** cannot be applied to it. Even were (1) bracketed differently so that **EI** could be applied, we could not move to '$Uv \supset (y)(z)(Syz \supset Uy)$': '$v$' already occurs earlier in our deduction, so proviso (i) on **EI** prohibits us from now instantiating to it. Well, if we can't change (1) so its antecedent looks like (9), what should we do? Change (9) so it looks like (1)'s antecedent!

10. $(\exists x)(Ux)$	9 EG
11. $(y)(z)(Syz \supset Uy)$	1, 10 MP

The left conjunct of (5), concerned with 'v', having now done its work, we turn our attention next to the right half of (5) and to 'v'':

12. $Rv' \& Sv'v$	5 Com, Simp
13. $Sv'v$	12 Com, Simp
14. $(z)(Sv'z \supset Uv')$	11 UI
15. $Sv'v \supset Uv'$	14 UI
16. Uv'	15, 13 MP
17. Rv'	12 Simp
18. $Rv' \& Uv'$	17, 16 Conj
19. $(\exists x)(Rx \& Ux)$	18 EG

II-3.

1. $(x)(Rx \supset \sim Gx)$	Premise
2. $(x)[(\sim Ex \& \sim Gx) \supset Hx]$	Premise

DEDUCE: $(x)(Ex \supset Gx) \supset (y)(Ry \supset Hy)$

Since the conclusion is a conditional, we should begin by assuming its antecedent. Needing '$(y)(Ry \supset Hy)$' then, we can assume 'Ru', hoping to show 'Hu' and then use **CP** and **UG** before dismissing our first assumption.

3. $(x)(Ex \supset Gx)$	Assumption
4. Ru	Assumption
5. $Ru \supset \sim Gu$	1 UI
6. $\sim Gu$	5, 4 MP
7. $Eu \supset Gu$	3 UI
8. $\sim Eu$	7, 6 MT
9. $\sim Eu \& \sim Gu$	8, 6 Conj
10. $(\sim Eu \& \sim Gu) \supset Hu$	2 UI
11. Hu	10, 9 MP
12. $Ru \supset Hu$	4-11 CP
13. $(y)(Ry \supset Hy)$	12 UG
14. $(x)(Ex \supset Gx) \supset (y)(Ry \supset Hy)$	3-13 CP

II-5. 1. $(x)(y)(z)(Sxy \supset Syz)$ Premise

DEDUCE: $(x)(y)[Sxy \supset (z)(Szx)]$

We'll assume 'Suu''' and then instantiate (1) three times to set up
MP:

$\rightarrow$ 2. Suu' Assumption
   3. $(y)(z)(Suy \supset Syz)$ 1 UI
   4. $(z)(Suu' \supset Su'z)$ 3 UI
   5. $Suu' \supset Su'u''$ 4 UI
   6. $Su'u''$ 5, 2 MP

Things are not looking good. We cannot **UG** on 'u''' since it occurs
in a **CP**-assumption we're still making. Even if we could,
'$(z)(Szu'')$' won't help. Crossing our fingers, let's instantiate (1)
again to set up **MP** with line (6):

   7. $(y)(z)(Su'y \supset Syz)$ 1 UI
   8. $(z)(Su'u'' \supset Su''z)$ 7 UI
   9. $Su'u'' \supset Su''u$ 8 UI
  10. $Su''u$ 9, 6 MP
  11. $(z)(Szu)$ 10 UG

12. $Suu' \supset (z)(Szu)$ 2-11 CP
13. $(y)[Suy \supset (z)(Szu)]$ 12 UG
14. $(x)(y)[Sxy \supset (z)(Szx)]$ 13 UG

For those who enjoyed this one, here's a new problem: use the
same premise, but let the conclusion be '$(x)(y)[Sxy \supset (z)(Sxz)]$'.

III-2. 1. $(x)(y)(Sxy \supset Syx)$ Premise
   2. $(x)(Sxd \supset {\sim}Tbx)$ Premise
   3. $(x)(y)(Rax \supset Txy)$ Premise   DEDUCE: $Rab \supset {\sim}Sdc$
$\rightarrow$ 4. Rab Assumption
   5. $(y)(Rab \supset Tby)$ 3 UI
   6. $Rab \supset Tbc$ 5 UI
   7. Tbc 6, 4 MP
   8. $Scd \supset {\sim}Tbc$ 2 UI
   9. ${\sim}Scd$ 8, 7 DN, MT
  10. $(y)(Sdy \supset Syd)$ 1 UI
  11. $Sdc \supset Scd$ 10 UI
  12. ${\sim}Sdc$ 11, 9 MT

13. $Rab \supset {\sim}Sdc$ 4-12 CP

III-5.   1. $(x)(\exists y)(Lxy)$ Premise

                                                        DEDUCE: $(y)(\exists z)(Lyz)$

      2. $(\exists y)(Luy)$ 1 UI

A simple idea is to remove the quantifiers by **UI** and **EI**, and then put back different ones. But if we apply **EI** to (2) and get 'Luv', 'u' will occur in the line obtained by **EI** and proviso (ii) on **UG** will prevent us from using **UG**. When stymied near the start like this, indirect proof can give us some maneuvering room so let's try:

    →3. $\sim(\exists z)(Luz)$ Assumption
      4. $(z)(\sim Luz)$ 3 QN
      5. $\sim Luu'$ 4 UI
      6. $(y)(\sim Luy)$ 5 UG
      7. $\sim(\exists y)(Luy)$ 6 QN

But now it's quicker to change plans yet again and close off the assumption at this point:

      8. $\sim(\exists z)(Luz) \supset \sim(\exists y)(Luy)$ 3-7 CP
      9. $(\exists y)(Luy) \supset (\exists z)(Luz)$ 8 Trans
    10. $(\exists z)(Luz)$ 9, 2 MP
    11. $(y)(\exists z)(Lyz)$ 10 UG

IV-2.   1. $(\exists x)(Sx \& Tx)$
      2. $(x)(Px \supset \sim Tx)$ ∴ $(\exists x)(Px \& \sim Sx)$

We could **EI** (1) to get '$Sv \& Tv$', pull 'Tv' out of that, and do **MT** on the result '$Pv \supset \sim Tv$' of **UI**-ing (2). But then we have '$\sim Pv$' and 'Sv', which do not help us to obtain the conclusion.

    So the argument is probably invalid. To show that it is, let the universe of discourse consist of *people*. If we interpret 'Px' as 'x is a baby' and 'Sx' as 'x is a child', our conclusion will be false since all babies are children. To make both premises true, then, we need a property for 'Tx' that some children have but that no babies do. So interpreting 'Tx' as 'x is six years old' will do the trick.

IV-5. 1. $(\exists x)(Bx \& Ux) \supset (y)(Ryi \supset Gy)$ Premise
      2. $(\exists z)(Czi \& Rzi)$ Premise
            DEDUCE: $(y)(Cyi \supset \sim Gy) \supset \sim(\exists x)(Bx \& Ux)$

>   3. $(y)(Cyi \supset \sim Gy)$ Assumption
     4. $Cvi \& Rvi$ 2 EI
     5. Cvi 4 Simp
     6. $Cvi \supset \sim Gv$ 3 UI
     7. $\sim Gv$ 6, 5 MP
     8. Rvi 4 Com, Simp
     9. $Rvi \& \sim Gv$ 8, 7 Conj
    10. $\sim(\sim Rvi \vee Gv)$ 9 DN, De M
    11. $\sim(Rvi \supset Gv)$ 10 Impl
    12. $(\exists y)[\sim(Ryi \supset Gy)]$ 11 EG
    13. $\sim(y)(Ryi \supset Gy)$ 12 QN
    14. $\sim(\exists x)(Bx \& Ux)$ 1, 13 MT

    15. $(y)(Cyi \supset \sim Gy) \supset \sim(\exists x)(Bx \& Ux)$ 3-14 CP

IV-8. 1. $(x)\{[(Vx \& Ax) \& Pjx] \supset \sim Sx\}$ Premise
      2. $(x)\{Ax \supset \{Sx \equiv [Vx \& (y)(Ryx \supset Ty)]\}\}$ Premise
            DEDUCE: $[(Va \& Aa) \& Pja] \supset (\exists y)(Rya \& \sim Ty)$

>   3. $(Va \& Aa) \& Pja$ Assumption
     4. $[(Va \& Aa) \& Pja] \supset \sim Sa$ 1 UI
     5. $\sim Sa$ 4, 3 MP
     6. $Va \& Aa$ 3 Simp

Lines (5) and (6) tell us that the Ontological Argument is a valid argument but is unsound. To conclude that it has a false premise, we must use (2):

     7. $Aa \supset \{Sa \equiv [Va \& (y)(Rya \supset Ty)]\}$ 2 UI
     8. Aa 6 Com, Simp
     9. $Sa \equiv [Va \& (y)(Rya \supset Ty)]$ 7, 8 MP
    10. $\{Sa \supset [Va \& (y)(Rya \supset Ty)]\} \&$
        $\{[Va \& (y)(Rya \supset Ty)] \supset Sa\}$ 9 Equiv
    11. $[Va \& (y)(Rya \supset Ty)] \supset Sa$ 10 Com, Simp
    12. $\sim[Va \& (y)(Rya \supset Ty)]$ 11, 5 MT
    13. $\sim Va \vee \sim(y)(Rya \supset Ty)$ 12 De M
    14. $\sim\sim Va$ 6 Simp, DN
    15. $\sim(y)(Rya \supset Ty)$ 13, 14 DS
    16. $(\exists y)[\sim(Rya \supset Ty)]$ 15 QN
    17. $\sim(Rva \supset Tv)$ 16 EI
    18. $\sim(\sim Rva \vee Tv)$ 17 Impl
    19. $Rva \& \sim Tv$ 18 De M, DN
    20. $(\exists y)(Rya \& \sim Ty)$ 19 EG

    21. $[(Va \& Aa) \& Pja] \supset (\exists y)(Rya \& \sim Ty)$ 3-20 CP

IV-14. 1. $(x)\{\{Ex \,\&\, (y)[(Ey \,\&\, Lyx) \supset Sy]\} \supset Sx\} \supset$
$(z)(Ez \supset Sz)$ Premise

DEDUCE: Line (18) below

2. $(x)\{(Ex \,\&\, \sim Sx) \supset (\exists y)[(Ey \,\&\,$
$Lyx) \,\&\, \sim Sy]\}$ Assumption
3. $Eu \,\&\, (y)[(Ey \,\&\, Lyu) \supset Sy]$ Assumption
4. $(y)[(Ey \,\&\, Lyu) \supset Sy]$ 3 Com, Simp
5. $(Eu' \,\&\, Lu'u) \supset Su'$ 4 UI
6. $\sim(Eu' \,\&\, Lu'u) \vee \sim\sim Su'$ 5 Impl, DN
7. $\sim[(Eu' \,\&\, Lu'u) \,\&\, \sim Su']$ 6 De M
8. $(y)\{\sim[(Ey \,\&\, Lyu) \,\&\, \sim Sy]\}$ 7 UG
9. $\sim(\exists y)[(Ey \,\&\, Lyu) \,\&\, \sim Sy]$ 8 QN
10. $(Eu \,\&\, \sim Su) \supset (\exists y)[(Ey \,\&\,$
$Lyu) \,\&\, \sim Sy]$ 2 UI
11. $\sim(Eu \,\&\, \sim Su)$ 10, 9 MT
12. $\sim Eu \vee Su$ 11 De M, DN
13. Eu 3 Simp
14. Su 12, 13 DN, DS

15. $\{Eu \,\&\, (y)[Ey \,\&\, Lyu] \supset Sy]\} \supset Su$ 3-14 CP
16. $(x)\{\{Ex \,\&\, (y)[(Ey \,\&\, Lyx) \supset Sy]\} \supset Sx]$ 15 UG
17. $(z)(Ez \supset Sz)$ 1, 16 MP

18. $(x)\{(Ex \,\&\, \sim Sx) \supset (\exists y)[(Ey \,\&\, Lyx) \,\&\, \sim Sy]\} \supset$
$(z)(Ez \supset Sz)$ 2-17 CP

If we take (18) as our premise and (1) as our conclusion, that argument is also deductively valid and yields to a similar approach.

Page 248

I-2.
1. $(z)(Guz)$ Assumption
2. Guu' 1 UI
3. $(y)(Guy)$ 2 UG

4. $(z)(Guz) \supset (y)(Guy)$ 1-3 CP
5. $\sim(y)(Guy) \supset \sim(z)(Guz)$ 4 Trans
6. $(\exists y)(\sim Guy) \supset \sim(z)(Guz)$ 5 QN
7. $(x)[(\exists y)(\sim Gxy) \supset \sim(z)(Gxz)]$ 6 UG

I-6.
1. $(x)(Fx \lor Ga)$ Assumption
2. $\sim(y)(Fy)$ Assumption
3. $(\exists y)(\sim Fy)$ 2 QN
4. $\sim Fv$ 3 EI
5. $Fv \lor Ga$ 1 UI
6. Ga 5, 4 DS

7. $\sim(y)(Fy) \supset Ga$ 2-6 CP
8. $(y)(Fy) \lor Ga$ 7 Impl, DN

9. $(x)(Fx \lor Ga) \supset [(y)(Fy) \lor Ga]$ 1-8 CP
10. $(y)(Fy) \lor Ga$ Assumption
11. $\sim Ga$ Assumption
12. $(y)(Fy)$ 10, 11 Com, DS
13. Fu 12 UI

14. $\sim Ga \supset Fu$ 11-13 CP
15. $Ga \lor Fu$ 14 Impl, DN
16. $(x)(Fx \lor Ga)$ 15 Com, UG

17. $[(y)(Fy) \lor Ga] \supset (x)(Fx \lor Ga)$ 10-16 CP
18. $\{(x)(Fx \lor Ga) \supset [(y)(Fy) \lor Ga]\}$ &
   $\{[(y)(Fy) \lor Ga] \supset (x)(Fx \lor Ga)\}$ 9, 17 Conj
19. $(x)(Fx \lor Ga) \equiv [(y)(Fy) \lor Ga]$ 18 Equiv

III-1. Again, each direction on the biconditional uses one **CP** nested inside another:

1. $\sim(\exists x)(Fx)$ Assumption
2. Fu Assumption
3. $(\exists x)(Fx)$ 2 EG

4. $Fu \supset (\exists x)(Fx)$ 2-3 CP
5. $\sim Fu$ 1, 4 MT
6. $(y)(\sim Fy)$ 5 UG

7. $\sim(\exists x)(Fx) \supset (y)(\sim Fy)$ 1-6 CP
8. $(y)(\sim Fy)$ Assumption
9. $(\exists x)(Fx)$ Assumption
10. Fv 9 EI

11. $(\exists x)(Fx) \supset Fv$ 9-10 CP
12. $\sim Fv$ 8 UI
13. $\sim(\exists x)(Fx)$ 11, 12 MT

14. $(y)(\sim Fy) \supset \sim(\exists x)(Fx)$ 8-13 CP

15. $[\sim(\exists x)(Fx) \supset (y)(\sim Fy)] \& [(y)(\sim Fy) \supset$
   $\sim(\exists x)(Fx)]$ 7, 14 Conj

16. $\sim(\exists x)(Fx) \equiv (y)(\sim Fy)$ 15 Equiv

Pages 253–254

I-2.
$\quad$1. $u = u'$ & $u = u''$ $\qquad$ Assumption
$\quad$2. $u = u'$ $\qquad$ 1 Simp
$\quad$3. $u = u''$ $\qquad$ 1 Com, Simp
$\quad$4. $u'' = u'$ $\qquad$ 2, 3 IE

$\quad$5. $(u = u'$ & $u = u'') \supset u'' = u'$ $\qquad$ 1-4 CP
$\quad$6. $(z)[(u = u'$ & $u = z) \supset z = u']$ $\qquad$ 5 UG
$\quad$7. $(y)(z)[(u = y$ & $u = z) \supset z = y]$ $\qquad$ 6 UG
$\quad$8. $(x)(y)(z)[(x = y$ & $x = z) \supset z = y]$ $\qquad$ 7 UG

II-2.
$\quad$1. Hc & $\sim Hm$ $\qquad$ Premise
$\quad$2. $c = t$ $\qquad$ Premise $\qquad$ DEDUCE: $m \neq t$
$\quad$3. $m = t$ $\qquad$ Assumption
$\quad$4. $\sim Hm$ $\qquad$ 1 Com, Simp
$\quad$5. $\sim Ht$ $\qquad$ 4, 3 IE

$\quad$6. $m = t \supset \sim Ht$ $\qquad$ 3-5 CP
$\quad$7. Hc $\qquad$ 1 Simp
$\quad$8. Ht $\qquad$ 7, 2 IE
$\quad$9. $m \neq t$ $\qquad$ 6, 8 DN, MT

II-7.
$\quad$1. $(x)(y)[(Rxy$ & $Ryx) \supset x = y]$ $\qquad$ Premise
$\quad$2. $(x)(y)(Rxy \supset Ryx)$ $\qquad$ Premise
$\qquad\qquad$ DEDUCE: $(x)[(\exists y)(Rxy$ v $Ryx) \supset Rxx]$

We want to deduce '$(\exists y)(Ruy$ v $Ryu) \supset Ruu$' by **CP** and then use **UG**. But if we assume the antecedent we'll be stuck: if we don't **EI** it it'll be useless, and if we do the later **UG** will be blocked. So instead we assume '$\sim Ruu$' and so need '$\sim (\exists y)(Ruy$ v $Ryu)$', that is, '$(y)[\sim(Ruy$ v $Ryu)]$'. An indirect approach establishes '$\sim Ruu$', and that and (2) get us '$\sim Ru'u$' as well:

3.	$\sim Ruu$	Assumption
4.	Ruu'	Assumption
5.	$(y)(Ruy \supset Ryu)$	2 UI
6.	$Ruu' \supset Ru'u$	5 UI
7.	$Ru'u$	6, 4 MP
8.	$Ruu' \& Ru'u$	4, 7 Conj
9.	$(y)[(Ruy \& Ryu) \supset u = y]$	1 UI
10.	$(Ruu' \& Ru'u) \supset u = u'$	9 UI
11.	$u = u'$	10, 8 MP
12.	Ruu	4, 11 IE
13.	$Ruu' \supset Ruu$	4-12 CP
14.	$\sim Ruu'$	13, 3 MT
15.	$(y)(Ru'y \supset Ryu')$	2 UI
16.	$Ru'u \supset Ruu'$	15 UI
17.	$\sim Ru'u$	16, 14 MT
18.	$\sim Ruu' \& \sim Ru'u$	14, 17 Conj
19.	$\sim(Ruu' \vee Ru'u)$	18 De M
20.	$(y)[\sim(Ruy \vee Ryu)]$	19 UG
21.	$\sim(\exists y)(Ruy \vee Ryu)$	20 QN
22.	$\sim Ruu \supset \sim(\exists y)(Ruy \vee Ryu)$	3-21 CP
23.	$(\exists y)(Ruy \vee Ryu) \supset Ruu$	22 Trans
24.	$(x)[(\exists y)(Rxy \vee Ryx) \supset Rxx]$	23 UG

III-3. 1. $(x)(y)\{[(Sx \,\&\, Sy) \,\&\, (z)(Mzx \equiv Mzy)] \supset x = y\}$ Premise

DEDUCE: Line (20) below

$\quad\quad$ 2. $Su \,\&\, Su'$ Assumption

$\quad\quad\quad$ 3. $u = u'$ Assumption

$\quad\quad\quad$ 4. $Mu''u \lor \sim Mu''u$ E-M I

$\quad\quad\quad$ 5. $Mu''u \supset Mu''u$ 4 Com, Impl

$\quad\quad\quad$ 6. $Mu''u \supset Mu''u$ 4 Com, Impl

$\quad\quad\quad$ 7. $(Mu''u \supset Mu''u) \,\&\, (Mu''u \supset Mu''u)$ 5, 6 Conj

$\quad\quad\quad$ 8. $Mu''u \equiv Mu''u$ 7 Equiv

$\quad\quad\quad$ 9. $(z)(Mzu \equiv Mzu)$ 8 UG

$\quad\quad\quad$ 10. $(z)(Mzu \equiv Mzu')$ 9, 3 IE

$\quad\quad$ 11. $u = u' \supset (z)(Mzu \equiv Mzu')$ 3-10 CP

$\quad\quad$ 12. $(y)\{[(Su \,\&\, Sy) \,\&\, (z)(Mzu \equiv Mzy)] \supset u = y\}$ 1 UI

$\quad\quad$ 13. $[(Su \,\&\, Su') \,\&\, (z)(Mzu \equiv Mzu')] \supset u = u'$ 12 UI

$\quad\quad$ 14. $(Su \,\&\, Su') \supset [(z)(Mzu \equiv Mzu') \supset u = u']$ 13 Exp

$\quad\quad$ 15. $(z)(Mzu \equiv Mzu') \supset u = u'$ 14, 2 MP

$\quad\quad$ 16. $[u = u' \supset (z)(Mzu \equiv Mzu')] \,\&$

$\quad\quad\quad\quad [(z)(Mzu \equiv Mzu') \supset u = u']$ 11, 15 Conj

$\quad\quad$ 17. $u = u' \equiv (z)(Mzu \equiv Mzu')$ 16 Equiv

$\quad$ 18. $(Su \,\&\, Su') \supset [u = u' \equiv (z)(Mzu \equiv Mzu')]$ 2-17 CP

$\quad$ 19. $(y)\{(Su \,\&\, Sy) \supset [u = y \equiv (z)(Mzu \equiv Mzy)]\}$ 18 UG

$\quad$ 20. $(x)(y)\{(Sx \,\&\, Sy) \supset [x = y \equiv (z)(Mzx \equiv Mzy)]\}$ 19 UG

APPENDIX A

Page 260

3. $\sim[D \lor (B \,\&\, \sim C)]$

$\quad \sim D \,\&\, \sim(B \,\&\, \sim C)$ De M

$\quad \sim D \,\&\, (\sim B \lor \sim\sim C)$ De M

$\quad \sim D \,\&\, (\sim B \lor C)$ DN

5. $[(A \supset B) \supset A] \supset A$

$\quad \sim[\sim(\sim A \lor B) \lor A] \lor A$ Impl

$\quad \sim[(\sim\sim A \,\&\, \sim B) \lor A] \lor A$ De M

$\quad [\sim(\sim\sim A \,\&\, \sim B) \,\&\, \sim A] \lor A$ De M

$\quad [(\sim\sim\sim A \lor \sim\sim B) \,\&\, \sim A] \lor A$ De M

$\quad [(\sim A \lor B) \,\&\, \sim A] \lor A$ DN

$\quad [A \lor (\sim A \lor B)] \,\&\, (A \lor \sim A)$ Com, Dist

$\quad (A \lor \sim A \lor B) \,\&\, (A \lor \sim A)$ Assoc

$\quad A \lor \sim A \lor B$ TE   Simpler form: $A \lor \sim A$.

Page 267

I-1.

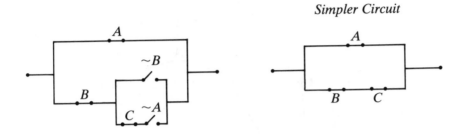

Simpler Circuit

Pages 275–276

2. '$(A \supset {\sim}B)$ & $(C \& D)$' is equivalent to '${\sim}[(A \& B) \vee {\sim}(C \& D)]$. From the latter we obtain:

APPENDIX B

Pages 289–290

I-3. Valid.

I-6. Invalid.

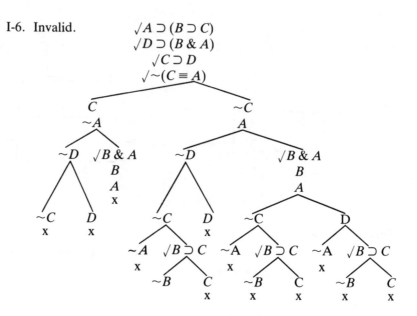

$$\sqrt{A \supset (B \supset C)}$$
$$\sqrt{D \supset (B \& A)}$$
$$\sqrt{C \supset D}$$
$$\sqrt{\sim(C \equiv A)}$$

III-2. Tautology. $\sqrt{\sim[(A \& \sim B) \vee \sim[\sim(B \& C) \& (A \& C)]]}$
$$\sqrt{\sim(A \& \sim B)}$$
$$\sqrt{\sim\sim[\sim(B \& C) \& (A \& C)]}$$
$$\sqrt{\sim(B \& C) \& (A \& C)}$$
$$\sqrt{\sim(B \& C)}$$
$$\sqrt{A \& C}$$
$$A$$
$$C$$

```
              ~B                        ~C
                                         x
        ~A          √~~B
         x           B
                     x
```

APPENDIX C

Pages 294–295

2. BASE CASE.

$$\frac{1\,(1+1)\,(2+1)}{6} = \frac{(2)\,(3)}{6} = \frac{6}{6} = 1 = 1^2.$$

INDUCTION STEP. Let m be any positive integer and assume that

$$1^2 + \ldots + m^2 = \frac{m\,(m+1)\,(2m+1)}{6}. \quad \text{Then}$$

$$\frac{(m+1)\,((m+1)+1)\,(2(m+1)+1)}{6} = \frac{(m+1)\,(m+2)\,(2m+3)}{6}$$

$$= \frac{(m^2+3m+2)\,(2m+3)}{6}$$

$$= \frac{2m^3+9m^2+13m+6}{6}$$

$$= \frac{(2m^3+3m^2+m)}{6} + \frac{(6m^2+12m+6)}{6}$$

$$= \frac{(m^2+m)\,(2m+1)}{6} + (m^2+2m+1)$$

$$= \frac{m(m+1)\,(2m+1)}{6} + (m+1)^2$$

$$= 1^2 + \ldots + m^2 + (m+1)^2.$$

APPENDIX D

Page 302

1. Yes. **N**'s axiom, '| — |', is true on this interpretation, since the number one is less than or equal to itself; and **N**'s rule preserves the feature of being true on this interpretation, for if one number is less than or equal to a second then the successor of the first is less than or equal to the successor of the second. By the general result (R), then, all theorems of **N** have the feature of being true on this interpretation.

INDEX